Andrzej Służalec

Theory of Metal Forming Plasticity

Springer

Berlin
Heidelberg
New York
Hong Kong
London
Milan
Paris
Tokyo

Andrzej Służalec

Theory of Metal Forming Plasticity

Classical and Advanced Topics

With 95 Figures

Springer

Professor Dr. Andrzej Służalec
Technical University of Czestochowa
ul. Akademicka 3
42-200 Czestochowa
Poland

Library of Congress Cataloging-in-Publication Data
Sluzalec, Andrzej.
Theory of metal forming plasticity: classical and advanced topics / Andrzej Sluzalec.
p.cm.
Includes bibliographical references and index.
ISBN 3-540-40648-4 (alk. paper)
1. Metal-work. 2. Metals--Plastic properties. I. Title.
TS205.S565 2003
671--dc22 2003059107

ISBN 3-540-40648-4 Springer-Verlag Berlin Heidelberg New York

Springer-Verlag Berlin Heidelberg New York
a member of BertelsmannSpringer Science + Business Media GmbH

http://www.springer.de

Typesetting: Digital data supplied by author
Cover-Design: deblik, Berlin
Printed on acid-free paper 62/3020GRw 5 4 3 2 1 0

Preface

The intention of this book is to reveal and discuss some aspects of the metal forming plasticity theory. The modern theory describes deformation of metallic bodies in cold and hot regimes under combined thermal and mechanical loadings. Thermal and deformation fields appear in metal forming in various forms. A thermal field influences the material properties, modifies the extent of plastic zones, etc. and the deformation of metallic body induces changes in temperature distribution. The thermal effects in metal forming plasticity can be studied at two levels, depending on whether uncoupled or coupled theories of thermo-plastic response have to be applied. A majority of metal forming processes can be satisfactorily studied within an uncoupled theory. In such an approach the temperature enters the stress-strain relation through the material constants and through the thermal dilatation. The description of thermo-plastic deformation in metal forming is carried out on the ground of thermodynamics.

The effective solutions of metal forming problems have become possible only recently. In the last two decades one observes the vigorous development of effective nonlinear methods in computational plasticity. Manufacturing industry is experiencing a rapidly-growing need for the analytical tools to handle complex problems of metal forming. Efficient analytical methods for combining geometrically, materially and thermally nonlinear problems are needed because experimental testing in such cases is often prohibitively expensive or physically impossible. In the book the theory of numerical solutions for metal forming plasticity is discussed because recent advances in computational plasticity make it possible to perform extensive calculations with great accuracy, at significantly reduced execution times and at reasonable cost.

The theory of sensitivity discussed in the book originates form purely mathematical studies of the influence of coefficient variations on differential equations. It was much later that the theory of sensitivity became the subject of studies in the field of metal forming plasticity. In metal forming a set of equations that defines the relationship between external loads, prescribed displacements, stress, etc. in metallic body are considered. A sensitivity of metal forming process to variations of its parameters is one of the most important aspects necessary for a proper understanding of the process.

Stochastic methods considered in the book have recently become an area of research in metal forming plasticity. As the name suggests, these methods combine two crucial methodologies developed to deal with problems of metal forming: analytical or numerical analysis with the stochastic one. The stochastic analysis in the broadest sense refers to the explicit treatment of uncertainty in any quantity entering the corresponding deterministic analysis. The exact values of these quantities are usually unknown because they cannot be precisely measured. Stochastic approach to metal forming problems is important not only because of random material parameters, but particularly because of boundary problems appearing in these processes. Contact problems die-workpiece have exceptional random character and lead to determine the boundary forces in the contact die-workpiece considering random character of friction between them. Existing uncertain variations in parameters may have significant effects on such fundamental final characteristics as strain and stress distributions, and they must affect the final design.

Standard analytical solutions evaluating plastic deformation have been included in the book. It describes the deformation in the simple way, which sometimes is very helpful.

This book is divided into eleven parts. It contains: deformation of metallic body (Part I), metal forming thermodynamics (Part II), plasticity (Part III), powder forming plasticity (Part IV), viscoplasticity (Part V), discontinuous fields (Part VI), numerical solution methods (Part VII), sensitivity in metal forming plasticity (Part VIII), stochastic metal forming process (Part IX), contact and friction (Part X) and simplified equations (Part XI).

Part VIII and Part IX provide new elements to the theory of metal forming plasticity not met yet in literature of the subject. The remaining parts are added to describe the theory comprehensively.

I would like to make an acknowledgement to Malgorzata Kowalik and my son Tom for their involvement in preparing the manuscript in its camera ready form.

Częstochowa
April 2003

Andrzej Służalec

Contents

Part I Deformation of Metallic Body

Part II Metal Forming Thermodynamics

Part III Plasticity

Part IV Powder Forming Plasticity

Part V Viscoplasticity

Part VI Discontinuous Fields

Part VII Numerical Solution Methods

Part VIII Sensitivity in Metal Forming Plasticity

Part IX Stochastic Metal Forming Process

Part X Contact and Friction

Part XI Simplified Equations

Part I

Deformation of Metallic Body

1 Description of Deformation

1.1 Introduction

A metallic body subjected to external loadings deforms. In the course of deformation the body changes its geometrical shape. The purpose of this chapter is to adapt the theoretical tools used in continuum mechanics to the description of the deformation and the motion of metallic bodies. The description of this deformation differs in no way from that of a standard continuum.

1.2 Description of Motion

Let B_{t_o} denote a domain occupied by the body at initial time t_o and B_t at current time t. Any elementary material particle in a reference configuration is determined by position vector $\mathbf{X}$ of components X_α in a Cartesian coordinate of orthogonal basis $\mathbf{e} = (\mathbf{e}_1, \mathbf{e}_2, \mathbf{e}_3)$. After deformation of the body, the material is in a new configuration, called the current configuration. In this configuration the material point is determined by the present position vector $\mathbf{x}$ of components x_i. The motion of the body can be described by the relation

$$\mathbf{x} = \mathbf{x}\,(\mathbf{X},t) \qquad x_i = x_i\,(X_\alpha,t) \qquad i, \alpha = 1, 2, 3 \tag{1.1}$$

The introduced X_α are called material coordinates, and x_α are called spatial coordinates. The notion of material coordinates (also called the Lagrange coordinates) was introduced by Euler, and the notion of spatial coordinates (sometimes called Euler coordinates) was introduced by d'Alembert (1752).

1.3 The Deformation Gradient

The tensor $\mathbf{F}$ defined by

$$\mathbf{F} = \operatorname{Grad} \mathbf{x} \qquad F_{i\alpha} = \frac{\partial x_i}{\partial X_\alpha} \qquad \mathbf{F} = F_{i\alpha}\, \mathbf{e}_i \otimes \mathbf{e}_\alpha \tag{1.2}$$

is called the deformation gradient.

The capital of the Grand operator, which appeared in Eq. (1.2) indicates a relation to the reference configuration and the symbol ⊗ stands for the tensorial product. This convention is used for all the introduced operators. The inverse of **F** denoted as $\mathbf{F}^{-1}$ and the transpose of **F** denoted as ${}^{T}\mathbf{F}$ are defined respectively by

$$ {}^{T}(F)_{i\alpha} = F_{\alpha i} \qquad \left(F^{-1}\right)_{\alpha i} = \frac{\partial X_{\alpha}}{\partial x_{i}} \tag{1.3}$$

The Jacobian of the mapping (1.2) is given by

$$ J = \det \mathbf{F} \tag{1.4}$$

1.4 The Polar Decomposition Theorem

According to the polar decomposition theorem the tensor **F** can be represented uniquely by

$$ \mathbf{F} = \mathbf{R}\,\mathbf{U} = \mathbf{V}\,\mathbf{R} \tag{1.5}$$

where **U**, **V** are positive-definite, symmetric tensors and **R** is an orthogonal tensor. Tensors **U** and **V** are called the right and left stretch tensors, respectively. They have common eigenvalues $\lambda_{(\alpha)}(\alpha = 1,2,3)$ called principal stretches and corresponding eigenvectors $\mathbf{N}^{(\alpha)}$, $\mathbf{n}^{(\alpha)}$ known as the Lagrangian and Eulerian trials, respectively. The following relation holds

$$ \mathbf{n}^{(\alpha)} = \mathbf{R}\,\mathbf{N}^{(\alpha)} \tag{1.6}$$

i.e. the deformation rotates eigenvectors of **U** into those of **V**.

1.5 Local Description of Deformation

The deformation can be described throughout the deformation gradient **F**. Two material points can be defined by their position vectors **X** and **X** + d**X** where d**X** is an infinitesimal material vector within the body in the reference configuration. Material vector d**X** after deformation becomes material vector d**x** joining the same two material points which are defined by the new position vectors **x** and **x** + d**x**.

Using the definition of deformation gradient **F**

$$ d\mathbf{x} = \mathbf{F} \cdot d\mathbf{X} \tag{1.7}$$

Vector $\mathbf{dx}$ is called the convective conveyor of vector $\mathbf{dX}$ with respect to the material body.

The deformation gradient $\mathbf{F}$ is sometimes called the tangent linear operator applied to the attached point $\mathbf{X}$. In the process of deformation the material vector $d\mathbf{X}$ is conveyed and transformed. The deformation gradient $\mathbf{F}$ links any vector $d\mathbf{X}$ belonging to the tangent vector space with point $\mathbf{X}$, to vector $d\mathbf{x}$ belonging to the tangent vector space with the transformed point $\mathbf{x}$.

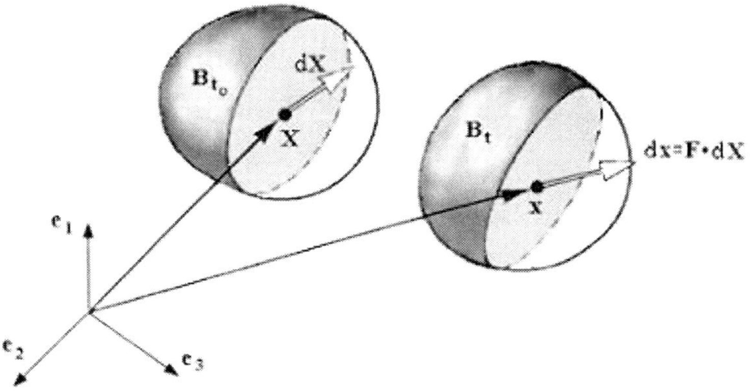

Fig. 1.1. Material vector $d\mathbf{X}$ in reference and current configurations

In the process of deformation, the elementary volume $dV_r = dX_1\, dX_2\, dX_3$ is transformed to the volume $dV = dx_1\, dx_2\, dx_3$ by the relation

$$dV = J\, dV_r \tag{1.8}$$

The volume dV_r is equal to the product $(d\mathbf{X}_1, d\mathbf{X}_2, d\mathbf{X}_3) = d\mathbf{X}_1 \cdot (d\mathbf{X}_2 \times d\mathbf{X}_3)$ of the vectors defined by $d\mathbf{X}_1 = dX_1 \cdot \mathbf{e}_1$, $d\mathbf{X}_2 = dX_2 \cdot \mathbf{e}_2$ and $d\mathbf{X}_3 = dX_3 \cdot \mathbf{e}_3$ and the volume dV after transformation is equal to the product $(d\mathbf{x}_1, d\mathbf{x}_2, d\mathbf{x}_3) = (\mathbf{F} \cdot d\mathbf{x}_1, \mathbf{F} \cdot d\mathbf{x}_2, \mathbf{F} \cdot d\mathbf{x}_3)$ of their convective conveyors, where the symbol $\times$ stands for the vector product.

The expression (1.8) is derived from Eq. (1.4) for the Jacobian J of transformation and from the relation $(\mathbf{F} \cdot d\mathbf{X}_1, \mathbf{F} \cdot d\mathbf{X}_2, \mathbf{F} \cdot d\mathbf{X}_3) = \det \mathbf{F}\, (d\mathbf{X}_1, d\mathbf{X}_2, d\mathbf{X}_3)$. Since the basis $\mathbf{e}_1$, $\mathbf{e}_2$, $\mathbf{e}_3$ can be arbitrary, the conveyance expression (1.8) applies to any infinitesimal material volume dV_r. An elementary material volume never becomes zero so the Jacobian J must remain strictly positive. This also ensures that the tangent linear operator is invertible.

After deformation the material surface $d\mathbf{A}$ of the infinitesimal area d(A) oriented by unit $\mathbf{N}$ i.e. $d\mathbf{A} = \mathbf{N} \cdot dA_r$ becomes the material surface $d\mathbf{a}$ of infinitesimal area da, oriented by unit normal $\mathbf{n}$ i.e. $d\mathbf{a} = \mathbf{n} \cdot da$. Material surface $d\mathbf{a} = \mathbf{n} \cdot da$ can be expressed as

$$\mathbf{n} \cdot da = J^T \mathbf{F}^{-1} \cdot \mathbf{N} \cdot dA_r \tag{1.9}$$

The unit vector **n** is not the convective conveyor of unit vector **N** because convective conveyance transforms the norm of vectors.

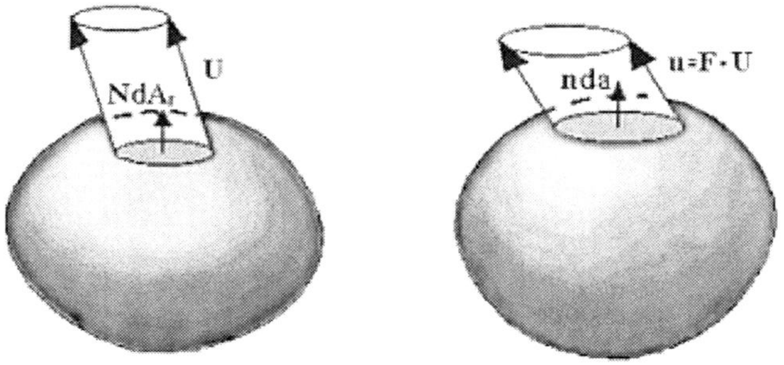

Fig. 1.2. Material surface in reference and current configurations

A geometrical interpretation of Eq. (1.9) is shown in Fig. 1.2. In current configuration the volume of the material generated by d**A** in a translation of any vector **U** is transported to the volume of the material generated by d**a** in a translation $\mathbf{u} = \mathbf{F} \cdot \mathbf{U}$.

Using expression (1.8), we have

$$\mathbf{n} \cdot \mathbf{F} \cdot \mathbf{U}\, da = J \cdot \mathbf{N} \cdot \mathbf{U}\, dA_r \tag{1.10}$$

for every **U**, which gives expression (1.9).

1.6 The Green-Lagrange Strain Tensor

Two infinitesimal material vectors $d\mathbf{X}^1$ and $d\mathbf{X}^2$ after deformation become vectors $d\mathbf{x}^1$ and $d\mathbf{x}^2$, respectively. By Eq. (1.7) we get

$$d\mathbf{x}^1 \cdot d\mathbf{x}^2 - d\mathbf{X}^1 \cdot d\mathbf{X}^2 = 2d\mathbf{X}^1 \cdot \mathbf{E} \cdot d\mathbf{X}^2 \tag{1.11}$$

where **E** is the so-called symmetric Green-Lagrange strain tensor defined as

$$2\mathbf{E} = {}^T\mathbf{F}\mathbf{F} - \mathbf{1} \tag{1.12}$$

The tensor **E** can be expressed as a function of displacement vector **u** defined as

$$\mathbf{u} = \mathbf{x} - \mathbf{X} \tag{1.13}$$

By Eqs. (1.2), (1.12) and (1.13) we get the following strain-displacement relation

$$\mathbf{E} = \mathrm{Grad}\,\mathbf{u} + {}^T\mathrm{Grad}\,\mathbf{u} + {}^T\mathrm{Grad}\,\mathbf{u} \cdot \mathrm{Grad}\,\mathbf{u} \tag{1.14}$$

The tensor **E** written by its Cartesian components $E_{\alpha\beta}$ reads

$$2E_{\alpha\beta} = \frac{\partial u_\alpha}{\partial X_\beta} + \frac{\partial u_\beta}{\partial X_\alpha} + \frac{\partial u_\gamma}{\partial X_\alpha}\frac{\partial u_\gamma}{\partial X_\beta} \tag{1.15}$$

1.7 The Logarithmic Strain Tensor

On the basis of finite strain tensors **E**, logarithmic strain tensor is defined as

$$\mathbf{H} = \frac{1}{2}\ln\left(\mathbf{1} + 2\,\mathbf{E}\right) \tag{1.16}$$

Definition (1.16) should be understood as expansions in infinite series of logarithms or their analytical extensions. In the general case of deformation, the use of definition (1.16) is complicated because of the necessity for repeated multiplication of the tensor $E_{\alpha\beta}$ by each other (with simultaneous contraction) and the use of infinite series.

1.8 Infinitesimal Transformation

In many engineering problems the hypothesis of infinitesimal transformation is assumed

$$\left\|\mathrm{Grad}\,\mathbf{u}\right\| << 1 \tag{1.17}$$

for every **X**. The displacement gradient Grad **u** belongs to finite dimensions vector space where the norm $\left\|\cdot\right\|$ needs to be specified.

The hypothesis given by (1.17) leads to the linearized strain tensor $\boldsymbol{\varepsilon}$ from the Green-Lagrange strain tensor **E**

$$2\boldsymbol{\varepsilon} = \mathrm{Grad}\,\mathbf{u} + {}^{T}\mathrm{Grad}\,\mathbf{u} \tag{1.18}$$

If Eqs. (1.13) and (1.17) apply, then both configurations i.e. reference and current configurations coincide

$$2\boldsymbol{\varepsilon} = \mathrm{grad}\,\mathbf{u} + {}^{T}\mathrm{grad}\,\mathbf{u} \tag{1.19}$$

$$2\boldsymbol{\varepsilon}_{ij} = \frac{\partial u_i}{\partial x_j} + \frac{\partial u_j}{\partial x_i} \tag{1.20}$$

It holds as far as spatial derivations are concerned. The grad operator lower-case letter appearing in Eq. (1.19) indicates that it is related to the current configuration. By expression (1.17), Eqs. (1.2), (1.4), (1.8), (1.13) and (1.19) we get

$$J = 1 + \text{div } \mathbf{u} \qquad \frac{dV - dV_r}{dV_r} = \text{div } \mathbf{u} = \text{tr}\,\boldsymbol{\varepsilon} = \varepsilon_{ii} \qquad (1.21)$$

The trace operator tr $\boldsymbol{\varepsilon}$ of the linearized stress tensor $\boldsymbol{\varepsilon}$ represents the volume change per unit volume in the deformation. It is called volume dilatation.

With the assumption (1.16), the non-diagonal term $2\varepsilon_{ij}$ ($i \neq j$) represents the decrease of the angle between the two vectors $\mathbf{e}_i$ and $\mathbf{e}_j$ of the material basis (Fig. 1.3). The diagonal term ($i = j$) represents the linear dilatation in the physical directions $\mathbf{e}_i$. The eigenvalues $\varepsilon_{(\alpha)}$ of the symmetric strain tensor $\boldsymbol{\varepsilon}$ are called the principal linear dilatations. The physical directions $\mathbf{e}_\alpha$ of the eigenvectors of $\boldsymbol{\varepsilon}$ are initially orthogonal and remain orthogonal throughout deformation.

$$\varepsilon_{ij} = \frac{1}{2}\left(\frac{\partial u_i}{\partial x_j} + \frac{\partial u_j}{\partial x_i}\right)$$

$$\gamma_{ij} \approx \frac{\partial u_i}{\partial x_j} \approx \frac{d(u_i)}{\partial x_j} = \text{tg}\,\gamma_{ij} \qquad \varepsilon_{ij} = \frac{\partial u_i}{\partial x_i} \approx \frac{d(u_i)}{\partial x_i} \qquad \omega_{ij} = \frac{1}{2}\left(\frac{\partial u_i}{\partial x_j} - \frac{\partial u_j}{\partial x_i}\right)$$

$$i \neq j \qquad\qquad i = j \qquad\qquad \omega_{ij} = -\omega_{ji}$$

general deformation   strain   rotation

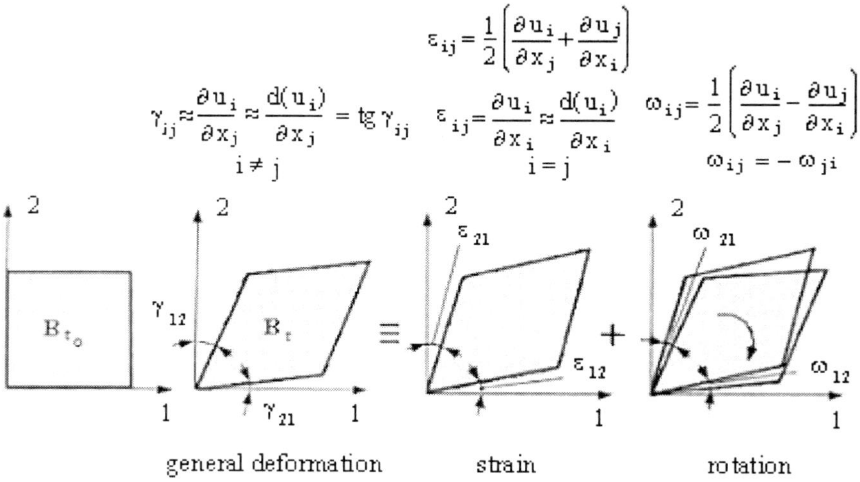

Fig. 1.3. Description of deformation on Euclidean plane

Infinitesimal transformation implies infinitesimal deformation, because $\|\mathbf{E}\|$ is of the same order of magnitude as $\|\text{Grad } \mathbf{u}\|$. But infinitesimal deformation does not imply the infinitesimal transformation. As an example, consider the rigid body motion, then $\mathbf{E}$ is equal to zero, and $\|\text{Grad } \mathbf{u}\|$ can be of any order of magnitude.

1.9 Lagrangian and Eulerian Strain Rate Tensors

In Lagrangian description, all physical quantities are functions of the position vector $\mathbf{X}$ and time t and are referred to the material in reference configuration. The partial derivation with respect to time gives the equations of the kinematics of deformation. This derivative is a total derivative, because the vector **X** is constant in transformation. In Eulerian description the derivative is carried out in the current configuration by using the velocity field **V**(**x**, t) of the material particle located at a point of the position vector **x**.

The Lagrangian strain rate tensor d**E**/dt involves a simple time derivation. Deriving Eq. (1.11) with respect to time yields

$$\frac{d}{dt}\left[d\,\mathbf{x}^1 \cdot d\,\mathbf{x}^2\right] = 2d\,\mathbf{X}^1 \cdot \frac{d\,\mathbf{E}}{dt} \cdot d\,\mathbf{X}^2 \tag{1.22}$$

In Eulerian formulation the kinematics of deformation is independently of reference configuration. The velocity field **V**(**x**, t) of the material particle is expressed as

$$\frac{d\,\mathbf{x}}{dt} = \mathbf{V}(\mathbf{x},t) \tag{1.23}$$

where **x** = **x** (**X**, t) is the position vector at time t in the current configuration. From Eq. (1.2) using the relation Grad (·) = grad (·) · **F** we get

$$\text{grad}\,\mathbf{V} = \frac{d\,\mathbf{F}}{dt} \cdot \mathbf{F}^{-1} \tag{1.24}$$

From Eq. (1.7) and Eq. (1.24) we have

$$\frac{d}{dt}\left[d\,\mathbf{x}\right] = \text{grad}\,\mathbf{V} \cdot d\,\mathbf{x} \tag{1.25}$$

Two points with position vectors **x** and **x** + d**x** posses the differences in velocities which can be obtained by the first – order differentiation

$$d\left[\frac{d\,\mathbf{x}}{dt}\right] = \mathbf{V}(\mathbf{x} + d\,\mathbf{x}, t) - \mathbf{V}(\mathbf{x},t) = \text{grad}\,\mathbf{V} \cdot d\,\mathbf{x} \tag{1.26}$$

By Eq. (1.25)

$$\frac{d}{dt}\left[d\,\mathbf{x}^1\, d\,\mathbf{x}^2\right] = 2d\,\mathbf{x}^1 \cdot \mathbf{d} \cdot d\,\mathbf{x}^2 \tag{1.27}$$

where **d** is the Eulerian strain rate tensor we arrive at

$$2\mathbf{d} = \text{grad}\,\mathbf{V} + {}^{T}\text{grad}\,\mathbf{V} \qquad 2d_{ij} = \frac{\partial V_i}{\partial x_j} + \frac{\partial V_j}{\partial x_i} \tag{1.28}$$

By Eqs. (1.7), (1.11), (1.12), (1.24) and (1.27) we get the relation for the tensor **d**

$$\mathbf{d} = {}^{\mathrm{T}}\mathbf{F}^{-1} \cdot \frac{\mathrm{d}\,\mathbf{E}}{\mathrm{dt}} \cdot \mathbf{F}^{-1} \tag{1.29}$$

The tensor **d** is independent of the reference configuration to which **E** refers. If the reference configuration is the current configuration then

$$\mathbf{x} = \mathbf{X} \qquad \mathbf{F} = \mathbf{1} \qquad \mathrm{J} = 1 \qquad \frac{\mathrm{dJ}}{\mathrm{dt}} = \operatorname{div} \mathbf{V} \qquad \frac{\mathrm{d}\,\mathbf{F}}{\mathrm{dt}} = \operatorname{grad} \mathbf{V} \tag{1.30}$$

From Eqs. (1.8) and (1.28) we have

$$\frac{\mathrm{d}}{\mathrm{dt}}[\mathrm{dV}]\Big/\mathrm{dV} = \operatorname{div} \mathbf{V} = \operatorname{tr} \mathbf{d} = \mathrm{d}_{\mathrm{ii}} \tag{1.33}$$

As we can see, the trace operator of tensor **d** represents the rate of volume dilatation. Assuming the hypothesis of infinitesimal transformation, from Eqs. (1.19) and (1.28) we get

$$\mathbf{d} = \frac{\mathrm{d}\varepsilon}{\mathrm{dt}} \tag{1.32}$$

1.10 Particulate and Material Derivatives

Consider a field G (**X**, t) in a Lagrange description. The particulate derivative of G (**X**, t) with respect to the material is equal to the variation between times t and t + dt of the function G (**X**, t). In terms of Euler variables the particulate derivative with respect to the material is the total time derivative of the field

$$\mathrm{g}\,[\mathbf{x}\,(\mathbf{X}, \mathrm{t})] = \mathrm{G}\,(\mathbf{X}, \mathrm{t}) \tag{1.33}$$

$$\frac{\mathrm{dg}}{\mathrm{dt}} = \frac{\partial \mathrm{g}}{\partial \mathrm{t}} + \operatorname{grad} \mathrm{g} \cdot \mathbf{V} \tag{1.34}$$

If g is the velocity vector **V**, then we get the expression for acceleration ***a*** in Euler variables

$$\boldsymbol{a} = \frac{\mathrm{d}\,\mathbf{V}}{\mathrm{dt}} = \frac{\partial\,\mathbf{V}}{\partial \mathrm{t}} + \operatorname{grad} \mathbf{V} \cdot \mathbf{V} \tag{1.35}$$

We construct the volume integral

$$\mathcal{G} = \int_{\mathrm{V}} \mathrm{g}(\mathbf{x}, \mathrm{t})\,\mathrm{dV} \tag{1.36}$$

where g (**x**, t) is the volume density in the current configuration.

The particulate derivative of the integral $\mathcal{G}$ with respect to the material denoted by $d\mathcal{G}/dt$ represents the time derivative in the material movement.
If $G(\mathbf{X}, t)$ is the volume Lagrangian density, then the corresponding Eulerian density is $g(\mathbf{x}, t)$

$$g(\mathbf{x},t)\,dV = G(\mathbf{X},t)\,dV_r \tag{1.37}$$

where the volume V_r refers to the reference configuration of the volume V in the current configuration.

The time derivation of Eq. (1.36) throughout the relation (1.37) has the form

$$\frac{d\mathcal{G}}{dt} = \int_{V_r} \frac{dG}{dt}\,dV_r \tag{1.38}$$

The expression (1.38) represents the particulate derivative of the volume integral $\mathcal{G}$ in terms of Lagrange variables.

In order to obtain the particulate time derivative in Eulerian variables we use Eq. (1.37).

By Eqs. (1.31) and (1.34) and the expression

$$\operatorname{div}(g \otimes \mathbf{V}) = g \operatorname{div} \mathbf{V} + \operatorname{grad} g \cdot \mathbf{V} \tag{1.39}$$

we get

$$\frac{dG}{dt}\,dV_r = \left[\frac{\partial g}{\partial t} + \operatorname{div}(g \otimes \mathbf{V})\right] dV \tag{1.40}$$

The particulate derivative of the volume integral $\mathcal{G}$ with respect to the material derived in Eulerian variables yields

$$\frac{d\mathcal{G}}{dt} = \int_{V} \left[\frac{\partial g}{\partial t} + \operatorname{div}(g \otimes \mathbf{V})\right] dV \tag{1.41}$$

The alternative form of Eq. (1.41) can be obtained by using the divergence theorem, then

$$\frac{d\mathcal{G}}{dt} = \int_{V} \frac{\partial g}{\partial t}\,dV_r + \int_{a} g\,\mathbf{V} \cdot \mathbf{n}\,da \tag{1.42}$$

where a is the surface of the volume V.

The first term of the right-hand side of Eq. (1.42) represents the variation of the field g between time t and t + dt in volume V, and the other one due to the movement of the same material volume.

The material derivative is used to determine the variation between time t and t + dt of any physical quantity attached to the whole material, which is contained at time t in volume V. The particulate derivative with respect to the material only partially takes into account this variation. It ignores any mass particles leaving the volume V, which is followed in the material movement.

In the infinitesimal time interval dt = Dt the variation of quantity g attached to the whole matter at time t in the volume V as the variation $D\mathcal{G}$ of the integral $\mathcal{G}$ given by Eq. (1.38) involves

$$\frac{D\mathcal{G}}{Dt} = \frac{d}{dt}\int_V g(\mathbf{x},t)dV \tag{1.43}$$

Substituting Eq. (1.42) of the particulate derivatives with respect to the material of the volume integral we obtain

$$\frac{D\mathcal{G}}{Dt} = \int_V \frac{\partial g}{\partial t}dV + \int_a g\,\mathbf{V}\cdot\mathbf{n}\,da \tag{1.44}$$

In the case of metal deformation the material derivative is equal to the particulate derivative.

$$\frac{D\mathcal{G}}{Dt} = \frac{d\mathcal{G}}{dt} \tag{1.45}$$

In the case of some classes of void-contained materials, this relation does not hold.

1.11 Mass Conservation

The overall mass density of the material considered as a whole will be denoted as ρ. The mass contained in the infinitesimal volume dV is equal to ρdV. We assume no overall mass creation, which implies the global mass balance

$$\frac{D}{Dt}\int_V \rho dV = 0 \tag{1.46}$$

Using Eqs. (1.44) and (1.46) the overall mass balance reads

$$\int_V \frac{\partial \rho}{dt}dV + \int_a \rho\,\mathbf{V}\cdot\mathbf{n}\,da = 0 \tag{1.47}$$

If we use the divergence theorem to Eq. (1.47) we get the local mass balance equation or continuity equation

$$\frac{\partial \rho}{\partial t} + \operatorname{div}(\rho\,\mathbf{V}) = 0 \tag{1.48}$$

or equivalently

$$\frac{\partial \rho}{\partial t} + \rho \operatorname{div}\mathbf{V} = 0 \tag{1.49}$$

The overall mass ρdV, which is contained in volume dV, may be written as

$$\rho \, dV = \rho_o \, dV_r \tag{1.50}$$

The overall mass conservation may now be written as

$$\frac{D}{Dt} \int_{V_r} \rho_o dV_r = 0 \tag{1.51}$$

From the transport formula we have

$$J\rho = \rho_o \tag{1.52}$$

which expresses the material mass conservation in Lagrange variables.

2 Stress Tensor

2.1 Momentum Balance

In the course of deformation of metals, due to volume and geometrical shape changes, interactions between molecules come into being and these oppose the changes. In any material domain V two kinds of external forces (Fig. 2.1.) are considered i.e. body and surface forces.

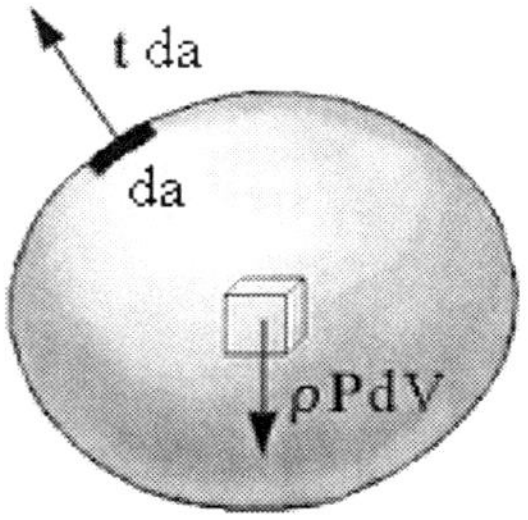

Fig. 2.1. External forces in material domain V

Body forces are defined throughout the density per mass unit **P** acting on the infinitesimal volume dV and are given as $\rho\mathbf{P}dV$. The density $\mathbf{P} = \mathbf{P}(\mathbf{x}, t)$ is dependent on the position vector associated with the volume dV considered. Surface forces acting on the surface a are represented by the density per surface of unit area $\mathbf{t}(\mathbf{x}, t, \mathbf{n})$ where **n** is the outward unit normal to the surface a at point **x**.

The elementary force **t**da acting on the infinitesimal material surface da depends on **x** and **n** with regard to space. This force is assumed to be local contact force, which will be essential in defining the stress tensor. In the present considerations the coupling of body and surface forces will not be analyzed.

The instantaneous momentum balance has the form

$$\frac{D}{Dt}\int_V \rho\,\mathbf{V}\,dV = \int_V \rho\,\mathbf{P}\,dV + \int_a \mathbf{t}\,da \tag{2.1}$$

$$\frac{D}{Dt}\int_V [\mathbf{x}\times\rho\,\mathbf{V}]dV = \int_V \mathbf{x}\times\rho\,\mathbf{P}\,dV + \int_a \mathbf{x}\times\mathbf{t}\,da \tag{2.2}$$

where $\rho\mathbf{V}$ dV is the linear momentum of the material in elementary material domain dV.

The expression (2.1) shows that the creation rate due to the external forces acting on material is the instantaneous time derivatives of the linear momentum of the material in the volume V.

The expression (2.2) states the same, but it concerns the angular momentum. In Eulerian expression of the material derivative of the volume integral, the left-hand side of Eq. (2.1) can be expressed as

$$\frac{D}{Dt}\int_V \rho\,\mathbf{V}\,dV = \int_V \frac{\partial}{\partial t}(\rho\,\mathbf{V})dV + \int_a [\rho\,\mathbf{V}\,\mathbf{V}\cdot\mathbf{n}]da \tag{2.3}$$

In Eq. (2.3) the volume integral corresponds to the time variation of linear momentum in elementary volume V, and the surface integral corresponds to the momentum carried away by the material leaving the same geometrical volume. Similarly, if $g = \mathbf{x}\times(\rho\mathbf{V})$ the right-hand side of (2.2) can be rewritten as

$$\frac{D}{Dt}\int_V [\mathbf{x}\times(\rho\,\mathbf{V})]dV = \int_V \left[\mathbf{x}\times\frac{\partial}{\partial t}(\rho\,\mathbf{V})\right]dV + \int_a [(\mathbf{x}\times\rho\,\mathbf{V})\mathbf{V}\cdot\mathbf{n}]da \tag{2.4}$$

The expressions (2.1) – (2.4) describe the Euler theorem, which says that the resultant and the overall moment of vectors

$$\frac{\partial}{\partial t}(\rho\,\mathbf{V})dV \tag{2.5}$$

distributed in the volume V, and the resultant and the overall moment of vectors

$$[(\rho\,\mathbf{V})\mathbf{V}\cdot\mathbf{n}]da \tag{2.6}$$

distributed on the surface a of the domain V is equal to the resultant of the elementary body and surface forces and the resultant of the corresponding elementary moments, respectively.

By the divergence theorem, Eqs. (2.3) and (2.4) are rewritten in the form

$$\frac{D}{Dt}\int_V \rho\,\mathbf{V}\,dV = \int_V \frac{\partial}{\partial t}(\rho\,\mathbf{V})dV + \int_V \mathrm{div}[\rho\,\mathbf{V}\otimes\mathbf{V}]dV \tag{2.7}$$

$$\frac{D}{Dt}\int_V \mathbf{x}\times\rho\,\mathbf{V}\,dV = \int_V \mathbf{x}\times\frac{\partial}{\partial t}(\rho\,\mathbf{V})dV + \int_V \mathbf{x}\times\mathrm{div}[\rho\,\mathbf{V}\otimes\mathbf{V}]dV \tag{2.8}$$

By the relation

$$\mathrm{div}(\mathbf{V}'\otimes\mathbf{V}) = \mathrm{grad}\,\mathbf{V}'\cdot\mathbf{V} + \mathbf{V}'\,\mathrm{div}\,\mathbf{V} \tag{2.9}$$

we get

$$\frac{\partial}{\partial t}(\rho\,\mathbf{V})+\operatorname{div}\left[\rho\,\mathbf{V}\otimes\mathbf{V}\right]=\rho\left[\frac{\partial\,\mathbf{V}}{\partial t}+\operatorname{grad}\mathbf{V}\cdot\mathbf{V}\right]+\mathbf{V}\left[\frac{\partial\rho}{\partial t}+\operatorname{div}(\rho\,\mathbf{V})\right] \tag{2.10}$$

By Eq. (1.34) we get

$$\frac{\partial}{\partial t}(\rho\,\mathbf{V})+\operatorname{div}\left[\rho\,\mathbf{V}\otimes\mathbf{V}\right]=\rho\boldsymbol{a} \tag{2.11}$$

The expressions (2.7), (2.8) and (2.11) give a new form of the momentum balance namely the dynamic theorem

$$\begin{aligned}&\int_V \rho\boldsymbol{a}\,dV=\int_V \rho\,\mathbf{P}\,dV+\int_a \mathbf{t}\,da\\&\int_V \mathbf{x}\times\rho\boldsymbol{a}\,dV=\int_V \mathbf{x}\times\rho\,\mathbf{P}\,dV+\int_a \mathbf{x}\times\mathbf{t}\,da\end{aligned} \tag{2.12}$$

which says that the resultant of the elementary body and surface forces and the resultant of the corresponding elementary moments are equal respectively to the resultant and the overall movement of the elementary dynamic forces.

2.2 The Cauchy Stress Tensor

We transform Eq. (2.12) to the form

$$\int_V\left[\rho\boldsymbol{a}-\rho\,\mathbf{P}\right]dV=\int_a \mathbf{t}\,da \tag{2.13}$$

We introduce the hypothesis of contact surface forces by the relation

$$\int_V \mathbf{f}\,(\mathbf{x},t)\,dV=-\int_a \mathbf{h}\,(\mathbf{x},t,\mathbf{n})\,da \tag{2.14}$$

By the tetrahedron Lemma which states that if a relation (2.14) holds for any volume V, then **h** may be written as a linear function of the components n_i of normal **n** as

$$\mathbf{h}\,(\mathbf{x},t,\mathbf{n})=h_i\,(\mathbf{x},t)\,n_i \tag{2.15}$$

From the above two expressions the existence of a stress tensor field $\boldsymbol{\sigma}$ so-called the Cauchy stress tensor is derived so that at every point

$$\boldsymbol{\sigma}\cdot\mathbf{n}=\mathbf{t} \tag{2.16}$$

The illustration of the physical meaning of the Cauchy stress tensor $\boldsymbol{\sigma}$ in the Euclidean space is given in Fig. 2.2.

Substituting Eq. (2.14) into (2.10) and applying the divergence theorem we get

$$\int_V [\operatorname{div}\boldsymbol{\sigma} + \rho(\mathbf{P} - \boldsymbol{a})]\,dV = 0 \tag{2.17}$$

The equation of motion is derived from (2.17). Since (2.17) holds for any V, we get

$$\operatorname{div}\boldsymbol{\sigma} + \rho\,(\mathbf{P} - \boldsymbol{a}) = 0 \qquad \frac{\partial \sigma_{ij}}{\partial x_j} + \rho(P_i - a_i) = 0 \tag{2.18}$$

which expresses the equation of motion.

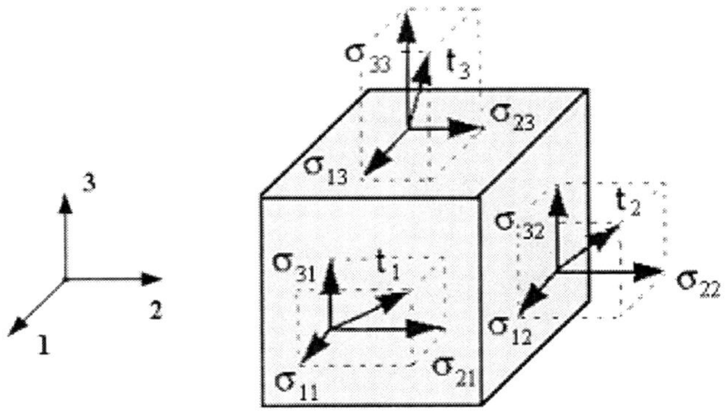

Fig. 2.2. Components of the stress vectors in Euclidean space

Substituting Eq. (2.14) into Eq. (2.11) gives

$$\int_V \{\mathbf{x} \times [\operatorname{div}\boldsymbol{\sigma} + \rho(\mathbf{P} - \boldsymbol{a}) + 2\,\mathbf{f}]\}\,dV = 0 \tag{2.19}$$

where **f** is the vector defined as

$$2\mathbf{f} = (\sigma_{23} - \sigma_{32})\,\mathbf{e}_1 + (\sigma_{31} - \sigma_{13})\,\mathbf{e}_2 + (\sigma_{12} - \sigma_{21})\,\mathbf{e}_3 \tag{2.20}$$

From the equation of motion (2.18) and Eq. (2.19) it follows that

$$\int_V \mathbf{f}\,dV = 0 \tag{2.21}$$

Since Eq. (2.21) holds for any V, it follows that

$$\mathbf{f} = 0 \tag{2.22}$$

By (2.20) the symmetry of the stress tensor is derived

$$\sigma_{ij} = \sigma_{ji} \qquad {}^{T}\boldsymbol{\sigma} = \boldsymbol{\sigma} \tag{2.23}$$

The symmetry of the stress tensor $\boldsymbol{\sigma}$ ensures the realness of its eigenvalues $\sigma_{(\alpha)}$ which are called the principal stresses, and the associated eigendirections are the principal stress directions.

The symmetry of the stress tensor is the dynamic moment theorem applied to the whole body, which forms at time t the material elementary parallelepiped dV.

2.3 The Virtual Work Rate

The strain work rate $R_{SR}(\mathbf{V}')$ associated with any velocity field $\mathbf{V}'$ and relative to material domain V is defined as

$$R_{SR}(\mathbf{V}') = \int_V \mathbf{V}' \cdot \rho \mathbf{P}\, dV + \int_a \mathbf{V}' \cdot \mathbf{t}\, da - \int_V \mathbf{V}' \cdot \rho \boldsymbol{a} dV \qquad (2.24)$$

The last term of Eq. (2.24) represents the work rate of inertia forces and the first two integrals of the right-hand side of Eq. (2.24) represent the work rate of the external body and surface forces.

The symmetry of the stress tensor $\boldsymbol{\sigma}$ and the divergence theorem applied yield the identity

$$\int_a \mathbf{V}' \cdot \boldsymbol{\sigma} \cdot \mathbf{n}\, da = \int_V \left(\boldsymbol{\sigma}\, \mathbf{d}^{v'} + \mathbf{V}' \cdot \operatorname{div} \boldsymbol{\sigma}\right) dV \qquad (2.25)$$

where $\boldsymbol{\sigma}\ \mathbf{d}^{v'} = \sigma_{ij}\, d_{ij}^{v'}$ and $\mathbf{d}^{v'}$ is the strain rate associated with velocity field $\mathbf{V}'$

$$2\mathbf{d}^{v'} = \text{grad } \mathbf{V}' + {}^{T}\text{grad } \mathbf{V}' \qquad (2.26)$$

By Eq. (2.14), the motion equation (2.18) and Eq. (2.25) the strain work rate $R_{SR}(\mathbf{V}')$ defined by (2.24) can be written as

$$R_{SR}(\mathbf{V}') = \int_V dr_{SR}(\mathbf{V}') \qquad dr_{SR}(\mathbf{V}') = \boldsymbol{\sigma}\ \mathbf{d}^{v'}\ dV \qquad (2.27)$$

where $dr_{SR}(\mathbf{V}')$ is the infinitesimal strain work rate of the elementary material domain dV.

If velocity field $\mathbf{V}'$ is equal to velocity $\mathbf{V}$ i.e. it is an actual velocity of the material particle, then

$$dr_{SR}(\mathbf{V}) = \boldsymbol{\sigma}\ \mathbf{d}\ dV \qquad (2.28)$$

The work rate of the internal forces denoted as R_{IF} is the opposite of the strain work rate R_{SR} i.e.

$$R_{IF}(\mathbf{V}') = -R_{SR}(\mathbf{V}') \qquad dr_{IF}(\mathbf{V}') = -dr_{SR} \qquad (2.29)$$

The strain work rate of external forces is defined as

$$R_{EF}(\mathbf{V}') = \int_V \mathbf{V}' \cdot \rho \mathbf{P}\, dV + \int_a \mathbf{V}' \cdot \mathbf{t}\, da \qquad (2.30)$$

and the work rate of inertia forces is

$$R_{IN}(\mathbf{V}') = -\int_V \mathbf{V}' \cdot \rho \boldsymbol{a} dV \tag{2.31}$$

The virtual work rate theorem is simply rewriting Eq. (2.24) in terms of (2.30) and (2.31)

$$R_{EF}(\mathbf{V}') + R_{IN}(\mathbf{V}') + R_{IF}(\mathbf{V}') = 0 \tag{2.32}$$

for every volume V and velocity $\mathbf{V}'$.

It states that for any actual or virtual velocity field $\mathbf{V}'$ and for any material domain V, the sum of the external forces $R_{EF}(\mathbf{V}')$, inertia forces $R_{IN}(\mathbf{V}')$ and internal forces $R_{IF}(\mathbf{V}')$ is equal to zero.

2.4 The Piola-Kirchhoff Stress Tensor

The strain work rate dr_{SR} defined by (2.28) is independent of the choice of the coordinate system used to describe the motion. The definition (2.28) corresponds to a Eulerian description.

By Eqs. (1.8) and (1.31) the expression (2.28) can be written as

$$dr_{SR}(\mathbf{V}) = \boldsymbol{\sigma}\, \mathbf{d}\, dV = J\left(\mathbf{F}^{-1} \cdot \boldsymbol{\sigma} \cdot {}^{T}\mathbf{F}^{-1}\right)\frac{d\mathbf{E}}{dt} dV_r \tag{2.33}$$

where $d\mathbf{E}/dt$ is the convective transporter of $\mathbf{d}$ in the reference configuration

$$\mathbf{d} = {}^{T}\mathbf{F}^{-1} \cdot \frac{\Delta \mathbf{E}}{dt} \cdot \mathbf{F}^{-1} \tag{2.34}$$

The expression (2.33) serves to introduce the symmetric Piola-Kirchhoff stress tensor $\mathbf{S}$ defined by

$$\mathbf{S} = J\,(\mathbf{F}^{-1} \cdot \boldsymbol{\sigma} \cdot {}^{T}\mathbf{F}^{-1}) \tag{2.35}$$

The tensors $\mathbf{d}$ and $\frac{d\mathbf{E}}{dt}$ represent the same material tensor in different configurations.

Using Eqs. (2.33) and (2.35), the Lagrangian description of the elementary and overall strain work rates $dr_{SR}(\mathbf{V})$ and $R_{SR}(\mathbf{V})$ are given, respectively, by

$$dr_{SR}(\mathbf{V}) = \mathbf{S}\,\frac{d\mathbf{E}}{dt} dV_r \qquad R_{SR}(\mathbf{V}) = \int_{V_r} dr_{SR}(\mathbf{V}) dV_r \tag{2.36}$$

The definition of Piola-Kirchhoff stress tensor $\mathbf{S}$, Eq. (2.35) and transport formula (1.9) give

$$\mathbf{F} \cdot \mathbf{S} \cdot \mathbf{N}\, dA_r = \boldsymbol{\sigma} \cdot \mathbf{n}\, da \tag{2.37}$$

The dynamic theorem Eq. (2.12) throughout Eqs. (2.16) and (2.37) expresses the equality between the dynamical resultant of external forces for the material domain V in a Lagrangian description as

$$\int_{A_r} \mathbf{F} \cdot \mathbf{S} \cdot \mathbf{N} \, dA_r + \int_{V_r} \rho_o (\mathbf{P} - \boldsymbol{a}) dV_r = 0 \tag{2.38}$$

where the domain V_r and the surface A_r enclosing this domain in reference configuration correspond to the domain V and the surface a enclosing it in the current configuration.

The equation of motion in a Lagrangian description is obtained by transformation of the surface integral into volume integral throughout the divergence theorem.

$$\mathrm{Div}\,(\mathbf{F} \cdot \mathbf{S}) + \rho_o (\mathbf{P} - \boldsymbol{a}) = 0 \tag{2.39}$$

$$\frac{\partial}{\partial X_\alpha}\left(\frac{\partial x_i}{\partial X_\beta} S_{\beta\alpha}\right) + \rho_o (P_i - a_i) = 0 \tag{2.40}$$

The body force **P** and the accelerations $\boldsymbol{a}$ are evaluated in the above equations at point **x** in the current configuration.

2.5 The Kinetic Energy Theorem

The kinetic energy K of the whole matter of the volume V is represented by the expression

$$K = \int_V \frac{1}{2} \rho \mathbf{V}^2 dV \tag{2.41}$$

The kinetic energy expression can be written in reference configuration as

$$K = \int_{V_r} \frac{1}{2} \rho_o \mathbf{V}^2 dV_r \tag{2.42}$$

The definition of the material derivative of the integral of an extensive quantity yields

$$\frac{DK}{Dt} = \frac{dK}{dt} \tag{2.43}$$

The conveyance formula (1.8), the continuity relation (1.50) and the definition of material acceleration $\boldsymbol{a}$ given by Eq. (1.35) lead to the following form of Eq. (2.41)

$$\frac{DK}{Dt} = \int \rho \mathbf{V} \cdot \boldsymbol{a} \, dV \tag{2.44}$$

The material derivative of the kinetic energy corresponds to the cpposite of the work rate of the inertia forces, which they develop in the actual movement. The inertia forces relative to the material develop their work rate in their own movement.

Let the velocity field **V'** be the actual material velocity field **V** in the virtual work rate theorem (2.32). Together with Eq. (2.44) it gives the kinetic energy theorem

$$\frac{DK}{Dt} + R_{SR}(\mathbf{V}) = R_{EF}(\mathbf{V}) \tag{2.45}$$

It says that in the actual movement and for any domain V, the work rate of the external forces is equal to the sum of the material derivative of the kinetic energy and of the strain work rate associated with the material strain rate.

3 Components of Stress and Strain Tensors

3.1 Components of the Green-Lagrange Strain Tensor

The components of the tensor **E** have the form

$$E_{11} = \frac{\partial u_1}{\partial X_1} + \frac{1}{2}\left[\left(\frac{\partial u_1}{\partial X_1}\right)^2 + \left(\frac{\partial u_2}{\partial X_1}\right)^2 + \left(\frac{\partial u_3}{\partial X_1}\right)^2\right] \tag{3.1}$$

$$E_{12} = \frac{1}{2}\left[\frac{\partial u_1}{\partial X_2} + \frac{\partial u_2}{\partial X_1} + \frac{\partial u_1}{\partial X_1}\frac{\partial u_1}{\partial X_2} + \frac{\partial u_2}{\partial X_1}\frac{\partial u_2}{\partial X_2} + \frac{\partial u_3}{\partial X_1}\frac{\partial u_3}{\partial X_2}\right] \tag{3.2}$$

$$E_{22} = \frac{\partial u_2}{\partial X_2} + \frac{1}{2}\left[\left(\frac{\partial u_1}{\partial X_2}\right)^2 + \left(\frac{\partial u_2}{\partial X_2}\right)^2 + \left(\frac{\partial u_3}{\partial X_2}\right)^2\right] \tag{3.3}$$

$$E_{23} = \frac{1}{2}\left[\frac{\partial u_2}{\partial X_3} + \frac{\partial u_3}{\partial X_2} + \frac{\partial u_1}{\partial X_2}\frac{\partial u_1}{\partial X_3} + \frac{\partial u_2}{\partial X_2}\frac{\partial u_2}{\partial X_3} + \frac{\partial u_3}{\partial X_2}\frac{\partial u_3}{\partial X_3}\right] \tag{3.4}$$

$$E_{33} = \frac{\partial u_3}{\partial X_3} + \frac{1}{2}\left[\left(\frac{\partial u_1}{\partial X_3}\right)^2 + \left(\frac{\partial u_2}{\partial X_3}\right)^2 + \left(\frac{\partial u_3}{\partial X_3}\right)^2\right] \tag{3.5}$$

$$E_{31} = \frac{1}{2}\left[\frac{\partial u_3}{\partial X_1} + \frac{\partial u_1}{\partial X_3} + \frac{\partial u_1}{\partial X_3}\frac{\partial u_1}{\partial X_1} + \frac{\partial u_2}{\partial X_3}\frac{\partial u_2}{\partial X_1} + \frac{\partial u_3}{\partial X_3}\frac{\partial u_3}{\partial X_1}\right] \tag{3.6}$$

The components of the strain tensor **E** in the cylindrical coordinate system $r = r(R, \vartheta, Z)$, $\vartheta = \vartheta(R, \vartheta, Z)$, $z = z(R, \vartheta, Z)$ are of the form

$$E_{RR} = \frac{1}{2}\left[\left(\frac{\partial r}{\partial R}\right)^2 + r^2\left(\frac{\partial \vartheta}{\partial R}\right)^2 + \left(\frac{\partial z}{\partial R}\right)^2 - 1\right] \tag{3.7}$$

$$E_{\vartheta\vartheta} = \frac{1}{2}\left[\left(\frac{\partial r}{\partial \vartheta}\right)^2 + r^2\left(\frac{\partial \vartheta}{\partial \vartheta}\right)^2 + \left(\frac{\partial z}{\partial \vartheta}\right)^2 - R^2\right] \tag{3.8}$$

$$E_{ZZ} = \frac{1}{2}\left[\left(\frac{\partial r}{\partial Z}\right)^2 + r^2\left(\frac{\partial \vartheta}{\partial Z}\right)^2 + \left(\frac{\partial z}{\partial Z}\right)^2 - 1\right] \tag{3.9}$$

$$E_{R\vartheta} = E_{\vartheta R} = \frac{1}{2}\left[\frac{\partial r}{\partial R}\frac{\partial r}{\partial \vartheta} + r^2\frac{\partial \vartheta}{\partial R}\frac{\partial \vartheta}{\partial \vartheta} + \frac{\partial z}{\partial R}\frac{\partial z}{\partial \vartheta}\right] \tag{3.10}$$

$$E_{\vartheta Z} = E_{Z\vartheta} = \frac{1}{2}\left[\frac{\partial r}{\partial \vartheta}\frac{\partial r}{\partial Z} + r^2\frac{\partial \vartheta}{\partial \vartheta}\frac{\partial \vartheta}{\partial Z} + \frac{\partial z}{\partial \vartheta}\frac{\partial z}{\partial Z}\right] \tag{3.11}$$

$$E_{ZR} = E_{RZ} = \frac{1}{2}\left[\frac{\partial r}{\partial Z}\frac{\partial r}{\partial R} + r^2\frac{\partial \vartheta}{\partial Z}\frac{\partial \vartheta}{\partial R} + \frac{\partial z}{\partial Z}\frac{\partial z}{\partial R}\right] \tag{3.12}$$

In the particular case of plane deformation where

$$r = R + u(R, \Theta) \qquad \vartheta = \Theta + \beta(R,\Theta) \qquad z = Z \tag{3.13}$$

non-zero components of the tensor **E** are

$$E_{RR} = \frac{\partial u}{\partial R} + \frac{1}{2}\left(\frac{\partial u}{\partial R}\right)^2 + \frac{1}{2}(R+u)^2\left(\frac{\partial \beta}{\partial R}\right)^2 \tag{3.14}$$

$$\begin{aligned} E_{\Theta\Theta} = {} & Ru + R^2\frac{\partial \beta}{\partial \Theta} + \frac{1}{2}u^2 + 2Ru\frac{\partial \beta}{\partial \Theta} + u^2\frac{\partial \beta}{\partial \Theta} + \frac{1}{2}\left(\frac{\partial u}{\partial \Theta}\right)^2 \\ & + \frac{1}{2}(R+u)^2\left(\frac{\partial \beta}{\partial \Theta}\right)^2 \end{aligned} \tag{3.15}$$

$$\begin{aligned} E_{R\Theta} = E_{\Theta R} = {} & \frac{1}{2}\left[\frac{\partial u}{\partial \Theta} + R^2\frac{\partial \beta}{\partial R} + \frac{\partial u}{\partial R}\frac{\partial u}{\partial \Theta} + 2Ru\frac{\partial \beta}{\partial R}\right. \\ & \left. + u^2\frac{\partial \beta}{\partial R} + (R+u)^2\frac{\partial \beta}{\partial R}\frac{\partial \beta}{\partial \Theta}\right] \end{aligned} \tag{3.16}$$

In another important case of axisymmetric deformation with respect to the Z axis, we get the following relations for the deformation describing the combined torsion and extension of a cylindrical tube

$$r = R + u\,(R) \qquad \vartheta = \Theta + \lambda\psi Z \qquad z = \lambda Z \tag{3.17}$$

and

$$E_{\alpha\beta} = \begin{Bmatrix} \frac{\partial u}{\partial R} + \frac{1}{2}\left(\frac{\partial u}{\partial R}\right)^2 & 0 & 0 \\ 0 & Ru + \frac{1}{2}u^2 & \frac{1}{2}(R+u)^2\lambda\psi \\ 0 & \frac{1}{2}(R+u)^2\lambda\psi & \frac{1}{2}\left[\lambda^2 - 1 + (R+u)^2\lambda^2\psi^2\right] \end{Bmatrix} \tag{3.18}$$

In the above relations ψ is the angle of torsion on unit length in the deformed state and λ is a nondimensional extension coefficient.

3.2 Components of Stress and Strain in Infinitesimal Transformation

3.2.1 Stress and Strain Tensors in Cartesian Coordinates

Consider infinitesimal transformation and rectangular parallelepiped element in Cartesian coordinates x_1, x_2, x_3. The components of body force per unit volume acting at the center of the rectangular element shown in Fig. 3.1 are denoted by P_1, P_2, P_3.

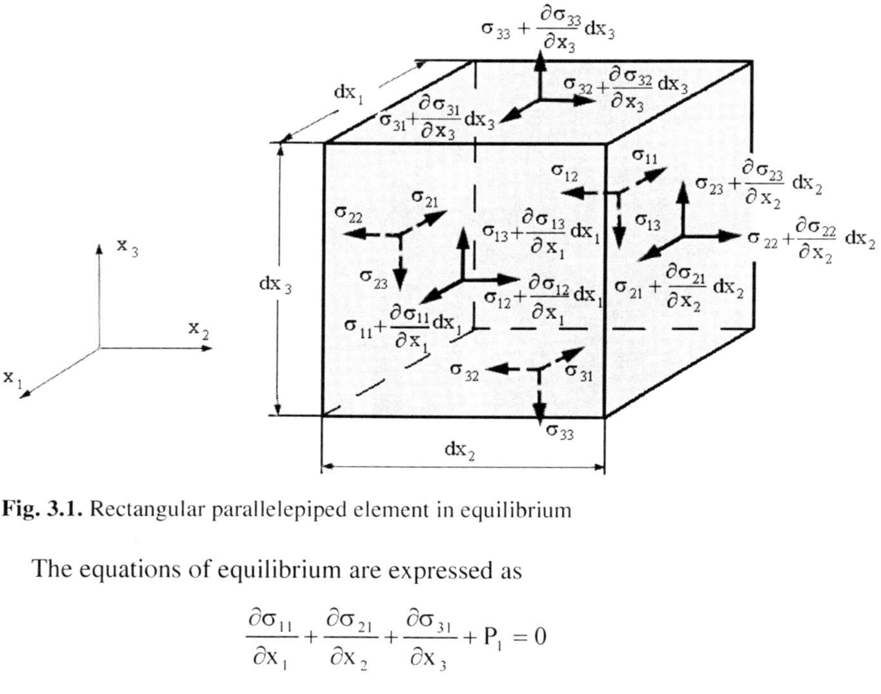

Fig. 3.1. Rectangular parallelepiped element in equilibrium

The equations of equilibrium are expressed as

$$
\begin{aligned}
&\frac{\partial\sigma_{11}}{\partial x_1}+\frac{\partial\sigma_{21}}{\partial x_2}+\frac{\partial\sigma_{31}}{\partial x_3}+P_1=0\\
&\frac{\partial\sigma_{12}}{\partial x_1}+\frac{\partial\sigma_{22}}{\partial x_2}+\frac{\partial\sigma_{32}}{\partial x_3}+P_2=0\\
&\frac{\partial\sigma_{13}}{\partial x_1}+\frac{\partial\sigma_{23}}{\partial x_2}+\frac{\partial\sigma_{33}}{\partial x_3}+P_3=0
\end{aligned}
\tag{3.19}
$$

If we define new shearing strains ε_{12}, ε_{23}, ε_{31} as half of the shearing strains γ_{12}, γ_{23} and γ_{31}

$$\varepsilon_{12}=\frac{\gamma_{12}}{2}\qquad \varepsilon_{23}=\frac{\gamma_{23}}{2}\qquad \varepsilon_{31}=\frac{\gamma_{31}}{2} \tag{3.20}$$

then the components of strain tensors are

$$\begin{aligned}
&\varepsilon_{11}=\frac{\partial u_1}{\partial x_1}\qquad \varepsilon_{22}=\frac{\partial u_2}{\partial x_2}\qquad \varepsilon_{33}=\frac{\partial u_3}{\partial x_3}\\
&\varepsilon_{12}=\frac{1}{2}\left(\frac{\partial u_2}{\partial x_1}+\frac{\partial u_1}{\partial x_2}\right)\quad \varepsilon_{23}=\frac{1}{2}\left(\frac{\partial u_3}{\partial x_2}+\frac{\partial u_2}{\partial x_3}\right)\quad \varepsilon_{31}=\frac{1}{2}\left(\frac{\partial u_1}{\partial x_3}+\frac{\partial u_3}{\partial x_1}\right)
\end{aligned} \tag{3.21}$$

In a material subjected to deformation, the three components of the displacement can be arbitrarily determined as a function of x_1, x_2, x_3 at each point. The six components of strain at each point are defined by differentiation of the three components of displacement with respect to the position coordinates. On the other hand, if the six components of strain are independently defined as functions of x_1, x_2, x_3, then we have six equations to determine the three unknown displacement components. This system of partial differential equations does not, in general, give a solution for the three unknown components of displacement unless the six components of strain relate to each other. Let us consider the relations among the six components of strain in this case. Differentiating the first of Eqs. (3.21) twice with respect to x_2, and the second one twice with respect to x_1, and adding them, we get

$$\frac{\partial^2\varepsilon_{11}}{\partial x_2^2}+\frac{\partial^2\varepsilon_{22}}{\partial x_1^2}=\frac{\partial^3 u_1}{\partial x_2^2\partial x_1}+\frac{\partial^3 u_2}{\partial x_1^2\partial x_2}=2\frac{\partial^2\varepsilon_{12}}{\partial x_1\partial x_2} \tag{3.22}$$

We can obtain five additional equations in the same way. Then, the set of set of six equations is

$$\begin{aligned}
&\frac{\partial^2\varepsilon_{11}}{\partial x_2^2}+\frac{\partial^2\varepsilon_{22}}{\partial x_1^2}=2\frac{\partial^2\varepsilon_{12}}{\partial x_1\partial x_2}\\
&\frac{\partial^2\varepsilon_{11}}{\partial x_2\partial x_3}=\frac{\partial}{\partial x_1}\left(-\frac{\partial\varepsilon_{23}}{\partial x_1}+\frac{\partial\varepsilon_{31}}{\partial x_2}+\frac{\partial\varepsilon_{12}}{\partial x_3}\right)\\
&\frac{\partial^2\varepsilon_{22}}{\partial x_3^2}+\frac{\partial^2\varepsilon_{33}}{\partial x_2^2}=2\frac{\partial^2\varepsilon_{23}}{\partial x_2\partial x_3}\\
&\frac{\partial^2\varepsilon_{22}}{\partial x_3\partial x_1}=\frac{\partial}{\partial x_2}\left(\frac{\partial\varepsilon_{23}}{\partial x_1}-\frac{\partial\varepsilon_{31}}{\partial x_2}+\frac{\partial\varepsilon_{12}}{\partial x_3}\right)
\end{aligned} \tag{3.23}$$

$$\frac{\partial^2 \varepsilon_{33}}{\partial x_1^2} + \frac{\partial^2 \varepsilon_{11}}{\partial x_3^2} = 2\frac{\partial^2 \varepsilon_{31}}{\partial x_3 \partial x_1}$$

$$\frac{\partial^2 \varepsilon_{33}}{\partial x_1 \partial x_2} = \frac{\partial}{\partial x_3}\left(\frac{\partial \varepsilon_{23}}{\partial x_1} + \frac{\partial \varepsilon_{31}}{\partial x_2} - \frac{\partial \varepsilon_{12}}{\partial x_3}\right)$$

or

$$\varepsilon_{ij,kl} + \varepsilon_{kl,ij} - \varepsilon_{ik,jl} - \varepsilon_{jl,ik} = 0 \tag{3.24}$$

These equations are called the Saint-Venant compatibility equations. The components of strain must satisfy these equations if the solution of the components of displacement under the known strain is to exist.

3.2.2 Stress and Strain Tensors in Cylindrical Coordinates

We consider the equilibrium of forces acting on a body in the cylindrical coordinate system (r, ϑ, z) shown in Fig. 3.2.

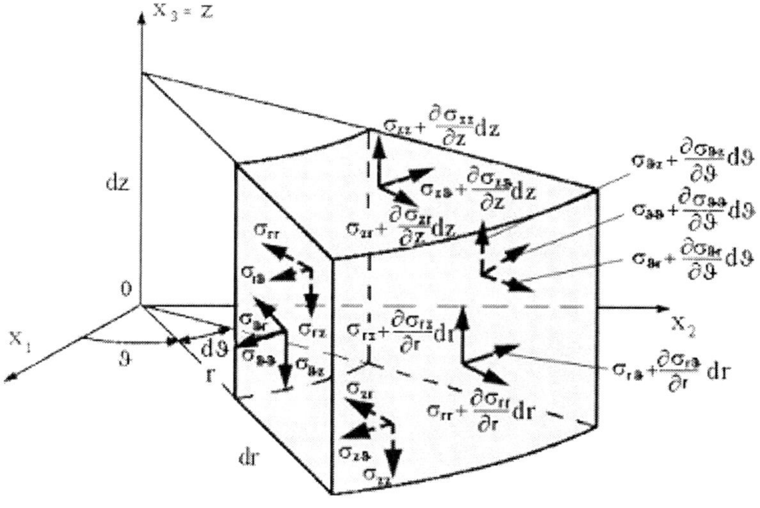

Fig. 3.2. Cylindrical element in equilibrium

When P_r, P_ϑ, P_z denote the components of the body force per unit volume acting on the cylindrical element, the equilibrium equations in the cylindrical coordinates are

$$\frac{\partial \sigma_{rr}}{\partial r} + \frac{1}{r}\frac{\partial \sigma_{\vartheta r}}{\partial \vartheta} + \frac{\partial \sigma_{zr}}{\partial z} + \frac{\sigma_{rr} - \sigma_{\vartheta\vartheta}}{r} + P_r = 0$$

$$\frac{\partial \sigma_{r\vartheta}}{\partial r} + \frac{1}{r}\frac{\partial \sigma_{\vartheta\vartheta}}{\partial \vartheta} + \frac{\partial \sigma_{z\vartheta}}{\partial z} + 2\frac{\sigma_{r\vartheta}}{r} + P_\vartheta = 0 \tag{3.25}$$

$$\frac{\partial \sigma_{rz}}{\partial r} + \frac{1}{r}\frac{\partial \sigma_{\vartheta z}}{\partial \vartheta} + \frac{\partial \sigma_{zz}}{\partial z} + \frac{\sigma_{rz}}{r} + P_z = 0$$

The components of strain tensor in the cylindrical coordinates in terms of the components of displacements are given as

$$\varepsilon_{rr} = \frac{\partial u_r}{\partial r} \qquad \varepsilon_{\vartheta\vartheta} = \frac{u_r}{r} + \frac{1}{r}\frac{\partial u_\theta}{\partial \vartheta} \qquad \varepsilon_{zz} = \frac{\partial u_z}{\partial z}$$

$$\varepsilon_{r\vartheta} = \frac{1}{2}\left(\frac{1}{r}\frac{\partial u_r}{\partial \vartheta} + \frac{\partial u_\vartheta}{\partial r} - \frac{u_\vartheta}{r}\right) \qquad \varepsilon_{\vartheta z} = \frac{1}{2}\left(\frac{\partial u_\vartheta}{\partial z} + \frac{1}{r}\frac{\partial u_z}{\partial \vartheta}\right) \tag{3.26}$$

$$\varepsilon_{zr} = \frac{1}{2}\left(\frac{\partial u_r}{\partial z} + \frac{\partial u_z}{\partial r}\right)$$

where u_r, u_ϑ, u_z are the components of displacement in the r, ϑ, z directions, respectively.

The compatibility conditions of strain tensor in the cylindrical coordinates can be obtained by eliminating the components of the displacement in Eqs. (3.26). We obtain

$$2\frac{\partial^2 (r\varepsilon_{r\vartheta})}{\partial r \partial \vartheta} = \frac{\partial^2 \varepsilon_{rr}}{\partial \vartheta^2} + r\frac{\partial^2 (r\varepsilon_{\vartheta\vartheta})}{\partial r^2} - r\frac{\partial \varepsilon_{rr}}{\partial r}$$

$$2\frac{\partial^2 \varepsilon_{zr}}{\partial r \partial z} = \frac{\partial^2 \varepsilon_{zz}}{\partial r^2} + \frac{\partial^2 \varepsilon_{rr}}{\partial z^2}$$

$$2\frac{\partial^2 (r\varepsilon_{\vartheta z})}{\partial \vartheta \partial z} = r^2\frac{\partial^2 \varepsilon_{\vartheta\vartheta}}{\partial z^2} - 2r\frac{\partial \varepsilon_{zr}}{\partial z} + r\frac{\partial \varepsilon_{zz}}{\partial r} + \frac{\partial^2 \varepsilon_{zz}}{\partial \vartheta^2}$$

$$\frac{\partial}{\partial z}\left(-\frac{\partial \varepsilon_{r\vartheta}}{\partial z} + \frac{1}{r}\frac{\partial \varepsilon_{zr}}{\partial \vartheta} + \frac{\partial \varepsilon_{\vartheta z}}{\partial r}\right) = \frac{\partial^2}{\partial r \partial \vartheta}\left(\frac{\varepsilon_{zz}}{r}\right) + \frac{1}{r}\frac{\partial \varepsilon_{\vartheta z}}{\partial z} \tag{3.27}$$

$$\frac{\partial}{\partial \vartheta}\left(\frac{\partial \varepsilon_{r\vartheta}}{\partial z} - \frac{1}{r}\frac{\partial \varepsilon_{zr}}{\partial \vartheta} + \frac{\partial \varepsilon_{\vartheta z}}{\partial r}\right) = \frac{\partial^2 (r\varepsilon_{\vartheta\vartheta})}{\partial r \partial z} - \frac{\partial \varepsilon_{rr}}{\partial z} - \frac{1}{r}\frac{\partial \varepsilon_{\vartheta z}}{\partial \vartheta}$$

$$\frac{\partial}{\partial r}\left(\frac{\partial \varepsilon_{r\vartheta}}{\partial z} + \frac{1}{r}\frac{\partial \varepsilon_{zr}}{\partial \vartheta} - \frac{\partial \varepsilon_{\vartheta z}}{\partial r}\right) = \frac{1}{r}\frac{\partial^2 \varepsilon_{rr}}{\partial \vartheta \partial z} - \frac{2}{r}\frac{\partial \varepsilon_{r\vartheta}}{\partial z} + \frac{\partial}{\partial r}\left(\frac{\varepsilon_{\vartheta z}}{r}\right)$$

3.2.3 Stress and Strain Tensors in Spherical Coordinates

The equilibrium of forces acting on a body in the spherical coordinates (r, ϑ, φ) is shown in Fig. 3.3. If P_r, P_ϑ, P_φ denote the components of the body force per unit volume acting on the spherical element, the equilibrium of forces in terms of r, ϑ, and φ are given by

$$\frac{\partial \sigma_{rr}}{\partial r} + \frac{1}{r}\frac{\partial \sigma_{\vartheta r}}{\partial \vartheta} + \frac{1}{r \sin \vartheta}\frac{\partial \sigma_{\varphi r}}{\partial \varphi} + \frac{1}{r}\left(2\sigma_{rr} - \sigma_{\vartheta\vartheta} - \sigma_{\varphi\varphi} + \sigma_{\vartheta r} \cot \vartheta\right) + P_r = 0$$

$$\frac{\partial \sigma_{r\vartheta}}{\partial r} + \frac{1}{r}\frac{\partial \sigma_{\vartheta\vartheta}}{\partial \vartheta} + \frac{1}{r \sin \vartheta}\frac{\partial \sigma_{\varphi\vartheta}}{\partial \varphi} + \frac{1}{r}\left[\left(\sigma_{\vartheta\vartheta} - \sigma_{\varphi\varphi}\right)\cot \vartheta + 3\sigma_{r\vartheta}\right] + P_\vartheta = 0 \qquad (3.28)$$

$$\frac{\partial \sigma_{r\varphi}}{\partial r} + \frac{1}{r}\frac{\partial \sigma_{\vartheta\varphi}}{\partial \vartheta} + \frac{1}{r \sin \vartheta}\frac{\partial \sigma_{\varphi\varphi}}{\partial \varphi} + \frac{1}{r}\left(3\sigma_{r\vartheta} + 2\sigma_{\vartheta\varphi} \cot \vartheta\right) + P_\varphi = 0$$

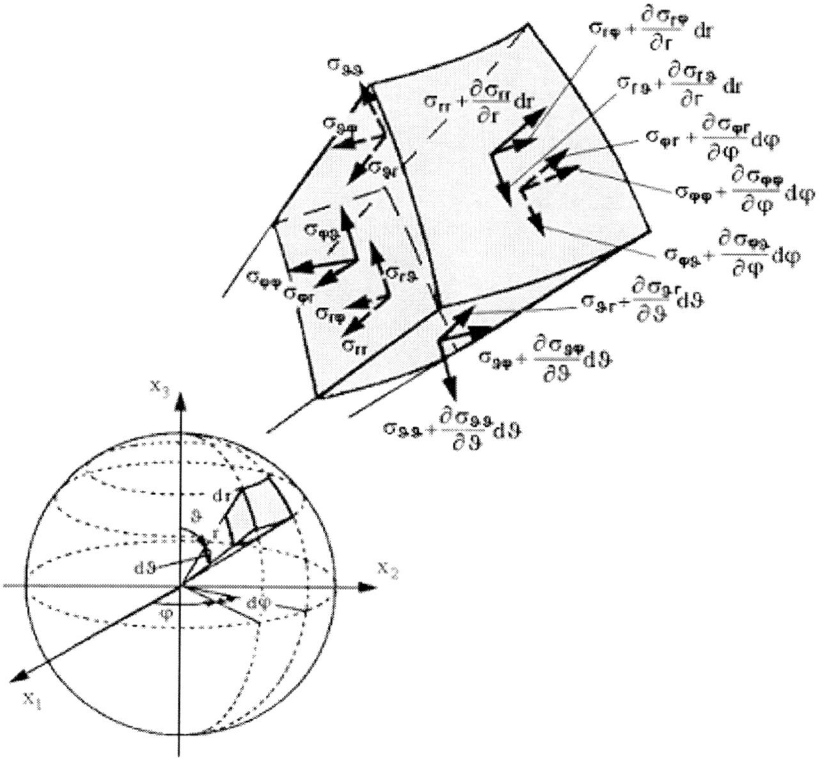

Fig. 3.3. Spherical element in equilibrium

The components of strain in spherical coordinates in terms of the components of displacement are

$$
\begin{aligned}
&\varepsilon_{rr} = \frac{\partial u_r}{\partial r}, \quad \varepsilon_{\vartheta\vartheta} = \frac{u_r}{r} + \frac{1}{r}\frac{\partial u_\vartheta}{\partial \vartheta} \\
&\varepsilon_{\varphi\varphi} = \frac{u_r}{r} + \cot\vartheta \frac{u_\vartheta}{r} + \frac{1}{r\sin\vartheta}\frac{\partial u_\varphi}{\partial \varphi} \\
&\varepsilon_{r\vartheta} = \frac{1}{2}\left(\frac{1}{r}\frac{\partial u_r}{\partial \vartheta} + \frac{\partial u_\vartheta}{\partial r} - \frac{u_\vartheta}{r}\right) \\
&\varepsilon_{\vartheta\varphi} = \frac{1}{2}\left(\frac{1}{r}\frac{\partial u_\varphi}{\partial \vartheta} - \cot\vartheta \frac{u_\varphi}{r} + \frac{1}{r\sin\vartheta}\frac{\partial u_\vartheta}{\partial \varphi}\right) \\
&\varepsilon_{\varphi r} = \frac{1}{2}\left(\frac{1}{r\sin\vartheta}\frac{\partial u_r}{\partial \varphi} + \frac{\partial u_\varphi}{\partial r} - \frac{u_\varphi}{r}\right)
\end{aligned}
\tag{3.29}
$$

where u_r, u_ϑ, u_φ are the components of displacement in the r, ϑ, φ directions, respectively.

The compatibility conditions of strain in the spherical coordinates may be obtained by eliminating the displacement components in Eqs. (3.29). We get

$$
\begin{aligned}
&-\frac{\varepsilon_{\vartheta\vartheta,\varphi\varphi}}{r^2\sin^2\vartheta} + \frac{\varepsilon_{\vartheta\vartheta,\vartheta}}{r^2\tan\vartheta} - \frac{\varepsilon_{\vartheta\vartheta,r}}{r} + \frac{2\varepsilon_{r\vartheta,\vartheta}}{r^2} + \frac{2\varepsilon_{r\vartheta}}{r^2\tan\vartheta} \\
&-\frac{2\varepsilon_{\vartheta\vartheta}}{r^2} - \frac{\varepsilon_{\varphi\varphi,\vartheta\vartheta}}{r^2} - \frac{\varepsilon_{\varphi\varphi,r}}{r} - \frac{2\varepsilon_{\varphi\varphi,\vartheta}}{r^2\tan\vartheta} + \frac{\varepsilon_{\varphi r}}{r^2\sin\vartheta} \\
&+\frac{2\cos\vartheta\,\varepsilon_{\vartheta\varphi,\varphi}}{r^2\sin^2\vartheta} + \frac{2\varepsilon_{\vartheta\varphi,\vartheta\varphi}}{r^2\sin^2\vartheta} + \frac{2\varepsilon_{rr}}{r^2} = 0
\end{aligned}
$$

$$
\begin{aligned}
&\frac{2\varepsilon_{\varphi r,\varphi r}}{r\sin\vartheta} + \frac{\varepsilon_{\varphi r,\varphi}}{r^2\sin\vartheta} - \frac{\varepsilon_{rr,\varphi\varphi}}{r^2\sin^2\vartheta} - \frac{2\varepsilon_{\varphi\varphi,r}}{r} - \varepsilon_{\varphi\varphi,rr} \\
&+\frac{2\varepsilon_{r\vartheta,r}}{r\tan\vartheta} + \frac{2\varepsilon_{r\vartheta}}{r^2\tan\vartheta} - \frac{\varepsilon_{rr,\vartheta}}{r^2\tan\vartheta} + \frac{\varepsilon_{rr,r}}{r} = 0
\end{aligned}
$$

$$
\frac{2\varepsilon_{r\vartheta,r\vartheta}}{r} - \frac{2\varepsilon_{\vartheta\vartheta,r}}{r} - \varepsilon_{\vartheta\vartheta,rr} - \frac{\varepsilon_{rr,r}}{r} - \frac{\varepsilon_{rr,\vartheta\vartheta}}{r^2} + \frac{2\varepsilon_{r\vartheta,\vartheta}}{r^2} = 0
$$

$$
\begin{aligned}
&\frac{\varepsilon_{rr,\vartheta\varphi}}{r^2\sin\vartheta} - \frac{\varepsilon_{r\vartheta,\varphi r}}{r\sin\vartheta} - \frac{\varepsilon_{r\vartheta,\varphi}}{r^2\sin\vartheta} + \frac{\varepsilon_{\varphi r}}{r^2\tan\vartheta} - \frac{\varepsilon_{\varphi r,\vartheta}}{r^2} \\
&-\frac{\varepsilon_{\varphi r,r\vartheta}}{r} + \varepsilon_{\vartheta\varphi,rr} + \frac{2\varepsilon_{\vartheta\varphi,r}}{r} + \frac{\varepsilon_{\varphi r,r}}{r\tan\vartheta} - \frac{\cos\vartheta\,\varepsilon_{rr,\varphi}}{r^2\sin^2\vartheta} = 0
\end{aligned}
$$

$$\frac{\varepsilon_{\vartheta\vartheta,\varphi r}}{r\sin\vartheta}-\frac{\varepsilon_{\vartheta\varphi,r\vartheta}}{r}-\frac{2\varepsilon_{\vartheta\varphi,r}}{r\tan\vartheta}+\frac{\cos\vartheta\varepsilon_{r\vartheta,\varphi}}{r^2\sin^2\vartheta}+\frac{\varepsilon_{\varphi r,\vartheta}}{r^2\tan\vartheta}$$
$$-\frac{\varepsilon_{r\vartheta,\vartheta\varphi}}{r^2\sin\vartheta}+\frac{\varepsilon_{\varphi r,\vartheta\vartheta}}{r^2}-\frac{\cos 2\vartheta\,\varepsilon_{\varphi r}}{r^2\sin^2\vartheta}-\frac{\varepsilon_{rr,\varphi}}{r^2\sin\vartheta}=0$$

$$\frac{\varepsilon_{\varphi\varphi,r\vartheta}}{r}+\frac{\varepsilon_{\varphi\varphi,r}}{r\tan\vartheta}-\frac{\varepsilon_{\varphi r,\vartheta\varphi}}{r^2\sin\vartheta}+\frac{2\varepsilon_{r\vartheta}}{r^2}-\frac{\varepsilon_{rr,\vartheta}}{r^2}$$
$$-\frac{\varepsilon_{\vartheta\varphi,\varphi r}}{r\sin\vartheta}+\frac{\varepsilon_{r\vartheta,\varphi\varphi}}{r^2\sin^2\vartheta}-\frac{\cos\vartheta\,\varepsilon_{\varphi r,\varphi}}{r^2\sin^2\vartheta}-\frac{\varepsilon_{\vartheta\vartheta,r}}{r\tan\vartheta}=0$$

Part II

Metal Forming Thermodynamics

4 Thermodynamical Considerations

4.1 Introduction

In order to understand a coupled thermo-mechanical process of metal forming it is necessary to describe metal forming as a thermodynamical process.

Any mechanical process starts in the initial state B_{t_0} of the material characterized by the initial configuration and the initial thermodynamical state of each material element. The initial state is described by the prescribed thermodynamical boundary conditions, the prescribed body forces and energy sources acting inside and on the surface of material. The process is governed by the field equations (balance equations) and by the constitutive law of the material.

Constitutive equations governing the thermodynamical process can be presented in a purely phenomenological way as the relation combining stress and strain and any relevant quantity suggested by experiment and by physical evidence. Experiments determine for any given material and specified conditions of use, the physical characteristics involved in the constitutive equations.

The aim of this section is to apply the two main laws of thermodynamics to the material deformed. This leads to the representation of material as a continuous and open thermodynamic system. The thermodynamic standpoints complete the kinematical and mechanical representation of a material considered.

Thermomechanics deals with the transformations which affect the energy of a system in evolution. The first law expresses the conservation of energy in all its forms. The second law is given through an inequality expressing that the quantity of energy i.e. its transformatility into efficient mechanical work, can only deteriorate. This fundamental inequality imposes some restrictions on the constitutive equations and it is not sufficient to achieve their specification. The identification of missing equations is carried out through the hypothesis of the normality of dissipative mechanisms associated with the dissipation of efficient energy. The hypothesis of normality will be useful to express a large class of constitutive equations, which are thermodynamically admissible. It should be noted that this hypothesis is not a law like for instance, the two first laws of thermodynamics.

4.2 The Local State Postulate

The thermodynamical state of each material system is uniquely characterized by the value of a finite set of state variables, even in an irreversible process. Such a phenomenological theory is restricted to some class of materials and to processes running not too far from thermodynamical equilibrium.

The characterization of a system means the characterization of its energy state, except kinetic energy. Thermostatics studies reversible and infinitely slow evolutions between two equilibrium states of homogeneous systems.

The state variable necessary for the description of the evolutions throughout the first and the second laws are directly observable and are linked by state equations through state functions characterizing the energy of the system.

Thermodynamics in contrast with thermostatics is the study of homogeneous systems in any evolution process. These evolutions are reversible, or not, and occur at any rate. The postulate of local state is to extend the concepts of thermostatics to these evolutions.

For a homogeneous system the postulate of local state is the following. The present state of a homogeneous system in any evolution can be characterized by the same variables as at equilibrium, and is independent of the rate of evolution.

The equilibrium fields of mechanical variables in a closed continuum, considered as a thermomechanical system are generally not homogeneous. The state of a continuum is characterized by fields of state variables. The local values of these variables characterize the state of the material particles, which constitute the continuum, and are considered as an elementary subsystem in homogeneous equilibrium.

The local state postulate for closed continua says that the elementary systems satisfy the local state postulate of a homogeneous system. The thermodynamics of closed continua is the study of material, which satisfies the extended local state postulate.

The reversible behaviour of the elementary system is necessary but not sufficient to ensure the reversibility of the evolution, since dissipative transport phenomena can occur. The local state postulate for closed continua is local in two aspects. It is local considering the time scale relative to the evolution rates and it refers to the space scale, which defines the dimensions of the system. The latter scale is supposed to be the same as the space scale used in the first sections to define the elementary systems.

The metal forming process is not strictly a reversible process. The evolutions of a system, which are not completely described by the state equations, are irreversible evolutions. The irreversible character of the evolutions can be due to the dependence of the constitutive equations with respect to the evolution rates, as it can be due to complementary equations of the system. This will be the case when the macroscopic characterization of a state requires a set of variables including not only external variables, but also internal variables. The equations, which give the evolutions of these internal variables, must be specified. The rates that the latter involve may be relative only to a chronological time.

4.3 The First Law of Thermodynamics

The conservation of energy is expressed by the first law of thermodynamics. It says that the material derivative of energy $\mathcal{E}$ of the material body contained in any domain V at any time is equal to the sum of the work rate P_{EF} of the external forces acting on this body, and of the rate Q^{o} of external heat supply.

The kinetic energy $\mathcal{K}$ of the body and the internal energy E give together the total energy of body $\mathcal{E}$.

The energy has an additive character. The internal energy E can be expressed by a volume density e such that edV is the internal energy of the whole body contained in the elementary domain dV.

The energy $\mathcal{E}$ of the body volume V is expressed as

$$\mathcal{E} = \mathcal{K} + \mathsf{E} = \int_V \frac{1}{2} \rho \mathbf{V}^2 \, dV + \int_V e dV \tag{4.1}$$

where e is a volume density, not a density per mass unit.

The hypothesis of external heat supply assumes that the external heat supply is due to contact effects, with the exterior through surface a limiting the material volume V which is the external heat provided by conduction. The external heat supply is also due to external volume heat sources. The rate Q^{o} can be written as

$$Q^{\circ} = \int_a q da + \int_V r dV \tag{4.2}$$

where q is the surface rate density of heat supply by conduction. The quantity q is assumed to be a function of position vector $\mathbf{x}$, time t and outward unit normal $\mathbf{n}$ to the surface a

$$q = q(\mathbf{x}, t, \mathbf{n}) \tag{4.3}$$

In Eq. (4.2) the density $r = r(\mathbf{x}, t)$ is a volume rate density of the heat provided to V.

The total work rate $R_{EF}(\mathbf{V})$ of the external forces in the whole body in volume V is

$$R_{EF}(\mathbf{V}) = \int_a \mathbf{V} \cdot \mathbf{t} \, da + \int_V \mathbf{V} \cdot \rho \mathbf{P} \, dV \tag{4.4}$$

then for any volume V the first law of thermodynamics reads

$$\frac{D\mathcal{E}}{Dt} = \frac{D\mathcal{K}}{Dt} + \frac{D\mathsf{E}}{Dt} = R_{EF} + Q^{\circ} \tag{4.5}$$

Combining the first law (4.5) and the kinetic energy theorem (2.45) yields

$$\frac{D\mathsf{E}}{Dt} = R_{SR} + Q^{\circ} \tag{4.6}$$

which expresses that the internal energy variation DE in time interval dt is due to the total strain work R_{SR}dt and the external heat supply Qdt.

4.4 The Energy Equation

If g = e in the Eulerian expression of the material derivative of a volume integral, then the material derivative of the internal energy gives

$$\frac{DE}{Dt} = \frac{D}{Dt}\int_V e\,dV = \int_V \left[\frac{\partial e}{\partial t} + \operatorname{div}(e\,\mathbf{V})\right] dV \tag{4.7}$$

Using (4.2), (2.36), (4.7) and the energy balance (4.6) we get the expression

$$\int_V \left[\frac{\partial e}{\partial t} + \operatorname{div}(e\,\mathbf{V}) - \boldsymbol{\sigma}\,\mathbf{d} - r\right] dV = \int_a q\,da \tag{4.8}$$

By the known relation

$$\int_V h(\mathbf{x}, t)\,dV = -\int_a f(\mathbf{x}, t, \mathbf{n})\,da \tag{4.9}$$

for f ($\mathbf{x}$, t, $\mathbf{n}$) = – q ($\mathbf{x}$, t, $\mathbf{n}$) we get

$$q = -\,\mathbf{q}\cdot\mathbf{n} \tag{4.10}$$

where $\mathbf{q}$ is the heat flux vector. Using the expression (1.33) of the particulate derivative for the internal energy e, the local expression for the first law in the Eulerian approach is in the form

$$\frac{de}{dt} + e\,\operatorname{div}\mathbf{V} = \boldsymbol{\sigma}\,\mathbf{d} + r - \operatorname{div}\mathbf{q} \tag{4.11}$$

The expression (4.11) is called the Eulerian energy equation. Multiplying Eq. (4.11) by dV and using Eqs. (1.30) and (4.8) we have

$$\frac{d}{dt}(e\,dV) = dr_{SR} + (r - \operatorname{div}\mathbf{q})\,dV \tag{4.12}$$

The expression (4.11) corresponds to a balance of internal energy for the elementary material system dV. In Eq. (4.12) the term d (e dV) represents the variation of internal energy of the body observed from the material particle between time t and t + dt. Equation (4.12) indicates that this variation is equal to the energy supplied to the open system during the same interval.

The energy supply is the sum of two terms: the elementary strain work dr_{SR}, that corresponds to the part of the external mechanical energy given to the system and not converted into kinetic energy and the external heat provided both by conduction given by the term – div $\mathbf{q}$ dt dV and by external volume heat sources given by the term r dt dV.

Define the Lagrangian vector **Q** by the relation

$$\mathbf{Q} \cdot \mathbf{N} \cdot \mathrm{dA_r} = \mathbf{q} \cdot \mathbf{n} \cdot \mathrm{da} \tag{4.13}$$

The expression (4.13) represents the heat flux $\mathbf{q} \cdot \mathbf{n} \cdot \mathrm{da}$ throughout the oriented material surface $\mathrm{d}\mathbf{a} = \mathbf{n} \cdot \mathrm{da}$ in terms of the oriented surface $\mathrm{d}\mathbf{A}_r = \mathbf{N} \cdot \mathrm{dA_r}$ which is the inverse convective conveyor of $\mathrm{d}\mathbf{a}$ in the reference configuration attached to the material.

By (1.9) we get

$$\mathbf{q} = \mathbf{F} \cdot \mathbf{Q} / \mathrm{J} \tag{4.14}$$

For any two vectors $\mathbf{V}^*$ and $\mathbf{v}^*$ in the relation $\mathbf{V}^* \cdot \mathbf{N} \cdot \mathrm{dA_r} = \mathbf{v}^* \cdot \mathbf{n}\,\mathrm{da}$ we have

$$\mathrm{Div}\,\mathbf{V}^*\,\mathrm{dV_r} = \mathrm{div}\,\mathbf{v}^*\,\mathrm{dV} \tag{4.15}$$

where dV is the convective conveyor of $\mathrm{dV_r}$. The Lagrangian volume density of internal energy is E = E ($\mathbf{x}$, t) given by

$$\mathrm{E\,dV_r} = \mathrm{e\,dV} \tag{4.16}$$

By using Eq. (4.16) together with the conveyance formulae

$$\boldsymbol{\sigma}\,\mathbf{d}\,\mathrm{dV} = \mathbf{S}\frac{\mathrm{d}\,\mathbf{E}}{\mathrm{dt}} \qquad \mathrm{r\,dV} = \mathrm{R\,dV_r} \tag{4.17}$$

and by applying identity (4.15) to the couples ($\mathbf{Q}$, $\mathbf{q}$) and the definition of the infinitesimal strain work rate $\mathrm{dr_{SR}}$, Eq. (4.12) finally gives

$$\frac{\mathrm{dE}}{\mathrm{dt}} = \mathbf{S}\frac{\mathrm{d}\,\mathbf{E}}{\mathrm{dt}} + \mathrm{R} - \mathrm{Div}\,\mathbf{Q} \tag{4.18}$$

The above equation corresponds to the Lagrangian formulation of the Eulerian energy equation (4.11).

4.5 The Second Law of Thermodynamics

The second law of thermodynamics says that the quality of energy can only deteriorate. The quantity of energy transformed to mechanical work can only decrease irreversibly. In the second law the new physical quantity is introduced, entropy, which represents a measure of this deterioration and which can increase when considering an isolated system. In a system that is no longer isolated, the second law defines a lower bound to the entropy increase, which takes into account the external entropy supply. The latter is defined in term of a new variable, the temperature. The second law can be formulated as below.

The material derivative of a thermodynamic function S, called entropy attached to any material system V is equal to or superior to the rate of entropy externally supplied to it. The external entropy rate supply can be defined in terms of a universal scale of absolute temperature denoted by T and positively defined.

The external entropy rate is then defined as the ratio of the heat supply rate and the absolute temperature at which the heat is provided to the considered subsystem. We denote by s the entropy volume density, such that the quantity sdV represents the entropy of all the matter presently contained in the open elementary system dV.

The total entropy S of all the matter contained in V is

$$\mathsf{S} = \int_V s\,dV \tag{4.19}$$

The second law reads

$$\frac{D\mathsf{S}}{Dt} \geq -\int_a \frac{\mathbf{q} \cdot \mathbf{n}}{T} da + \int_V \frac{r}{T} dV \tag{4.20}$$

The right-hand side of inequality (4.20) represents the rate of external entropy supply. This external rate is composed of both the entropy influx associated with the heat provided by conduction through surface a enclosing the considered material volume V, and of the volume entropy rate associated with the external heat sources distributed within the same volume.

Eq. (4.20) implies that the internal entropy production rate cannot be negative in real evolutions. The entropy S of a material system V at $\mathbf{q} = 0$ and $r = 0$ cannot spontaneously decrease.

The material derivative of a volume integral in Eulerian expression for g = s reads

$$\frac{D\mathsf{S}}{Dt} = \frac{D}{Dt}\int_V s\,dV = \int_V \left[\frac{\partial s}{\partial t} + \operatorname{div}\left(s\,\mathbf{V}\right)\right] dV \tag{4.21}$$

By Eqs. (4.20) and (4.21) it follows that the volume integral must be non-negative for any system V,

$$\frac{\partial s}{\partial t} + \operatorname{div}\left(s\,\mathbf{V}\right) + \operatorname{div}\frac{\mathbf{q}}{T} - \frac{r}{T} \geq 0 \tag{4.22}$$

where the surface integral has been transformed to the volume integral. The expression (4.22) can be rewritten as

$$\frac{\partial s}{\partial t} + s\operatorname{div}\mathbf{V} + \operatorname{div}\frac{\mathbf{q}}{T} - \frac{r}{T} \geq 0 \tag{4.23}$$

Multiplying Eq. (4.23) by dV and using Eq. (1.30) the above inequality becomes

$$\frac{d}{dt}\left(s\,dV\right) + \left(\operatorname{div}\frac{\mathbf{q}}{T} - \frac{r}{T}\right) dV \geq 0 \tag{4.24}$$

In Eq. (4.24) the term d(s dV) represents the variation in entropy of this open system, during the infinitesimal time interval dt observed from any material point.

By energy equation (4.11), the fundamental inequality (4.23) can be written as

$$\boldsymbol{\sigma}\,\mathbf{d} + T\frac{ds}{dt} - \frac{de}{dt} - (e - T\,s)\,\mathrm{div}\,\mathbf{V} - \frac{\mathbf{q}}{T}\cdot\mathrm{grad}\,T \geq 0 \tag{4.25}$$

Now we define the free volume energy of the open system dV

$$\Psi = e - T\,s \tag{4.26}$$

By Eqs. (4.26) and (4.25) we get

$$\boldsymbol{\sigma}\,\mathbf{d} + s\frac{dT}{dt} - \frac{d\Psi}{dt} - \Psi\,\mathrm{div}\,\mathbf{V} - \frac{\mathbf{q}}{T}\cdot\mathrm{grad}\,T \geq 0 \tag{4.27}$$

The Eq. (4.27) is the fundamental inequality or so-called Clausius-Duhem inequality. It corresponds to a Eulerian approach.

The Lagrangian approach of the fundamental inequality leads to expressing DS/Dt in terms of the Lagrangian entropy density S defined by

$$S\,dV_r = s\,dV \tag{4.28}$$

By transport formula (4.13) and (4.17) together with a Lagrangian expression of the material derivative of the integral of an extensive variable, the inequality (4.21) can be rewritten as

$$\frac{DS}{Dt} = \frac{D}{Dt}\int_{V_r} S\,dV_r = \int_{V_r}\frac{dS}{dt}\,dV_r \geq -\int\frac{\mathbf{Q}\cdot\mathbf{N}}{T}\,dA_r + \int_{V_r}\frac{R}{T}\,dV_r \tag{4.29}$$

The expression (4.29), which holds for any domain V_r can be written in Lagrangian formulation of the Eulerian inequality in the form

$$\frac{dS}{dt} + \mathrm{Div}\frac{\mathbf{Q}}{T} - \frac{R}{T} \geq 0 \tag{4.30}$$

By the Lagrangian energy equation (4.18) and the positiveness at absolute temperature T, from the inequality (4.30) it follows that

$$\mathbf{S}\frac{d\,\mathbf{E}}{dt} + T\frac{dS}{dt} - \frac{dE}{dt} - \frac{\mathbf{Q}}{T}\cdot\mathrm{Grad}\,T \geq 0 \tag{4.31}$$

We define the free Lagrangian energy density Ψ as

$$\Psi\,dV_r = \psi\,dV \qquad \Psi = E - T\,S \tag{4.32}$$

By Eqs. (4.31) and (4.32) we get

$$\mathbf{S}\frac{d\,\mathbf{E}}{dt} - S\frac{dT}{dt} - \frac{d\Psi}{dt} - \frac{dE}{dt} - \frac{\mathbf{Q}}{T}\cdot\mathrm{Grad}\,T \geq 0 \tag{4.33}$$

The expression (4.33) is the Lagrangian formulation of the Clausius-Duhem inequality.

4.6 Dissipations

The left-hand side of Eq. (4.33) we denote by Π. It is called the dissipation per unit of initial volume dV_r. The second law requires the dissipation Π and the associated internal entropy production Π / T to be non-negative. The dissipation Π is the sum of two terms

$$\Pi = \Pi_1 + \Pi_2 \tag{4.34}$$

where

$$\Pi_1 = \mathbf{S}\frac{d\,\mathbf{E}}{dt} - S\frac{dT}{dt} - \frac{d\Psi}{dt} \tag{4.35}$$

The local state postulate assumes that all the quantities appearing in Eq. (4.35) depend only on the state variables characterizing the free energy ΨdV_r of the open elementary system dV_r.

The second dissipation Π_2 is defined as

$$\Pi_2 = -\frac{\mathbf{Q}}{T}\cdot \text{Grad } T \tag{4.36}$$

and is called the thermal dissipation associated with heat conduction. By definition (4.32) of Ψ and definition of Π_1 (4.35), the Lagrangian energy equation (4.18) is rewritten as

$$T\frac{dS}{dt} = R - \text{Div}\,\mathbf{Q} + \Pi_1 \tag{4.37}$$

The expression (4.37) is called the Lagrangian thermal equation. Using Eqs. (4.36) (4.37) we get

$$\frac{d}{dt}\left(S dV_r\right) = \left[\frac{R}{T} - \text{Div}\left(\frac{\mathbf{Q}}{T}\right)\right] dV_r + \frac{\Pi}{T} dV_r \tag{4.38}$$

The thermal equation (4.37) or equivalently Eq. (4.38) corresponds to a balance in entropy for the elementary system dV_r. The term $d(SdV_r)$ is the entropy variation, during the time interval dt, observed from the point of the open system dV_r.

The internal source of entropy $(\Pi / T)\, dt\, dV_r$ is the sum of the production $(\Pi_2 / T)\, dt\, dV_r$ associated with heat conduction and the production $(\Pi_1 / T)\, dt\, dV_r$ associated with the mechanical dissipation $\Pi_1\, dV_r\, dt$. As it will be seen later, $\Pi_2\, dV_r\, dt$ corresponds to a mechanical energy converted into heat by mechanical dissipation. Thus, the mechanical dissipation Π_1 appears as a heat source in thermal equation (4.37).

A Lagrangian approach to thermal equation is introduced below. If Π_i is the Eulerian dissipation volume densities, then

$$\pi_i\, dV = \Pi_i\, dV_r \quad J\,\pi_i = \Pi_i \tag{4.39}$$

The respective expressions can be written as

$$\pi_1 = \boldsymbol{\sigma}\,\mathbf{d} - s\frac{dT}{dt} - \frac{d\Psi}{dt} - \Psi \operatorname{div} \mathbf{V}$$
$$\pi_2 = -\frac{\mathbf{q}}{T}\cdot \operatorname{grad} T \tag{4.40}$$

By Eqs. (4.34) and (4.39), for the densities π_i the following relation holds

$$\pi_1 + \pi_2 \geq 0 \tag{4.41}$$

The Eulerian thermal equation takes the form

$$T\left[\frac{ds}{dt} + s \operatorname{div} \mathbf{V}\right] = r - \operatorname{div} \mathbf{q} + \pi_1 \tag{4.42}$$

4.7 Equations of State for an Elementary System

For any evolution of the continuum the energy states of the elementary systems depend on the same state variables as at equilibrium. The free energy volume density Ψ depends locally on the state variables, but not on their rates nor on their spatial gradients. We assume that it is also true for the Piola-Kirchhoff stress tensor **S**. In Eq. (4.34) the dissipation Π_2 depends on Grad T. Let Grad T = 0 ($\Pi_2 = 0$), then the non-negativeness of intrinsic dissipation Π_1 is derived independently of the non-negativeness of total dissipation Π

$$\Pi_1 = \mathbf{S}\frac{d\mathbf{E}}{dt} - S\frac{dT}{dt} - \frac{d\Psi}{dt} \geq 0 \tag{4.43}$$

The expression (4.43) is derived from the second law and the local state postulate.

In Eq. (4.43) the energy **S** d**E** represents the energy supplied to the elementary system dV_r, not in the form of heat and not converted into kinetic energy during the time interval dt. It is actually supplied if positive, extracted if negative. The energy $d\Psi + SdT = dE - TdS$ is the part of the previous energy that the system effectively stores in the time interval dt in any other form but heat. It is actually stored if positive, extracted if negative.

The Lagrangian expression of the particulate derivatives used here do not involve gradients, contrary to the corresponding Eulerian expression, which has importance in this reasoning.

The free energy Ψ can be written as

$$\Psi = \Psi\,(T, E_{\alpha\beta}, m_1, \ldots, m_n) \tag{4.44}$$

based on the local state postulate.

The variables T, $E_{\alpha\beta}$, m_1, ..., m_n constitute a set of state variables characterizing the state of the open system dV_r. By the local state postulate, these state variables are macroscopic variables. The non-negativeness of the intrinsic dissipation expressed by (4.43) gives

$$\left(\mathbf{S}-\frac{\partial\Psi}{\partial\mathbf{E}}\right)\frac{d\mathbf{E}}{dt}-\frac{\partial\Psi}{\partial\mathbf{m}}\cdot\frac{d\mathbf{m}}{dt}-\left(S+\frac{\partial\Psi}{\partial T}\right)\frac{dT}{dt}\geq 0 \tag{4.45}$$

In Eq. (4.45) $\partial\Psi/\partial\mathbf{E}$ stands for the tensor of components $\partial\Psi/\partial E_{\alpha\beta}$ and

$$\frac{\partial\Psi}{\partial\mathbf{m}}\cdot\frac{d\mathbf{m}}{dt}=\frac{\partial\boldsymbol{\Psi}}{\partial\mathbf{m}}\cdot\dot{\mathbf{m}}=\frac{\partial\Psi}{\partial m_I}\cdot\dot{m}_I \tag{4.46}$$

where

$$\dot{\mathbf{m}}=\frac{d\mathbf{m}}{dt} \tag{4.47}$$

The inequality (4.45) is relative only to this elementary open system dV_r under consideration.

We say that a set of state variables characterizing an elementary system is normal if variations of this particular state can occur independently of the variations of the other variables of the set.

The hypothesis introduced by Helmholtz states that it is possible to find a set of variables, which is normal with respect to absolute temperature T. In Eq. (4.45) with such a set of variables, real evolution can occur with arbitrary variations dT and zero variations for the other variables of the set.

From Eq. (4.45) it follows that

$$S=-\frac{\partial\Psi}{\partial T} \tag{4.48}$$

because the inequality (4.45) must remain satisfied for these particular real evolutions, and assuming that the present value of entropy S is independent of the temperature rate dT/dt.

Assume that the variable set is normal with respect to the state variables $E_{\alpha\beta}$, and that the present value of the Piola-Kirchoff stress tensor **S** is independent of the rates d**E**/dt. Then we have

$$\mathbf{S}=\frac{\partial\Psi}{\partial\mathbf{E}} \tag{4.49}$$

The expression (4.49) associates the state variables T, $E_{\alpha\beta}$ with their dual thermodynamic variables -S, **S**. They are called the state equations of the system. They still hold for a system of evolution in holding at equilibrium.

The Maxwell symmetry relations are given by Eqs. (4.48) and (4.49)

$$\frac{\partial S}{\partial E_{\alpha\beta}} = -\frac{\partial S_{\alpha\beta}}{\partial T} \qquad \frac{\partial S_{\alpha\beta}}{\partial E_{\gamma\lambda}} = \frac{\partial S_{\gamma\lambda}}{\partial E_{\alpha\beta}} \tag{4.50}$$

By Eq. (4.44) and (4.49) the dissipation Π_1 fulfills the expression

$$\Pi_1 = -\frac{\partial \Psi}{\partial \mathbf{m}} \cdot \dot{\mathbf{m}} \geq 0 \tag{4.51}$$

The above equation can be written in the form

$$\Pi_1 = \mathbf{B}_{\dot{\mathbf{m}}} \cdot \dot{\mathbf{m}} \geq 0 \qquad \mathbf{B}_{\dot{\mathbf{m}}} = -\frac{\partial \Psi}{\partial \mathbf{m}} \tag{4.52}$$

The rates $\dot{\mathbf{m}}_I$ of the internal variables are associated with the thermodynamical forces $\mathbf{B}_{\dot{\mathbf{m}}_I}$.

The hypothesis of normality of a dissipative mechanism consists of introducing the existence of both a set of internal variables $\mathbf{m}_I$ and a function H of their rates

$$\mathbf{B}_{\dot{\mathbf{m}}} = \frac{\partial H}{\partial \dot{\mathbf{m}}} \qquad B_{\dot{m}_I} = \frac{\partial H}{\partial \dot{m}_I} \tag{4.53}$$

The function H is called the dissipation potential. H is a convex function with respect to arguments $\dot{m}_I$ and has a minimum value when all its arguments are zero. By Eqs. (4.53) the Maxwell symmetric relations (4.50) hold

$$\frac{\partial B_{\dot{m}_I}}{\partial \dot{m}_J} = \frac{\partial B_{\dot{m}_J}}{\partial \dot{m}_I} \tag{4.54}$$

From Eq. (4.53) the dissipation Π_1 defined by (4.51) can be expressed as a function of the arguments of H.

If H is a quadratic form of its arguments $\dot{m}_I$ then $\Pi_1 = 2H$. The relation combining $\mathbf{B}_{\dot{\mathbf{m}}}$ and $\dot{\mathbf{m}}$ given by (4.53) is linear and the reversible process corresponding to thermodynamic forces is also linear.

4.8 The Heat Conduction Law

The second law and the local state postulate give the relation

$$\pi_1 + \pi_2 \geq 0 \qquad \pi_1 \geq 0 \tag{4.55}$$

where dissipations π_1, π_2 are Eulerian intrinsic dissipations and Eulerian dissipation associated with transport phenomena, per unit volume dV in the current configuration expressed by (4.39). The internal entropy production rate π_2/T is due to the assembly of adjacent elementary systems that ensures the continuity of the medium, contrary to the intrinsic internal entropy production rate π_1/T related to the elementary system, which is considered independently of the other system. The hypothesis of dissipation decoupling assumes

$$\pi_1 \geq 0 \qquad \pi_2 \geq 0 \tag{4.56}$$

We formulated the heat conduction law on the basis of hypothesis (4.56). This hypothesis requires the non-negativeness of thermal dissipation π_2

$$\mathbf{B}_{q/T} \cdot \frac{\mathbf{q}}{T} \geq 0 \qquad \mathbf{B}_{q/T} = -\operatorname{grad} T \tag{4.57}$$

From Eq. (4.57) the entropy vector $\mathbf{q}/T$ and the thermodynamic force – grad T are associated as in Eq. (4.57). The decoupling hypothesis (4.56) and the resulting inequality (4.57) states that the heat flows from high temperatures. The positiveness of the associated internal entropy production rate π_2/T expresses that it is not-less evident to find a cold source to extract efficient mechanical work from this heat. Let H_2 ($\mathbf{q}/T$) be the dissipation potential. Assume the normality of the associated dissipative mechanism. From Eq. (4.57) we get the heat conduction law

$$-\operatorname{grad} T = \frac{\partial H_2}{\partial(\mathbf{q}/T)} \tag{4.58}$$

If we define H_2 as

$$H_2\left(\frac{\mathbf{q}}{T}\right) = \frac{1}{2T}\,\mathbf{q} \cdot \mathbf{k}^{-1} \cdot \mathbf{q} \tag{4.59}$$

where $\mathbf{k}$ is a symmetric tensor, Eqs. (4.58) and (4.59) give the linear heat conduction law called the Fourier law

$$\mathbf{q} = -\,\mathbf{k} \operatorname{grad} T \tag{4.60}$$

where $\mathbf{k}$ is the so-called thermal conductivity tensor, relative to the current configuration. Since H_2 defined by Eq. (4.59) is a quadratic function, the corresponding irreversible process is linear and the thermal dissipation associated with heat transport is $\Pi_2 = H_2$ per unit volume dV.

In Lagrangian approach we have

$$H_2\left(\frac{\mathbf{q}}{T}\right) dV = H_2\left(\frac{\mathbf{Q}}{T}\right) dV_r \tag{4.61}$$

$$H_2\left(\frac{\mathbf{Q}}{T}\right) = \frac{1}{2T}\,\mathbf{Q} \cdot \mathbf{K}^{-1} \cdot \mathbf{Q} \tag{4.62}$$

$$- \text{Grad } T = \frac{\partial H_2}{\partial \left(\frac{\mathbf{Q}}{T} \right)} \tag{4.63}$$

and finally we get

$$\mathbf{Q} = - \mathbf{K} \cdot \text{Grad } T \tag{4.64}$$

where **K** is the convective conveyor of **k**

$$\mathbf{K} = J\, \mathbf{F}^{-1} \cdot \mathbf{k} \cdot {}^{T}\mathbf{F}^{-1} \tag{4.65}$$

For isotropic material in reference configuration the tensor K is written as

$$\mathbf{K} = K\, \mathbf{1} \qquad \mathbf{k} = (K / J)\, \mathbf{F} \cdot {}^{T}\, \mathbf{F} \tag{4.66}$$

4.9 Equations of the Thermodynamical Process

In order to solve the equations of the thermodynamical process the number of equations should be equal to the number of unknown parameters, which have to be determined. The detailed account with the use of Lagrange variables is made in Table 4.1. and 4.2. The functions in space and time have arguments **X** and t. The number of unknowns in the thermodynamical process is given in Table 4.1. and is equal to 20 + n, where n is the number of internal variables. The number of equations governing the process is 20 + n as shown in Table 4.2. To describe completely the thermodynamical process the boundary conditions should be specified, describing the values of the variables at the limit of the considered material. These boundary conditions concern the components of the stress tensor **t**, and in Lagrangian approach they concern the stress vector $B \cdot \mathbf{N} = \mathbf{F} \cdot \mathbf{S} \cdot \mathbf{N}$ as $B \cdot \mathbf{N}\, dA_r = \mathbf{t} \cdot da$, or the components of the displacement vector **u**. The boundary conditions concern the absolute temperature T or the heat flow $\mathbf{Q} \cdot \mathbf{N}$. If the expressions of the free energy Ψ are specified and once the evolution laws governing the internal variables m_l are given then the system of equations is complete.

Table 4.1. Unknown parameters in the thermodynamical process

Unknowns	Number of unknowns
Temperature T	1
Heat flow vector **Q**	3
Entropy S	1
Displacement **u**	3
Strain tensor **E**	6
Stress tensor **S**	6
Internal variables m_l	n

Table 4.2. Governing equations in the thermodynamical process

Equation	Number of equations
Thermal equation	
$T\left[\dfrac{dS}{dt}\right] = R - \mathrm{Div}\mathbf{Q} + \Pi_2$	1
Conduction law $\mathbf{q} = - \mathbf{K} \cdot \mathrm{Grad}\, T$	3
State equation	
$\mathbf{S} = \dfrac{d\Psi}{d\mathbf{E}} \qquad S = -\dfrac{d\Psi}{dT}$	7
Strain-displacement relation	
$2\,\mathbf{E} = \mathrm{Grad}\,\mathbf{u} + {}^{T}\mathrm{Grad}\,\mathbf{u} + {}^{T}\mathrm{Grad}\,\mathbf{u} \cdot \mathrm{Grad}\,\mathbf{u}$	6
Equations of motion Div $\mathbf{F} \cdot \mathbf{S} + \rho_o\,(\mathbf{P} - \boldsymbol{a})$	3
Evolution laws	
$\mathbf{B}_{\dot{m}} = \dfrac{\partial H}{\partial \dot{\mathbf{m}}}$	n

5 Temperature Field in Material

5.1 Introduction

Deformation of materials is carried out at room temperature and at elevated temperatures. At elevated temperatures, recovery and recrystallization take place, and the flow curves depend strongly on such material parameters as temperature or strain rate.

In the metal forming process heat generation phenomena are important. If dies are at a considerably lower temperature than a workpiece, the heat losses by conduction to the dies and by radiation and convection to the environment can result in severe temperature gradients within the workpiece. The inclusion of temperature effects in the analysis of the metal forming problem is important. Furthermore, at elevated temperatures, plastic deformation can induce phase transformations and alterations in grain structures, which in turn can modify the flow resistance of the material as well as other mechanical properties.

The method for the full theoretical analysis of metal forming requires the coupling of metal deformation and heat transfer. In order to solve a coupled thermo-plastic deformation problem, it is necessary to solve simultaneously the flow problem for a given temperature distribution and to solve the thermal equations.

5.2 Thermal Behaviour

The thermal behaviour of material is defined by assuming the dissipation being equal to zero in any evolution, and thus by the absence of internal variables. The constitutive equations of the thermal system reduce the state equation to

$$S = -\frac{\partial \Psi}{\partial T} \tag{5.1}$$

where the energy Ψ depends on the external variable T

$$\Psi = \Psi\,(T) \tag{5.2}$$

The expression (5.2) has to be specified to get the constitutive equations. The linearization presented here assumes small temperature variations $\theta = T - T_o$. Under this assumption the expression of the free energy $\Psi = \Psi\,(T)$ as a second-order expansion with regard to argument θ is assumed as

$$\Psi = -S_o\theta - \frac{1}{2}b\theta^2 \tag{5.3}$$

By Eq. (5.3) we get

$$S = S_o + b\theta \tag{5.4}$$

The time differentiation of Eq. (5.4) gives

$$T_o\frac{dS}{dt} = T_o b\frac{d\theta}{dt} \tag{5.5}$$

In the physical linearization limit, the thermal Eq. (4.37) reads

$$T_o\frac{dS}{dt} = R - \mathrm{Div}\mathbf{Q} \tag{5.6}$$

In Eq. (5.6) $R - \mathrm{Div}\,\mathbf{Q}$ represents the external rate of heat supply to the elementary open system and $C_\Delta = T_o\, b$ is the volume heat capacity per unit of initial volume, so that the heat needed to produce a temperature variation θ in an iso-deformation ($\mathbf{E} = 0$) is equal to $C_\Delta\theta$.

5.3 Heat Transfer in Cartesian Coordinates

In the case of infinitesimal transformation in Cartesian coordinates the point of the space is described by the position vector $\mathbf{x} = \mathbf{x}\,(x_1, x_2, x_3)$. Under the assumption of infinitesimal transformation the reference configuration is equal to the current configuration, and the following relations hold $\mathbf{x} = \mathbf{x}$, $\mathbf{F} = \mathbf{1}$, $J = 1$ and $\mathbf{r} = \mathbf{R}$, $\mathbf{q} = \mathbf{Q}$. Moreover for simplicity let $T_o = 0$.

By Eqs. (5.5) and (5.6) we get

$$C_\Delta\frac{d\theta}{dt} = r - \mathrm{Div}\mathbf{q} \tag{5.7}$$

By Eq. (4.60) we get from (5.7)

$$C_\Delta\frac{d\theta}{dt} = r + \mathrm{div}\left(\mathbf{k}\,\mathrm{grad}\,\theta\right) \qquad C_\Delta\frac{d\theta}{dt} = r + \frac{\partial}{\partial x_i}\left(k_{ij}\frac{\partial\theta}{\partial x_j}\right) \tag{5.8}$$

Using the definition of particulate derivative, Eq. (5.8) becomes

$$\begin{gathered} C_\Delta\left(\frac{\partial\theta}{\partial t} + \mathbf{V}\,\mathrm{grad}\,\theta\right) = r + \mathrm{div}\left(\mathbf{k}\,\mathrm{grad}\,\theta\right) \\ C_\Delta\left(\frac{\partial\theta}{\partial t} + V_i\frac{\partial\theta}{\partial x_i}\right) = r + \frac{\partial}{\partial x_i}\left(k_{ij}\frac{\partial\theta}{\partial x_j}\right) \end{gathered} \tag{5.9}$$

which is a form of the Navier-Stokes equations. If $\mathbf{V} = 0$, then we get the following form of heat transfer equation

$$C_{\Delta} \frac{\partial \theta}{\partial t} = r + \operatorname{div}\left(\mathbf{k} \operatorname{grad} \theta\right) \tag{5.10}$$

5.4 The Heat Conduction Equation

The heat conduction equation given in the form (5.8) may also be derived throughout the analysis of thermal energy in an elementary volume. By the Fourier law the heat flux vector is parallel to the temperature gradient. In one-dimensional situations this law can be expressed as

$$\mathbf{q} = -k \frac{\partial \theta}{\partial x} \tag{5.11}$$

where k is the heat conduction coefficient which depends upon the properties of the material. The minus sign in Eq. (5.11) indicates that the heat always flows from hot to cold.

Derive the Fourier heat conduction equation for material based on the Fourier law of the heat conduction (5.11). The temperature at an arbitrary point $\mathbf{x}$ and at time t is denoted by $\theta\ (x_1, x_2, x_3, t)$. Consider the thermal energy balance in an element volume (dx_1, dx_2, dx_3) of the material with thermal conductivity $\mathbf{k}$, shown in Fig. 5.1. Assume that the thermal conductivity is different in all directions, so the components of heat flux along the x_1, x_2 and x_3 directions are expressed, respectively, by

$$q_1 = -k_1 \frac{\partial \theta}{\partial x_1} \qquad q_2 = -k_2 \frac{\partial \theta}{\partial x_2} \qquad q_3 = -k_3 \frac{\partial \theta}{\partial x_3} \tag{5.12}$$

where k_1, k_2, and k_3 denote the thermal conductivities along the axes x_1, x_2, and x_3 directions, respectively. The amount of heat $(dq_1)_{in}$ flowing into the element volume through the surface A_1 in the time interval dt is

$$\left(dq_1\right)_{in} = q_1 dx_2 dx_3 dt = -k_1 \frac{\partial \theta}{\partial x_1} dx_2 dx_3 dt \tag{5.13}$$

The heat $(dq_1)_{out}$ flowing out of the element volume through the surface A_2 in the time interval dt is

$$\begin{aligned}\left(dq_1\right)_{out} &= \left(q_1 + \frac{\partial q_1}{\partial x_1} dx_1\right) dx_2 dx_3 dt \\ &= -\left[k_1 \frac{\partial \theta}{\partial x_1} + \frac{\partial}{\partial x_1}\left(k_1 \frac{\partial \theta}{\partial x_1}\right) dx_1\right] dx_2 dx_3 dt\end{aligned} \tag{5.14}$$

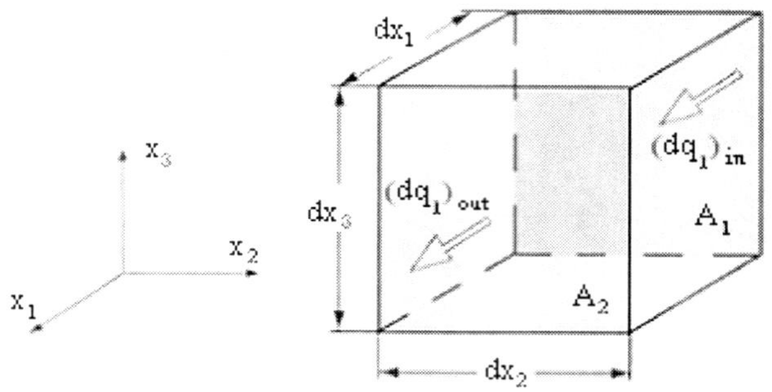

Fig. 5.1. An element of volume

The amount of heat which remains in the element volume due to the heat flow along the x_1 direction is

$$\begin{aligned}(dq_1)_{in} - (dq_1)_{out} &= -\frac{\partial q_1}{\partial x_1} dx_1 dx_2 dx_3 dt \\ &= \frac{\partial}{\partial x_1}\left(k_1 \frac{\partial \theta}{\partial x_1}\right) dx_1 dx_2 dx_3 dt\end{aligned} \tag{5.15}$$

Similar expressions hold for the directions of x_2 and x_3

$$\begin{aligned}(dq_2)_{in} - (dq_2)_{out} &= -\frac{\partial q_2}{\partial x_2} dx_1 dx_2 dx_3 dt \\ &= \frac{\partial}{\partial x_2}\left(k_2 \frac{\partial \theta}{\partial x_2}\right) dx_1 dx_2 dx_3 dt\end{aligned} \tag{5.16}$$

$$\begin{aligned}(dq_3)_{in} - (dq_3)_{out} &= -\frac{\partial q_3}{\partial x_3} dx_1 dx_2 dx_3 dt \\ &= \frac{\partial}{\partial x_3}\left(k_3 \frac{\partial \theta}{\partial x_3}\right) dx_1 dx_2 dx_3 dt\end{aligned} \tag{5.17}$$

The amount of heat dq^1 which remains in the element volume due to the heat flow through the surfaces is

$$\begin{aligned}dq^1 &= -\left(\frac{\partial q_1}{\partial x_1} + \frac{\partial q_2}{\partial x_2} + \frac{\partial q_3}{\partial x_3}\right) dx_1 dx_2 dx_3 dt \\ &= -\operatorname{div} q \, dx_1 dx_2 dx_3 dt = \left[\frac{\partial}{\partial x_1}\left(k_1 \frac{\partial \theta}{\partial x_1}\right)\right. \\ &\left. + \frac{\partial}{\partial x_2}\left(k_2 \frac{\partial \theta}{\partial x_2}\right) + \frac{\partial}{\partial x_3}\left(k_3 \frac{\partial \theta}{\partial x_3}\right)\right] dx_1 dx_2 dx_3 dt\end{aligned} \tag{5.18}$$

In the case of an internal heat generation r per unit volume per unit time, the amount of heat dq generated in the element volume in the time interval dt is equal to

$$dq^2 = r\ dx_1 dx_2 dx_3 dt \tag{5.19}$$

The temperature rise $d\theta$ in the element volume occurs due to the amount of heat dq^1 and dq^2. The amount of heat dq^3 accumulated due to the temperature rise $d\theta$ in the element volume is

$$dq^3 = C_\Delta\, d\theta\, dx_1 dx_2 dx_3 \tag{5.20}$$

The thermal energy balance gives

$$C_\Delta \frac{\partial\theta}{\partial t} = \frac{\partial}{\partial x_1}\left(k_1 \frac{\partial\theta}{\partial x_1}\right) + \frac{\partial}{\partial x_2}\left(k_2 \frac{\partial\theta}{\partial x_2}\right) + \frac{\partial}{\partial x_3}\left(k_3 \frac{\partial\theta}{\partial x_3}\right) + r \tag{5.21}$$

The expression (5.21) is a partial differential equation of the second order and it is called the Fourier heat conduction equation. This equation holds for a non-homogeneous anisotropic material.

In a homogeneous material these constants are the same at all points. In an anisotropic body the material constants at any point may change in each direction. In an isotropic material these constants at any point in a material are the same in all directions. If the thermal conductivities have different values at various points, but have the same value in all directions at each particular point, then the material is a nonhomogeneous isotropic and $k_1 = k_2 = k_3 = k\,(\mathbf{x})$.

Similarly, if the thermal conductivities have the same value at any point, but have a different value in each direction x_i this material is an homogeneous anisotropic material.

For the homogeneous isotropic material in which the thermal conductivity k in Eq. (5.21) is constant, the heat conduction equation reduces to

$$C_\Delta \frac{\partial\theta}{\partial t} = k\left(\frac{\partial^2\theta}{\partial x_1^2} + \frac{\partial^2\theta}{\partial x_2^2} + \frac{\partial^2\theta}{\partial x_3^2}\right) + r \tag{5.22}$$

The expression $k/C_\Delta = a$ is called the coefficient of thermal diffusion or temperature conduction.

For many practical problems it is convenient to use a system of coordinates other than Cartesian. The heat conduction in cylindrical coordinates r, ϑ, z, has the form

$$a\left(\frac{\partial^2\theta}{\partial r^2} + \frac{1}{r}\frac{\partial\theta}{\partial r} + \frac{1}{r^2}\frac{\partial^2\theta}{\partial\vartheta^2} + \frac{\partial^2\theta}{\partial z^2}\right) + \frac{r}{C_\Delta} = \frac{\partial\theta}{\partial t} \tag{5.23}$$

and in spherical coordinates r, ϑ, ϕ

$$\frac{1}{r}\frac{\partial^2(r\theta)}{\partial r^2} + \frac{1}{r^2\sin\vartheta}\frac{\partial}{\partial\vartheta}\left(\sin\vartheta\frac{\partial\theta}{\partial\vartheta}\right) + \frac{1}{r^2\sin\vartheta}\frac{\partial^2\theta}{\partial\phi^2} + \frac{r}{k} = \frac{1}{a}\frac{\partial\theta}{\partial t} \tag{5.24}$$

5.5 Heat Convection

A frequently encountered case of practical significance in metal forming is the heat exchange between a material wall and an adjacent gas or liquid. Heat exchange in fluids takes place by convection, but near a wall there exists a very thin layer in which heat exchange takes place by conduction. In the case where heat exchange is stationary, the heat is transferred from the wall towards the center of the fluid. If the intensity of heat transferred is higher, then the drop in temperature per unit length in a direction perpendicular to the wall is lower. Near the wall one observes a significant drop of temperature because in the thin boundary layer conduction plays a decisive role and heat exchange is less intensive in the boundary layer than in areas remote from the wall, where convection also takes place.

The phenomenon described above is known as heat convection. Mathematically it is described by the Newton equation

$$q = \alpha(\theta_w - \theta_f) \tag{5.29}$$

where θ_w is the wall temperature, θ_f is the fluid temperature at a sufficiently great distance from the wall, and the method of determination of θ_f is usually precisely laid down. The magnitude α determining the heat exchange intensity is called the heat convection coefficient.

5.6 Heat Radiation

The concept of a black body is essential to radiation theory. A black body is a hypothetical body which absorbs all radiant energy falling on it, transmitting and reflecting nothing. Heat radiation occurs in accordance with the Stefan-Boltzmann law, which says that the energy radiated by a black body is proportional to the fourth power of the absolute temperature of that body. Mathematically this law is expressed by the formula

$$q = C_o \left[\frac{T^a}{100} \right]^4 \tag{5.30}$$

where C_o is the so-called radiation coefficient of the black body and T^a is the absolute temperature. The heat radiated through the surface A per unit time is

$$q_h = C_o A \left[\frac{T^a}{100} \right]^4 \tag{5.31}$$

Real bodies are not black bodies and at a given temperature will radiate less energy than a black body. If the ratio of the energy radiated by the real body to the energy radiated by the black body in the same conditions does not depend on radiation wavelength, then this body is called a gray body. The heat exchange between gray bodies is described by the equation

$$q_{1-2} = C_o A_1 \phi_{1-2} \left[\left(\frac{T_1^a}{100} \right)^4 - \left(\frac{T_2^a}{100} \right)^4 \right] \tag{5.32}$$

where T_1^a and T_2^a are the absolute temperatures of the bodies radiating the heat, A_1 is the surface of the body at temperature T_1, and ϕ_{1-2} is the coefficient taking into consideration the deviation of the properties of the analyzed body from the properties of the black body and the geometrical system of the two bodies.

Heat exchange based on pure conduction, convection or radiation very rarely holds in practice. These three fundamental kinds of heat exchange normally appear in various combinations. A common case is the exchange of heat through a solid wall by a combination of radiation and convection. In this case one introduces a substitute coefficient of heat exchange by radiation α_r, which is defined as follows

$$\alpha_r = \frac{q_{1-2}}{A_1 (T_1 - T_0)} \tag{5.33}$$

where q_{1-2} is the heat exchanged by radiation, given by Eq. (5.32), T_1 is the wall temperature, and T_0 is the reference temperature. The reference temperature T_0 does not have to be equal to the temperature T_2 appearing in Eq. (5.32), and in the case of convection and radiation one puts it equal to the fluid temperature T_f. The expression (5.33) can then be rewritten in the form

$$\alpha_r = \frac{C_o \phi_{1-2} \left[\left(\frac{T_1^a}{100} \right)^4 - \left(\frac{T_2^a}{100} \right)^4 \right]}{T_1 - T_f} \tag{5.34}$$

Heat exchange by both convection and radiation can be described by the relation

$$q = (\alpha + \alpha_r)\,(\theta_w - \theta_f) \tag{5.35}$$

where α_r is the heat radiation coefficient described by Eq. (5.33) or (5.34), and T_w and T_f are wall and fluid temperatures respectively.

5.7 Initial and Boundary Conditions

In order to solve the heat conduction equation it is necessary to know the initial and boundary conditions. The initial condition prescribes the temperature at time $t = 0$, that is

$$\theta\,(\mathbf{x}, 0) = f\,(\mathbf{x}) \tag{5.36}$$

The boundary conditions describe the heat exchange at the boundary of the body and are given by one of three possible cases.

1. The temperature distribution on the boundary of the body at any time t

$$\theta_w\,(t) = g\,(t) \tag{5.37}$$

where θ_w (t) is the body surface temperature. This condition is called a boundary condition of the first kind.

2. The heat flux is determined at each point of the body surface

$$q_w (t) = h (t) \tag{5.38}$$

This condition is called a boundary condition of the second kind.

3. The temperature of the surrounding medium and the relation describing the heat exchange between the heat-conducting material and the surroundings are known. The heat exchange between the heat-conducting body and its surroundings takes place by convection, radiation or by both of these phenomena and is most conveniently described by the Newton equation (5.29).

The Newton equation written for the surface element dA

$$dq_h = \alpha (\theta_w - \theta_f) \, dA \tag{5.39}$$

shows the amount of heat exchanged by the element with the surroundings. On the other hand the same amount of heat has to be conducted at the boundary of the body, i.e.

$$dq_h = - k \, (\text{grad } \theta)_w \, dA \tag{5.40}$$

where $(\text{grad } \theta)_w$ denotes the magnitude of the temperature gradient between the boundary of the body and the surroundings. Comparison of the above two expressions for dq_h gives

$$(\text{grad } \theta)_w = -\frac{\alpha}{k}\left(\theta_w - \theta_f\right) \tag{5.41}$$

The above condition is called the boundary condition of the third kind. In the mathematical theory of heat conduction the ratio α/k is often denoted by h and is called the heat exchange coefficient.

A fourth kind of boundary condition covers heat exchange with surroundings by conduction. In heat exchange with surroundings by conduction the material surface temperature θ'_w and the surroundings material temperature θ''_w are identical

$$\theta'_w(t) = \theta''_w(t) \tag{5.42}$$

Moreover magnitudes of heat fluxes on the surface separating the materials considered are identical

$$-k'\left(\frac{\partial\theta}{\partial n}\right)'_w = -k''\left(\frac{\partial\theta}{\partial n}\right)''_w \tag{5.43}$$

5.8 Thermomechanical Behaviour

At this point we describe thermo-mechanical coupling in thermo-elastic material. The constitutive equations of thermo-mechanical elementary systems reduce to the state equations

$$S = -\frac{\partial \Psi}{\partial T} \quad \mathbf{S} = \frac{\partial \Psi}{\partial \mathbf{E}} \tag{5.44}$$

where the free energy Ψ depends only on the external variables T and **E** as

$$\Psi = \Psi\,(T, \mathbf{E}) \tag{5.45}$$

Under the assumptions of small temperature variations $\theta = T - T_o$, the free energy is written as

$$\Psi = \mathbf{S}^o \mathbf{E} - S_o \theta + \frac{1}{2} \mathbf{E}\,\mathbf{C}\,\mathbf{E} - \theta\,\mathbf{A}\,E - \frac{1}{2} b\theta^2 \tag{5.46}$$

where **C**, **A**, b are material characteristics. Tensor **C,** which is a tensor of the fourth order, is called the elasticity tensor

$$\mathbf{C} = C_{\alpha\beta\gamma\delta}\, \mathbf{e}_\alpha \otimes \mathbf{e}_\beta \otimes \mathbf{e}_\gamma \otimes \mathbf{e}_\delta \tag{5.47}$$

Tensor **C** is symmetric

$$C_{\alpha\beta\gamma\delta} = C_{\beta\alpha\gamma\delta} = C_{\alpha\beta\delta\gamma} = C_{\beta\alpha\delta\gamma} = C_{\gamma\delta\alpha\beta} \tag{5.48}$$

The tensor **A** is a symmetric second-order tensor characterizing the properties of the material. Through thermo-mechanical coupling, $-\mathbf{A}\,\theta$ represents the stress variation $S - S_o$ produced by the temperature variation θ in deformation when $\mathbf{E} = 0$.

Using Eqs. (5.44) and (5.46), the constitutive equations are given as

$$\mathbf{S} = \mathbf{S}^o + \mathbf{C}\,\mathbf{E} - \mathbf{A}\,\theta \tag{5.49}$$

$$S = S_o + \mathbf{A}\,\mathbf{E} - b\theta \tag{5.50}$$

A time differentiation of Eq. (5.50) gives

$$T_o \frac{dS}{dt} = T_o\,\mathbf{A} \frac{d\,\mathbf{E}}{dt} + T_o b \frac{d\theta}{dt} \tag{5.51}$$

The thermal equation (4.37) reads

$$T_o \frac{dS}{dt} = R - \mathrm{Div}\,Q + \Pi^1 \tag{5.52}$$

where

$$\Pi^1 = \mathbf{S} \frac{dE}{dt} - S \frac{dT}{dt} - \frac{d\Psi}{dt} \tag{5.53}$$

is the intrinsic volume dissipation associated with the open system dV_r. The dissipation Π_1 due to the non-dissipative character of thermo-elastic behaviour is equal to zero.

Part III

Plasticity

6 Plastic Behaviour of Material

6.1 Introduction

A property of various metals that describes their ability to undergo permanent stains is called plasticity. The uniaxial tension test is a convenient method to indicate plastic behaviours of a material. A curve for typical material - soft steel in the system nominal stress-Cauchy strain presented in Fig. 6.1 indicates the following characteristic points: 1: proportional limit; 2: elastic limit; 3: yield point; 3-4: platform of ideal plasticity; 4-5: plastic hardening; 5: necking point; 6: rupture point. The typical curve shown does not describe the character of stress-strain curves for other materials.

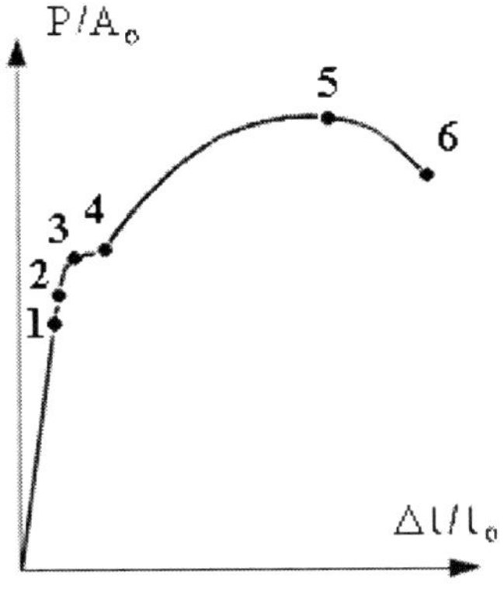

Fig. 6.1. The stress-strain curve for soft steel

Many materials do not give evidence of the existence of a linear segment (annealed aluminium for example). The stress-strain relations for chosen materials are presented in Fig. 6.2. In order to describe the behaviour of material in the best possible way the modeling process is necessary by introducing thermodynamical notions and relations connecting these notions.

The behaviour of an elementary system is said to be elasto-plastic, if in the present state and in the space of stress tensors $\{\boldsymbol{\sigma}\}$, a domain exists that entirely contains a loading path to which the elastic evolutions are associated.

In general, all materials and their mechanical properties are sensitive to temperature and strain rate. Fig. 6.3 illustrates the influence of temperature on flow stress for aluminium.

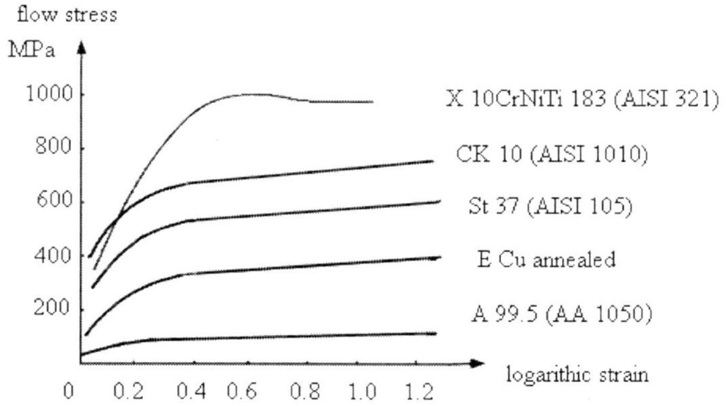

Fig. 6.2. Stress-strain curves of different materials at room temperature [Krause, 1962]

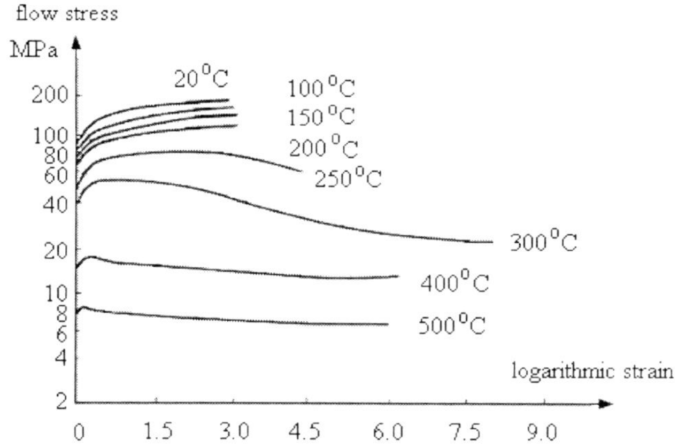

Fig. 6.3. Stress-strain curves of aluminium at different temperatures [Stüwe, 1965]

In this chapter we assume that the variation of strain depends on the loading history and is independent of the loading rate. The plastic models with the viscosity effects are not considered here and are analyzed in the next section.

The plasticity modeling is carried out in the framework of physical linearization and infinitesimal transformations. Since permanent strain is involved, the plastic evolution of an elementary system is irreversible. This evolution can be viewed as the superposition of thermodynamic equilibrium states.

6.2 Plastic Strain

In order to characterize the state of material and to describe its irreversible evolutions we have to know both external variables i.e. the strain tensor $\boldsymbol{\varepsilon}$, temperature T and internal variables.

Consider an elementary system in a present state, characterized by the stress tensor $\boldsymbol{\sigma}$ and the temperature T. From this present state let $d\boldsymbol{\sigma}$ and dT be incremental loading variations in stress and temperature, respectively, and $d\boldsymbol{\varepsilon}$ the incremental strain.

In the elementary unloading process defined by the opposite increments $-\,d\boldsymbol{\sigma}$, $-\,dT$, the reversible elastic strain increments $-\,d\boldsymbol{\varepsilon}^e$ is observed.
The plastic strain increment $d\boldsymbol{\varepsilon}^p$ is defined by

$$d\boldsymbol{\varepsilon} = d\boldsymbol{\varepsilon}^e + d\boldsymbol{\varepsilon}^p \tag{6.1}$$

In the initial configuration, for which $\boldsymbol{\varepsilon} = 0$ the initial stress $\boldsymbol{\sigma}^o$ and the temperature T_o are given. The plastic strain is defined as the integrals of increments $d\boldsymbol{\varepsilon}^p$

$$\boldsymbol{\varepsilon} = \boldsymbol{\varepsilon}^e + \boldsymbol{\varepsilon}^p \tag{6.2}$$

The plastic strain $\boldsymbol{\varepsilon}^p$ can be measurable in the unloading process from the current state. This state restores the initial stress denoted by $\boldsymbol{\sigma}^o$ and the temperature T_o. According to its definition and to the experiments needed for its determination, plastic variable $\boldsymbol{\varepsilon}^p$ is an internal state variable, since it is not accessible to direct observation. The increment $d\boldsymbol{\varepsilon}$ of the external variable can be measured in experiments. The increment $d\boldsymbol{\varepsilon}^e$ of the internal variable is obtained by subtracting the successive value of the external variable increment in experiments that correspond to the opposite variation of the loading. The incremental nature of plasticity is observed as the internal plastic variable $\boldsymbol{\varepsilon}^p$ measured only as integrals of increments, contrary to the external variable $\boldsymbol{\varepsilon}$, which is directly measurable.

6.3 State Equations

Assume the infinitesimal transformation hypothesis defined by the relation

$$\|\text{Grad}\,\mathbf{u}\| << 1 \tag{6.3}$$

where $\mathbf{u} = \mathbf{x} - \mathbf{X}$ is the displacement vector. Then we use the linearized approximation $\boldsymbol{\varepsilon}$ of the Green-Lagrange strain tensor $\mathbf{E}$

$$2\mathbf{E} \cong 2\boldsymbol{\varepsilon} = \text{grad}\,\mathbf{u} + {}^{T}\text{grad}\,\mathbf{u} \tag{6.4}$$

The Jacobian J of the transformation is

$$J \cong 1 + \text{tr}\,\boldsymbol{\varepsilon} \tag{6.5}$$

The relation combining the Cauchy stress tensor $\boldsymbol{\sigma}$ and the Piola-Kirchhoff stress tensor $\mathbf{S}$ is

$$\boldsymbol{\sigma} = J^{-1}\,\mathbf{F} \cdot \mathbf{S} \cdot {}^{T}\mathbf{F} \tag{6.6}$$

where $\mathbf{F} = \mathbf{1} + \text{Grad}\,\mathbf{u}$.

By the infinitesimal transformation hypothesis

$$\boldsymbol{\sigma} = (-\,\text{tr}\,\boldsymbol{\varepsilon})\,\mathbf{S} + \text{grad}\,\mathbf{u} \cdot \mathbf{S} + \mathbf{S}\,{}^{T}\text{grad}\,\mathbf{u} \tag{6.7}$$

The thermodynamic states of a plastic material in infinitesimal transformation are characterized by the external variables T, $\boldsymbol{\varepsilon}$ and the internal variables $\boldsymbol{\varepsilon}^{p}$ and m_I (I = 1, ..., N). The free energy is a function of the external and internal variables

$$\Psi = \Psi\,(T, \boldsymbol{\varepsilon}, \boldsymbol{\varepsilon}^{p}, \mathbf{m}) \tag{6.8}$$

The variables **m** in Eq. (6.8) characterize the hardening state. The state equations are expressed in the form

$$S = -\frac{\partial \Psi}{\partial T} \qquad \boldsymbol{\sigma} = \frac{\partial \Psi}{\partial \boldsymbol{\varepsilon}} \tag{6.9}$$

The expressions (6.9) are based on the normality of external variables T, $\boldsymbol{\varepsilon}$ with regard to the whole set of state variables. In plastic behaviour actual evolutions of these external variables varies independently of the other and must correspond to infinitesimal elastic evolutions from the present state in which $d\boldsymbol{\varepsilon}^{p} = 0$. The validity of state equations (6.9) on the border of the elasticity domain is ensured so long as free energy Ψ is continuously differentiable with respect to the external variables. The expression of free energy Ψ is presented as the second-order expansion with respect to the variables T, $\boldsymbol{\varepsilon}$, $\boldsymbol{\varepsilon}^{p}$ close to the initial state.

$$\begin{aligned} \Psi &= \boldsymbol{\sigma}^{o}\left(\boldsymbol{\varepsilon} - \boldsymbol{\varepsilon}^{p}\right) - S_{o}\theta + \frac{1}{2}\left(\boldsymbol{\varepsilon} - \boldsymbol{\varepsilon}^{p}\right)C\left(\boldsymbol{\varepsilon} - \boldsymbol{\varepsilon}^{p}\right) \\ &- \theta\,\mathbf{A}\left(\boldsymbol{\varepsilon} - \boldsymbol{\varepsilon}^{p}\right) - \frac{1}{2}\frac{C_{\varepsilon}}{T_{o}}\theta^{2} + U(\mathbf{m}) \end{aligned} \tag{6.10}$$

with $\theta = T - T_o$. From Eqs. (6.9) and (6.10) the elastic state equations are

$$\boldsymbol{\sigma} = \boldsymbol{\sigma}^{o} + \mathbf{C}\left(\boldsymbol{\varepsilon} - \boldsymbol{\varepsilon}^{p}\right) - \mathbf{A}\,\theta \tag{6.11}$$

$$S = S_{o} + \mathbf{A}\left(\boldsymbol{\varepsilon} - \boldsymbol{\varepsilon}^{p}\right) + C_{\varepsilon}\theta/T_{o} \tag{6.12}$$

In an isotropic case Eq. (6.11) reads

$$\boldsymbol{\sigma} = \boldsymbol{\sigma}^{o} + (K - 2/3\mu)\,\text{tr}\left(\boldsymbol{\varepsilon} - \boldsymbol{\varepsilon}^{p}\right)\mathbf{1} + 2\mu\left(\boldsymbol{\varepsilon} - \boldsymbol{\varepsilon}^{p}\right) - 3\alpha K\theta\mathbf{1} \tag{6.13}$$

$$S = S_{o} + 3\alpha\,K\text{tr}\left(\boldsymbol{\varepsilon} - \boldsymbol{\varepsilon}^{p}\right) + C_{\varepsilon}\theta/T_{o} \tag{6.14}$$

Through thermo-mechanical coupling, $-\mathbf{A}\,\theta$ represents the stress variation $\boldsymbol{\sigma}-\boldsymbol{\sigma}^{o}$ produced by temperature variation θ in deformation when $\boldsymbol{\varepsilon}-\boldsymbol{\varepsilon}^{p}=0$. K appearing in Eq. (6.13) is the bulk modulus and C_{ε} is the volume heat capacity per unit of initial volume, α is the thermal dilatation coefficient and μ is the Lame constant. The above state equations can be obtained for an elastic case by assuming $\boldsymbol{\varepsilon}^{p}=0$. The frozen energy U (**m**) appearing in Eq. (6.10) is the energy due to hardening and is assumed to be independent of the external state variable $\boldsymbol{\varepsilon}$.

Assume that the function U (**m**) is independent of temperature T thus, the evolution of entropy S is decoupled from the hardening evolution and the hardening latent heat is assumed to be zero. The case when the function U depends on temperature (i.e. for a non-zero hardening latent heat) is discussed in Section 6.8. Expression (6.10) of free energy Ψ is based on a hypothesis of separation of energies. The sum of an energy depends only on reversible variables and frozen energy.

Introduce the definition of the reduced potential in the form

$$W=\boldsymbol{\sigma}^{o}\left(\boldsymbol{\varepsilon}-\boldsymbol{\varepsilon}^{p}\right)-\theta\,\mathbf{A}\left(\boldsymbol{\varepsilon}-\boldsymbol{\varepsilon}^{p}\right)+\frac{1}{2}\left(\boldsymbol{\varepsilon}-\boldsymbol{\varepsilon}^{p}\right)\mathbf{C}_{o}\left(\boldsymbol{\varepsilon}-\boldsymbol{\varepsilon}^{p}\right) \tag{6.15}$$

which leads to rewriting the state equations (6.11) and (6.12) in the form

$$\boldsymbol{\sigma}=\frac{\partial W}{\partial\left(\boldsymbol{\varepsilon}-\boldsymbol{\varepsilon}^{p}\right)}=\frac{\partial W}{\partial\boldsymbol{\varepsilon}^{e}} \tag{6.16}$$

Let $W^{*}(\boldsymbol{\sigma})$ be the Legendre-Fenchel transform of W $(\boldsymbol{\varepsilon}^{e})$ defined as

$$W^{*}=\boldsymbol{\sigma}\;\boldsymbol{\varepsilon}^{e}-W \tag{6.17}$$

where $\boldsymbol{\sigma}$ and $\boldsymbol{\varepsilon}^{e}$ are linked by relation (6.16). Equations (6.15) – (6.17) yield

$$W^{1}=\frac{1}{2}\boldsymbol{\alpha}_{o}\,\mathbf{A}_{o}\,\theta^{2}+\left(\boldsymbol{\sigma}-\boldsymbol{\sigma}^{o}\right)\boldsymbol{\alpha}_{o}\theta+\frac{1}{2}\left(\boldsymbol{\sigma}-\boldsymbol{\sigma}^{o}\right)\mathbf{C}_{o}^{-1}\left(\boldsymbol{\sigma}-\boldsymbol{\sigma}^{o}\right) \tag{6.18}$$

where

$$\boldsymbol{\alpha}_{o}=\mathbf{C}_{o}^{-1}\,\mathbf{A}_{o} \tag{6.19}$$

By the properties of the Legendre-Fenchel transform

$$\boldsymbol{\varepsilon}-\boldsymbol{\varepsilon}^{p}=\boldsymbol{\varepsilon}^{e}=\frac{\partial W^{*}}{\partial\boldsymbol{\sigma}} \tag{6.20}$$

The expressions (6.18) and (6.19) yield

$$\boldsymbol{\varepsilon}-\boldsymbol{\varepsilon}^{p}=\mathbf{C}_{o}^{-1}\left(\boldsymbol{\sigma}-\boldsymbol{\sigma}^{o}\right)+\boldsymbol{\alpha}_{o}\theta \tag{6.21}$$

which in the isotropic case give

$$\boldsymbol{\varepsilon}-\boldsymbol{\varepsilon}^{p}=\frac{1+\nu_{o}}{E_{o}}\left(\boldsymbol{\sigma}-\boldsymbol{\sigma}^{o}\right)-\frac{\nu_{o}}{E_{o}}\mathrm{tr}\left(\boldsymbol{\sigma}-\boldsymbol{\sigma}^{o}\right)\mathbf{1}+\alpha_{o}\theta\,\mathbf{1} \tag{6.22}$$

where ν is the Poisson ratio and E is the Young modulus with the subscript o referring to the initial state.

6.4 Evolution Relations

6.4.1 The Plastic Work Rate

Assume the hypothesis of infinitesimal transformation. If **E** and **S** are replaced by $\boldsymbol{\varepsilon}$ and $\boldsymbol{\sigma}$ in Eq. (4.43), then the intrinsic dissipation Π_1 is

$$\Pi_1 = \boldsymbol{\sigma}\frac{d\boldsymbol{\varepsilon}}{dt} - \frac{d\Psi}{dt} - S\frac{dT}{dt} \geq 0 \tag{6.23}$$

The intrinsic dissipation Π_1 can be rewritten in the form

$$\Pi_1 = \boldsymbol{\sigma}\frac{d\boldsymbol{\varepsilon}^p}{dt} - \frac{\partial U}{\partial \mathbf{m}} \cdot \frac{d\mathbf{m}}{dt} \geq 0 \tag{6.24}$$

The above is obtained by Eq. (6.10) of the free energy Ψ and the state equations (6.11) and (6.12) retaining the terms of the same order of magnitude. The thermodynamic hardening force $\boldsymbol{\eta}$ is defined as

$$\boldsymbol{\eta} = -\frac{\partial U}{\partial \mathbf{m}} \tag{6.25}$$

The expression (6.25) can be rewritten as

$$\mathbf{m} = -\frac{\partial U^*}{\partial \boldsymbol{\eta}} \tag{6.26}$$

In Eq. (6.26) $U^*(\boldsymbol{\eta})$ is the Legendre-Fenchel transform of $U(\mathbf{m})$

$$U^* = -\boldsymbol{\eta}\mathbf{m} - U \tag{6.27}$$

The intrinsic dissipation Π_1 is

$$\Pi_1 = \Pi^p - \frac{dU}{dt} = \Pi^p + \boldsymbol{\eta} \cdot \frac{d\mathbf{m}}{dt} \tag{6.28}$$

where Π^p is the plastic work defined by

$$\Pi^p = \boldsymbol{\sigma}\frac{d\boldsymbol{\varepsilon}^p}{dt} \tag{6.29}$$

The name of the plastic dissipation is used for the plastic work rate. It should be noted that the plastic dissipation is the energy $\Pi_1 dt = \Pi^p dt - dU$ and not the energy $\Pi^p dt$ dissipated into heat during the time interval dt of the loading, but which cannot be immediately recovered as efficient work during load reversal. The energy denoted by U in Eq. (6.10) is called frozen energy due to hardening. The evolution of U(**m**) relies on the evolution of variables **m** defining the hardening state of the material. The frozen energy is met during reversed loading. Then a certain amount of energy is not recovered nor transported into efficient work during reversed loading, although having not been converted into heat during loading.

6.4.2 Plasticity Criterion

The stress $\boldsymbol{\sigma}$ at any loading state characterizes any open elementary system. The loading point ($\boldsymbol{\sigma}$) in the stress space {$\boldsymbol{\sigma}$} represents the present loading state. Denote the domain of elasticity in initial state by $\mathcal{E}_D$. It contains the zero loading point ($\boldsymbol{\sigma}$) = (**0**). In the elasticity domain the strain increase remains reversible or elastic, for any path of the loading point ($\boldsymbol{\sigma}$) starting from the origin of space and lying inside this domain (Fig. 6.4.).

A hardening frozen energy is absent in ideal plastic material without any hardening effect. The initial domain of elasticity for this material is not changed by the appearance of plastic strain. The elasticity domain is identical to the initial domain, and the loading point ($\boldsymbol{\sigma}$) cannot leave this domain (Fig. 6.4.). If the loading point is and remains on the boundary of the elasticity domain $\mathcal{E}_D$ as illustrated by the loading path **12** in Fig. 6.4. then the evolutions of plastic strain occur. Consider a loading path leaving the boundary towards the interior of domain $\mathcal{E}_D$. It can be for instance the path **23** in Fig. 6.4, corresponding to a purely elastic evolution of the elementary system. It corresponds to an elastic unloading.

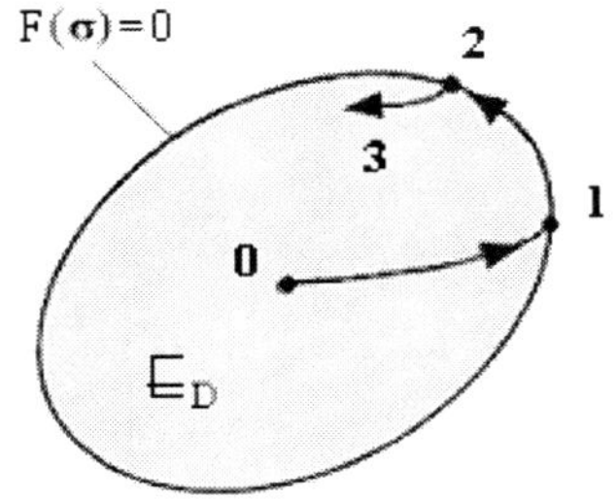

Fig. 6.4. Elasticity domain of ideal plastic material

The elasticity domain is defined by a scalar function F. It is called the loading function and has $\boldsymbol{\sigma}$ as its arguments. It is such that F ($\boldsymbol{\sigma}$) < 0 represents the interior of domain E_D, F ($\boldsymbol{\sigma}$) = 0 represents the boundary of domain E_D and F ($\boldsymbol{\sigma}$) > 0 represents the exterior of domain E_D. The criterion F ($\boldsymbol{\sigma}$) < 0 is the elasticity criterion. The criterion F ($\boldsymbol{\sigma}$) = 0 is the plasticity criterion. The surface in the space of loading points {$\boldsymbol{\sigma}$}, defined by F ($\boldsymbol{\sigma}$) = 0, represents the boundary of domain E_D and is called the yield locus. The plastically admissible loading state ($\boldsymbol{\sigma}$) satisfies the criterion F ($\boldsymbol{\sigma}$) ≤ 0.

The elasticity domain for hardening materials is altered by the appearance of plastic strain. In the space {$\boldsymbol{\sigma}$} of loading points, the present elasticity domain is defined as the domain arisen by the set of elastic unloading paths, or reversible loading paths, which issue from a present loading point **2**, as path **23** in Fig. 6.5.

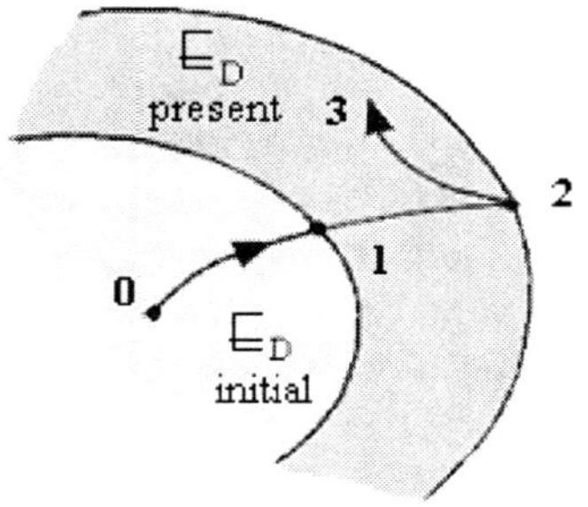

Fig. 6.5. Elasticity domains of hardening material

The present loading point is not necessarily on the boundary of the present elasticity domain, such as point **3** in Fig. 6.5. There still exists an initial elasticity domain but, as soon as the loading point ($\boldsymbol{\sigma}$) reaches for the first time the boundary of the initial elasticity domain (point **1**), further loading can deform this domain while carrying it along (loading path **12**). This is the phenomenon of hardening. The present elasticity domain depends not only on the present loading point ($\boldsymbol{\sigma}$), but also on the loading path followed before, and thus on the hardening state.

Consider the domain of elasticity in the present state E_D. It is defined by a scalar loading function F, with arguments $\boldsymbol{\sigma}$ and with some hardening parameters represented by hardening force $\boldsymbol{\eta}$. For the hardening material, it is such that F ($\boldsymbol{\sigma}$, $\boldsymbol{\eta}$) < 0 represents the interior of domain E_D, F ($\boldsymbol{\sigma}$, $\boldsymbol{\eta}$) = 0 represents the boundary of domain E_D, F ($\boldsymbol{\sigma}$, $\boldsymbol{\eta}$) > 0 represents to the exterior of domain E_D. The criterion of elasticity is expressed by F ($\boldsymbol{\sigma}$, $\boldsymbol{\eta}$) < 0. The plasticity threshold or criterion is expressed by F ($\boldsymbol{\sigma}$, $\boldsymbol{\eta}$) = 0. The surface defined by F ($\boldsymbol{\sigma}$, $\boldsymbol{\eta}$) = 0, in the space of loading points {$\boldsymbol{\sigma}$}, representing the boundary of the present domain E_D is called the present yield locus.

We say that a loading state ($\boldsymbol{\sigma}$) the plastically admissible in the present state if it satisfies the criterion F ($\boldsymbol{\sigma}$, $\boldsymbol{\eta}$) ≤ 0.

The loading function F considered in the space {$\boldsymbol{\sigma} \times \boldsymbol{\eta}$} by the equation F ($\boldsymbol{\sigma}$, $\boldsymbol{\eta}$) ≤ 0 defines a generalized elasticity domain E, which is now fixed as in ideal plasticity and which the generalized loading point ($\boldsymbol{\sigma}$, $\boldsymbol{\eta}$) cannot escape.

The present elasticity domain $\mathbb{E}_D$, as previously defined in the space $\{\boldsymbol{\sigma}\}$, appears in the extended space $\{\boldsymbol{\sigma} \times \boldsymbol{\eta}\}$ as the intersection of the fixed domain $\mathbb{E}$ with the hyperplane $\boldsymbol{\eta} = \eta_{present}$, where $\eta_{present}$ is the present value of hardening force $\boldsymbol{\eta}$. Note that, owing to hardening phenomena, the origin $O = (\mathbf{0})$ of space $\{\boldsymbol{\sigma}\}$ may become outside the present elasticity domain, as illustrated in Figure 6.7.
Consider hardening parameters. A zero hardening force ($\boldsymbol{\eta} = 0$) corresponds to a material state without any hardening history. The origin $O = (\mathbf{0}, \mathbf{0})$ in space $\{\boldsymbol{\sigma} \times \boldsymbol{\eta}\}$ necessarily belongs to the extended elasticity domain $\mathbb{E}$, since it corresponds to a material without any loading history. The loading paths **01**, **12** and **23** in space $\{\boldsymbol{\sigma}\}$, as represented in Figure 6.6 can be simply interpreted in the space $\{\boldsymbol{\sigma} \times \boldsymbol{\eta}\}$. The loading path **01** corresponds to an elastic evolution from the virgin state ($\boldsymbol{\sigma} = \boldsymbol{\eta} = 0$), without evolution of the hardening state ($d\boldsymbol{\eta} = 0$). Therefore, the loading path **01** in the space $\{\boldsymbol{\sigma}\}$ corresponds to the loading path **01** in the space $\{\boldsymbol{\sigma} \times \boldsymbol{\eta}\}$. At point **1** the plasticity criterion is satisfied, and loading path **12** corresponds to a plastic evolution during which the loading point carries along the elasticity domain $\mathbb{E}_D$, while deforming it. The hardening state is modified and the plasticity criterion $F = 0$ is constantly satisfied during this evolution. Hence, in the space $\{\boldsymbol{\sigma} \times \boldsymbol{\eta}\}$ the loading point moves on the boundary of the fixed domain $\mathbb{E}$ from point **1** to point **2**. The loading path **23** corresponds to an elastic unloading without evolution of the hardening state ($d\boldsymbol{\eta} = 0$). Therefore, the hardening force $\boldsymbol{\eta}$ keeps the value $\boldsymbol{\eta}_2$ reached at point **2**. The loading path **23** in the space $\{\boldsymbol{\sigma}\}$ corresponds to the loading path **23** in the space $\{\boldsymbol{\sigma} \times \boldsymbol{\eta}\}$, as illustrated in Figure 6.6.

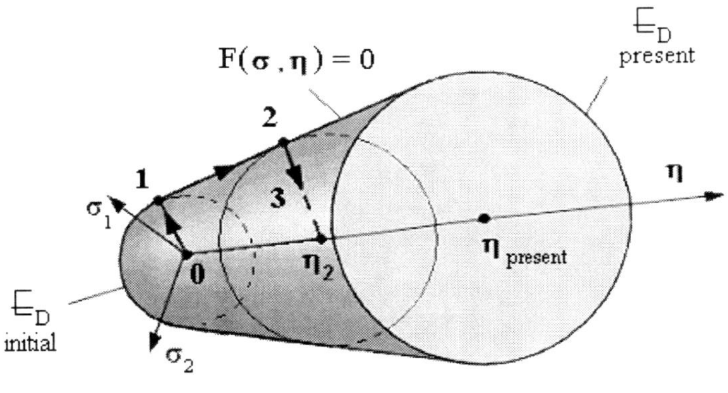

Fig. 6.6. Elasticity domains in the space $\{\boldsymbol{\sigma} \times \boldsymbol{\eta}\}$

The present elasticity domain $\mathbb{E}_D$ depends on the present value of the hardening force $\boldsymbol{\eta}$. This dependency is the basis of their experimental identification. To be useful in practice, models must involve a few hardening variables, which correspond to a few components for vector $\boldsymbol{\eta}$. For this purpose, simple hardening models have been designed.

The first one is the isotropic hardening model. In this model the elasticity domain in space {**σ**} is transformed by a homothety centred at the origin, as illustrated in Figure 6.7. The hardening force is reduced to a single scalar parameter **η** required to characterize this homothety.

The second one is the kinematic hardening model. In this model the boundaries in space {**σ**} of the elasticity domain are obtained through a translation of the boundary of the initial domain. The hardening variables are the variables characterizing this translation. They reduce to a tensor parameter **η** relative to the translation with respect to the stress tensor (Fig. 6.7). The two previous hardening models can also be combined to yield an isotropic and kinematic hardening model, as illustrated in Figure 6.7. As defined in this section, the hardening force **η** represents only a set of variables well suited for mathematical description of the observed evolution of the elasticity domain, and thus may not yet be considered as a thermodynamic force.

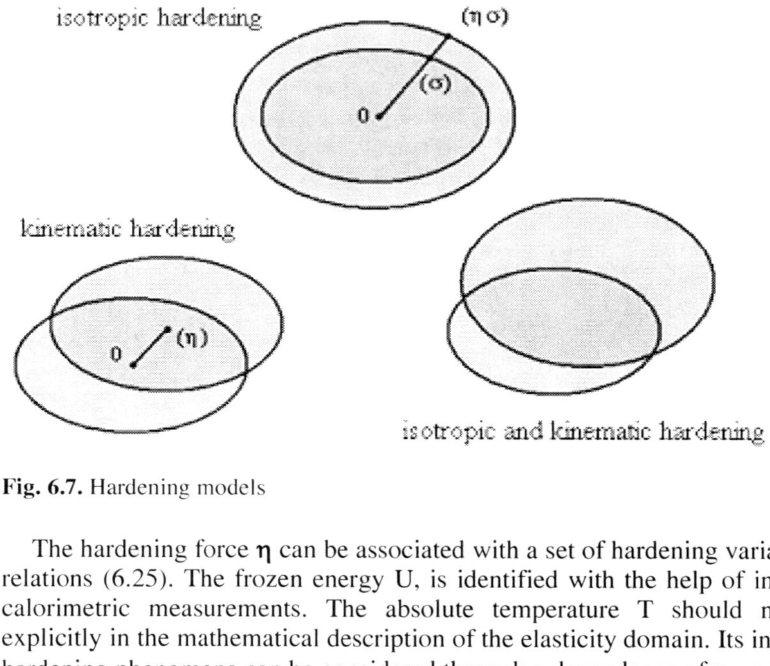

Fig. 6.7. Hardening models

The hardening force **η** can be associated with a set of hardening variables **m** by relations (6.25). The frozen energy U, is identified with the help of independent calorimetric measurements. The absolute temperature T should not appear explicitly in the mathematical description of the elasticity domain. Its influence on hardening phenomena can be considered through a dependence of **η** upon T.

According to relation (6.25) the temperature T can be an argument of energy U. The influence of T on hardening phenomena will not be considered in the following. According to relation (6.25), the virgin hardening state, which has been previously defined by **η** = 0 as illustrated in Fig. 6.6, now corresponds to the relation dU = 0, which is irrespective of any particular choice of hardening parameters.

The initial and present elasticity domains are convex. This property of convexity constitutes one of the sufficient criterions for the stability of plastic materials. In the loading point space $\{\boldsymbol{\sigma} \times \boldsymbol{\eta}\}$, the fundamental geometrical property of a convex domain is that all points of a segment of a line that joins two points on the boundary of the domain lie inside this domain.

6.4.3 The Plastic Flow Rule

The plastic flow rule describes how the plastic phenomena occur. If the loading point $(\boldsymbol{\sigma})$ lies within the elasticity domain $\mathcal{E}_D$, $F(\boldsymbol{\sigma}) < 0$, then the strain increments $d\boldsymbol{\varepsilon}$ are elastic or reversible. If the loading point (σ) is on the boundary of $\mathcal{E}_D$ but leaves it during elastic unloading the strain increment is also elastic. The elastic unloading criterion is $F = 0$ and $dF < 0$.

In the case when the loading point $(\boldsymbol{\sigma})$ is on the boundary of the domain $\mathcal{E}_D$, i.e. $F = dF = 0$ we say that the corresponding infinitesimal loading increment $d\boldsymbol{\sigma}$ is a neutral loading increment. Then the increment $d\boldsymbol{\varepsilon}$ may not be elastic. This is expressed by

$$d\boldsymbol{\varepsilon}^p = 0 \quad \text{if} \quad F(\boldsymbol{\sigma}) < 0 \quad \text{or if} \quad F(\boldsymbol{\sigma}) = 0 \quad \text{and} \quad dF = \frac{\partial F}{\partial \boldsymbol{\sigma}} d\boldsymbol{\sigma} < 0 \tag{6.30}$$

$$d\boldsymbol{\varepsilon} = d\boldsymbol{\varepsilon}^e + d\boldsymbol{\varepsilon}^p \quad \text{if} \quad F = dF = 0$$

The notation dF says that the function F is differentiable with respect to its arguments, at the considered loading point $(\boldsymbol{\sigma})$ on the boundary of the domain $\mathcal{E}_D$.

Let ∂F be the subdifferential of F with respect to $(\boldsymbol{\sigma})$ at the present loading point $(\boldsymbol{\sigma})$. The subdifferential ∂F at a regular point on the boundary of domain $\mathcal{E}_D$ defined by $F(\boldsymbol{\sigma}) = 0$ reduces to the gradient of F with respect to $(\boldsymbol{\sigma})$ and thus corresponds geometrically at this point to the outward normal $\partial F/\partial \boldsymbol{\sigma}$ to the boundary. At a singular point in the boundary of domain $\mathcal{E}_D$ a convex loading function shows that the set $\alpha\partial F$ with $\alpha \geq 0$ is constituted by the cone F of outward normals to the boundary of domain $\mathcal{E}_D$.

The expression (6.30) indicates when an evolution of the plastic strain occurs. All values for $d\boldsymbol{\varepsilon}^p$ are not admissible since they must ensure the non-negativeness of intrinsic dissipation Π_1. Since the intrinsic dissipation reduces to the plastic work rate for ideal plastic materials, the non-negativeness of intrinsic dissipation, or equivalenty of the infinitesimal plastic work rate requires the relation for the plastic increments $d\boldsymbol{\varepsilon}^p$

$$\boldsymbol{\sigma}\, d\boldsymbol{\varepsilon}^p \geq 0 \tag{6.31}$$

Write the relation

$$d\boldsymbol{\varepsilon}^p \in d\lambda\, \mathcal{G}(\boldsymbol{\sigma}) \qquad d\lambda \geq 0 \tag{6.32}$$

where $d\lambda$ is a so-called the plastic multiplier. The plastic multiplier is a non-negative scalar factor. The set of thermodynamically admissible directions in the loading point space $\{\boldsymbol{\sigma}\}$ for the vector $(d\boldsymbol{\varepsilon}^p)$ ensuring the non-negativeness of its scalar product $\boldsymbol{\sigma}\, d\boldsymbol{\varepsilon}^p$ is represented by the set $\mathcal{G}\,(\boldsymbol{\sigma})$. Fig. 6.8 illustrates the set $\mathcal{G}\,(\boldsymbol{\sigma})$ in the loading point space $\{\boldsymbol{\sigma}\}$. The admissible directions for the vector $(d\boldsymbol{\varepsilon}^p)$ are independent of the plastic criterion.

The plastic flow rule derived from the non-negativeness of intrinsic dissipation for an ideal plastic material can be summarized by

$$d\boldsymbol{\varepsilon}^p \in d\lambda\ \mathcal{G}\,(\boldsymbol{\sigma}) \qquad d\lambda \geq 0 \qquad \text{where} \quad \begin{array}{l} d\lambda \geq 0 \text{ if } F = 0 \text{ and } dF = 0 \\ d\lambda = 0 \text{ if } F < 0 \text{ or } dF < 0 \end{array} \tag{6.33}$$

If the plastic increments $d\boldsymbol{\varepsilon}^p$ have non-zero values, then we say that the loading is plastic.

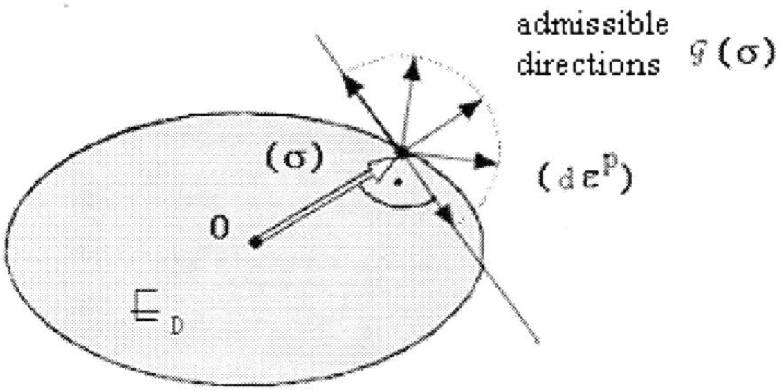

Fig. 6.8. Admissible directions of plastic increments for ideal plastic material

The evolution for hardening materials is elastic with no change of the hardening state, if in the space $\{\boldsymbol{\sigma} \times \boldsymbol{\eta}\}$ the present extended loading point $(\boldsymbol{\sigma}, \boldsymbol{\eta})$ lies inside the fixed domain E $(F < 0)$ or on its boundary $F = 0$ and leaves it $(dF < 0)$. The above can be written in the form

$$(d\boldsymbol{\varepsilon}^p, d\mathbf{m}) \in d\lambda\ \mathcal{G}\,(\boldsymbol{\sigma}, \boldsymbol{\eta}) \qquad d\lambda \geq 0 \quad \text{where} \quad \begin{array}{l} d\lambda \geq 0 \text{ if } F = 0 \text{ and } dF = 0 \\ d\lambda = 0 \text{ if } F = 0 \text{ or } dF = 0 \end{array} \tag{6.34}$$

If the increments $(d\boldsymbol{\varepsilon}^p, d\mathbf{m})$ have non-zero values, then we say that the loading is plastic. The scalar $d\lambda$ in Eq. (6.34) is the plastic multiplier, and $\mathcal{G}\,(\boldsymbol{\sigma}, \boldsymbol{\eta})$ are the set of thermodynamically admissible directions for the increments $(d\boldsymbol{\varepsilon}^p, d\mathbf{m})$ satisfying the relation of the non-negativeness of the intrinsic dissipation

$$\boldsymbol{\sigma}\, d\boldsymbol{\varepsilon}^p + \boldsymbol{\eta} \cdot d\mathbf{m} \geq 0 \tag{6.35}$$

The illustration of the criterion (6.35) is given in Fig. 6.9. The increment $d\mathbf{m}$ of the hardening variable is written in the form

$$d\mathbf{m} = d\lambda\ \mathcal{G}_m\,(\boldsymbol{\sigma}, \boldsymbol{\eta}) \tag{6.36}$$

In the above $\mathcal{G}_m\,(\boldsymbol{\sigma}, \boldsymbol{\eta})$ are the actual directions of the permissible directions for $d\mathbf{m}$.

In the case of hardening plasticity the loading function depends on the hardening force $\boldsymbol{\eta}$. Then the expression for dF is

$$dF = d_{\eta}F + d_{\sigma}F \tag{6.37}$$

with

$$d_{\eta}F = \frac{\partial F}{\partial \boldsymbol{\sigma}} d\boldsymbol{\sigma} \tag{6.38}$$

$$d_{\sigma}F = \frac{\partial F}{\partial \boldsymbol{\eta}} d\boldsymbol{\eta} \tag{6.39}$$

If the plastic increments $d\boldsymbol{\varepsilon}^p$ have non-zero values, then we say that the loading is plastic.

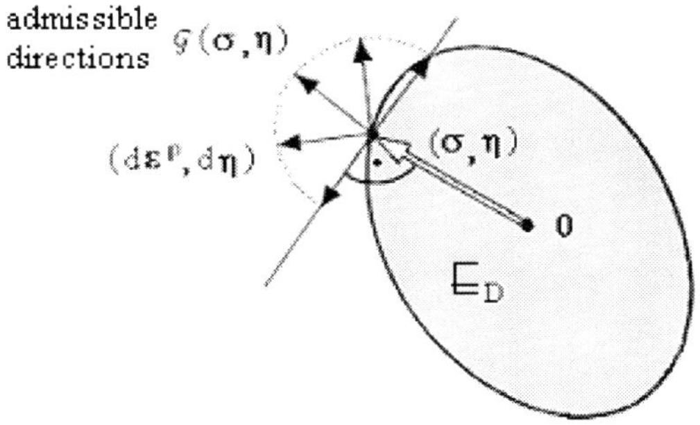

Fig. 6.9. Admissible directions for the flow rule in the case of hardening material

In the above $d_{\eta}F$ is the differential of function F with the hardening force $\boldsymbol{\eta}$ being held constant. The expression (6.39) can be written as $d_{\sigma}F = -H\,d\lambda$, where H is the hardening modulus defined by

$$H = -\frac{\partial F}{\partial \boldsymbol{\eta}} \cdot \frac{\partial \boldsymbol{\eta}}{\partial \mathbf{m}} \cdot \frac{\partial \mathbf{m}}{\partial \lambda} = \frac{\partial F}{\partial \boldsymbol{\eta}} \cdot \frac{\partial^2 U}{\partial \mathbf{m}^2} \cdot \mathcal{G}_{m}(\boldsymbol{\sigma}, \boldsymbol{\eta}) \tag{6.40}$$

The definition (6.39) of $d_{\sigma}F$ and the relation (6.40) show that H has a stress unit. In order to specify the plastic flow rule, we need to examine the specific case of hardening ($H > 0$) and the case of softening ($H < 0$).

In the case of hardening

$$\begin{aligned} &\left(d\boldsymbol{\varepsilon}^p, d\mathbf{m}\right) \in \frac{\partial_{\eta}F}{H}\, \mathcal{G}(\boldsymbol{\sigma}, \boldsymbol{\eta}) \quad \text{if } F = 0 \quad \text{and } d_{\eta}F > 0 \\ &d\boldsymbol{\varepsilon}^p = d\mathbf{m} = 0 \quad \text{if } F < 0 \quad \text{or if } F = 0 \quad \text{and } d_{\eta}F \le 0 \end{aligned} \tag{6.41}$$

In the case of softening

$$\left(d\boldsymbol{\varepsilon}^{p}, d\mathbf{m}\right) \in \frac{\partial_{\eta} F}{H}\, \mathcal{G}(\boldsymbol{\sigma}, \boldsymbol{\eta}) \quad \text{or } d\boldsymbol{\varepsilon}^{p} = d\mathbf{m} = 0 \quad \text{if } F = 0 \quad \text{and } d_{\eta}F < 0 \tag{6.42}$$

$$d\boldsymbol{\varepsilon}^{p} = d\mathbf{m} = 0 \quad \text{if } F < 0 \quad \text{or if } F = d_{\eta}F = 0$$

The relations (6.41) and (6.42) for hardening or softening materials show that if the loading point ($\boldsymbol{\sigma}$) is and remains on the boundary of the present elasticity domain E_D (i.e. if $F = d_{\eta}F = 0$), the evolution is purely elastic without a change in the hardening state (i.e. $d\boldsymbol{\varepsilon}^{p} = d\mathbf{m} = d\boldsymbol{\eta} = 0$). Then we say that the loading increment $d\boldsymbol{\sigma}$ is neutral. This situation is illustrated in Fig. 6.10.

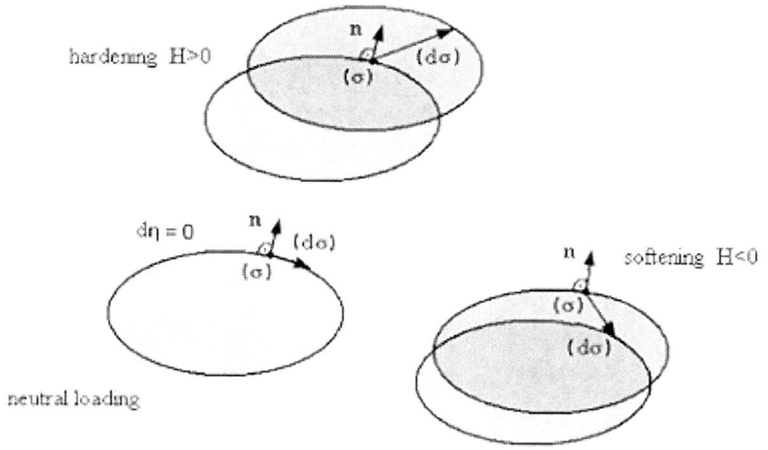

Fig. 6.10. Interpretation of hardening in the space $\{\boldsymbol{\sigma}\}$

The interpretation of the hardening sign can be explained as follows. In the case of hardening, expression (6.41) says that $d_{\eta}F$ is strictly positive for a plastic evolution to occur. The geometrical meaning of $d_{\eta}F$ (the scalar product of the present loading increment ($d\boldsymbol{\sigma}$) with outward unit normal ($\partial F/\partial \boldsymbol{\sigma}$)) shows that the vector of present loading increment $d\boldsymbol{\sigma}$ in the space $\{\boldsymbol{\sigma}\}$ must be oriented outwards with regard to the present elasticity domain E_D at the present loading point ($d\boldsymbol{\sigma}$). In fact, in the case of a regular point, its scalar product with unit outward normal **n** must be positive. In other words, the new loading point ($\boldsymbol{\sigma} + d\boldsymbol{\sigma}$) escapes from the present elasticity domain E_D, while carrying it along. Fig. 6.10 illustrates this geometrical interpretation in the space $\{\boldsymbol{\sigma}\}$ and in Fig. 6.11 in the space $\{\boldsymbol{\sigma} \times \boldsymbol{\eta}\}$.

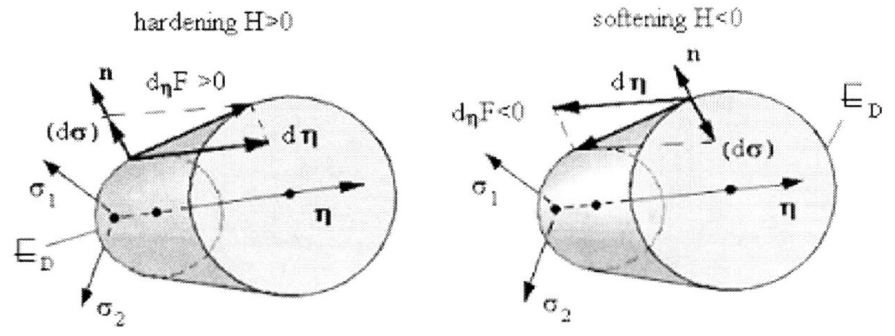

Fig. 6.11. Hardening sign in the space $\{\boldsymbol{\sigma} \times \boldsymbol{\eta}\}$

In the case of softening, expression (6.42) says that $d_\eta F$ is strictly negative for a plastic evolution to occur. Consider the loading point $(\boldsymbol{\sigma} + d\boldsymbol{\sigma})$. This loading point carries the elasticity domain inside E_D when softening occurs. Taking into account the hardening sign, we say that softening is negative hardening.

6.4.4 Thermal Hardening

Consider the case where the hardening force $\boldsymbol{\eta}$ representing the evolution of the elasticity domain is independent of the temperature. The expression of free energy Ψ in this case is of the form

$$\Psi = \Psi\,(T, \boldsymbol{\varepsilon}, \boldsymbol{\varepsilon}^p, \mathbf{m}) = \varphi\,(\boldsymbol{\varepsilon} - \boldsymbol{\varepsilon}^p) + U\,(\mathbf{m}) \tag{6.43}$$

and

$$\boldsymbol{\eta} = -\frac{\partial \Psi}{\partial \mathbf{m}} \qquad S = -\frac{\partial \Psi}{\partial T} \tag{6.44}$$

In the case of thermal hardening effects, in the expression of free energy Ψ the temperature $\theta = T - T_o$ is included

$$\Psi = \Psi\,(\theta = T - T_o, \boldsymbol{\varepsilon}, \boldsymbol{\varepsilon}^p, \mathbf{m}) = \varphi\,(\boldsymbol{\varepsilon} - \boldsymbol{\varepsilon}^p, \theta\,) + U^*\,(\mathbf{m}, \theta) \tag{6.45}$$

and

$$\boldsymbol{\eta} = -\frac{\partial \Psi}{\partial \mathbf{m}} \qquad S = -\frac{\partial \Psi}{\partial \theta} \tag{6.46}$$

Substituting expression (6.45) into (6.46) shows that the hardening force $\boldsymbol{\eta}$ depends on temperature variation $\theta = T - T_0$.

Assume that the temperature variation is small. The function $U^*\,(\mathbf{m}, \theta)$ can be expressed as

$$U^*(\mathbf{m}, \theta) = U(\mathbf{m}) - \theta S^*(\mathbf{m}) \tag{6.47}$$

The function $U(\mathbf{m})$ is interpreted as the frozen free energy. By (6.45), (6.46) and (6.47) we get

$$\eta = -\frac{\partial U}{\partial \mathbf{m}} + \theta \frac{\partial S^*}{\partial \mathbf{m}} \qquad S = -\frac{\partial \varphi}{\partial \theta} + S^*(\mathbf{m}) \tag{6.48}$$

Consider the elasticity domain $\mathcal{E}_D$ in the loading space $\{\boldsymbol{\sigma}\}$ defined by

$$F = F(\boldsymbol{\sigma}, \eta) \leq 0 \tag{6.49}$$

According to Eq. (6.48), due to the temperature variation $d\theta$, the hardening force η and the elasticity domain $\mathcal{E}_D$ may change. Thermal hardening occurs for a zero frozen free energy $U(\mathbf{m}) = 0$. An inverse temperature variation $-d\theta$ restores the previous elasticity domain. The second relation (6.48) shows also that there is an unrecovered change in entropy, yielding the frozen entropy term $S^*(\mathbf{m})$. With respect to the linearized thermal equation (5.6) the quality $T_o S^*(\mathbf{m})$ can easily be related to a hardening latent heat effect. The flow rule can be expressed as

$$(d\boldsymbol{\varepsilon}^p, d\mathbf{m}) \in d\lambda\, \mathcal{G}(\boldsymbol{\sigma}, \eta) \quad d\lambda \geq 0 \quad \text{where} \quad \begin{array}{lll} d\lambda \geq 0 & \text{if } F = 0 & \text{and } dF = 0 \\ d\lambda = 0 & \text{if } F < 0 & \text{or } dF < 0 \end{array} \tag{6.50}$$

where the plastic multiplier $d\lambda$ and the hardening modulus H are now expressed in the form

$$d\lambda = \frac{d_m F}{H} \quad H = -\frac{\partial F}{\partial \eta} \cdot \frac{\partial \eta}{\partial \mathbf{m}} \cdot \frac{\partial \mathbf{m}}{\partial \lambda} = \frac{\partial F}{\partial \eta} \cdot \frac{\partial^2 U}{\partial \mathbf{m}^2} \cdot \mathcal{G}_m(\boldsymbol{\sigma}, \eta) \tag{6.51}$$

Substituting (6.48) in to (6.49) $d_\eta F$ can be expressed in the form

$$d_m F = \frac{\partial F}{\partial \boldsymbol{\sigma}} d\boldsymbol{\sigma} + \frac{\partial F}{\partial \eta} \cdot \frac{dS^*}{\partial \mathbf{m}} d\theta \tag{6.52}$$

In order to express the loading function F in terms of θ and $\mathbf{m}$ we use Eq. (6.48) and Eq. (6.49). Assume $dF = 0$, then Eqs. (6.50) and (6.51) give

$$F(\boldsymbol{\sigma}, \theta, \mathbf{m} + d\mathbf{m}) = F(\boldsymbol{\sigma}, \theta, \mathbf{m}) + \frac{\partial F}{\partial \mathbf{m}} \cdot d\mathbf{m} = F(\boldsymbol{\sigma}, \theta, \mathbf{m}) - H\, d\lambda \tag{6.53}$$

6.5 Plastic Behaviours

6.5.1 The Hypothesis of Maximal Plastic Work

The admissible directions of the plastic strain increment $d\boldsymbol{\varepsilon}^p$ are defined by the hypothesis of maximal plastic work. This hypothesis is formulated as follows.

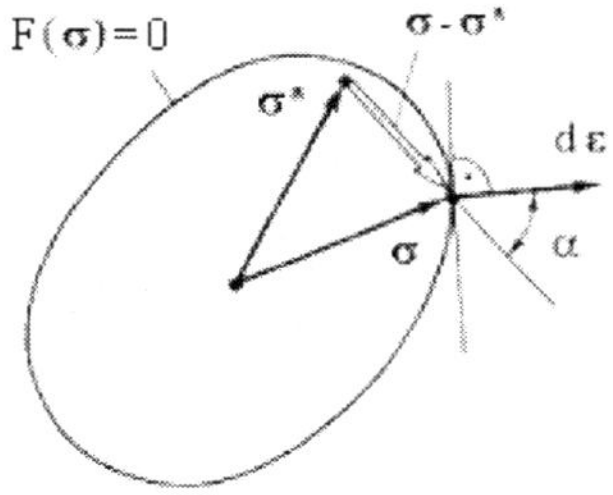

Fig. 6.12. Geometrical interpretation of the hypothesis of maximal plastic work

Let $(\boldsymbol{\sigma})$ be the present loading state in an elementary material system being non-exterior to the present elasticity domain i.e. $(\boldsymbol{\sigma}) \in E_D$ and $d\boldsymbol{\varepsilon}^p$ be the associated plastic increment of strain. Let $(\boldsymbol{\sigma}^*)$ be another loading state being non-exterior to the present elasticity domain i.e. $(\boldsymbol{\sigma}^*) \in E_D$. The hypothesis of maximal plastic work is expressed by

$$(\boldsymbol{\sigma} - \boldsymbol{\sigma}^*)\, d\boldsymbol{\varepsilon}^p \geq 0 \tag{6.54}$$

for every $(\boldsymbol{\sigma}^*) \in E_D$.

This means that the loading state $(\boldsymbol{\sigma})$ to be associated with the present plastic increment of strain $(d\boldsymbol{\varepsilon}^p)$ is the one among those admitted by the present plastic criterion, maximizing the infinitesimal plastic work.

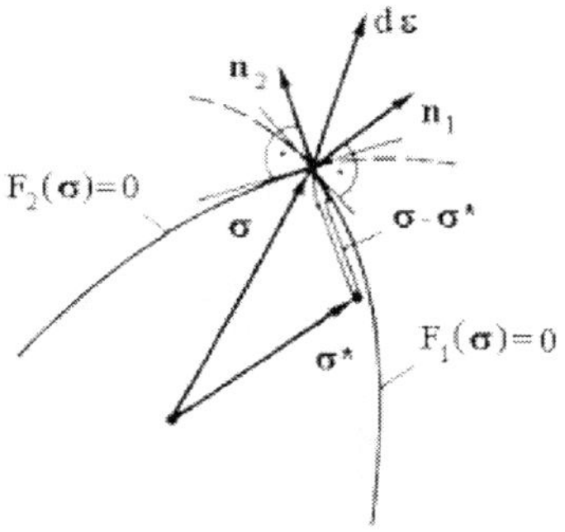

Fig. 6.13. Geometrical interpretation of the hypothesis of maximal plastic work at a singular point

Figure 6.12 represents the geometrical interpretation of the hypothesis of maximal plastic work. The directions of vectors $(\boldsymbol{\sigma} - \boldsymbol{\sigma}^*)$ and $d\boldsymbol{\varepsilon}^p$ form the non-obtuse angle. The geometrical interpretation of the hypothesis of maximal plastic work for a singular point is given in Fig. 6.13. For directions of strain increment $(d\boldsymbol{\varepsilon}^p)$ laying between directions n_1 and n_2 the hypothesis of maximal plastic work is satisfied. If the loading function is not convex then the hypothesis of maximal plastic work is not satisfied (Fig. 6.14).

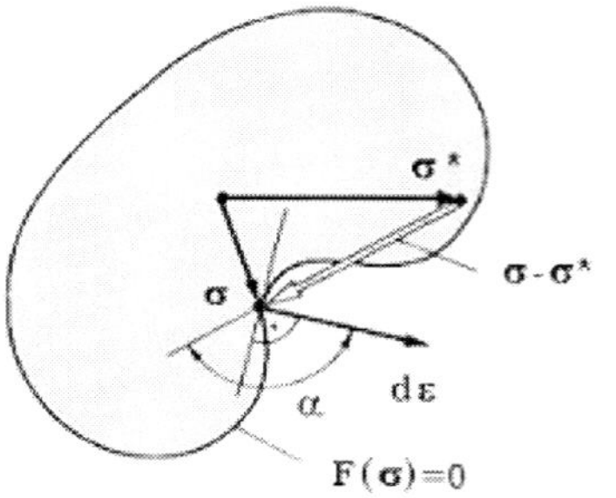

Fig. 6.14. The hypothesis of maximal plastic work for non-convex loading function is not satisfied.

The hypothesis of maximal plastic work can be geometrically represented in the loading point space $\{\boldsymbol{\sigma}\}$ by the convexity of the elasticity domain E_D and by the directions permissible for the plastic increment ($d\boldsymbol{\varepsilon}^p$).

The flow rule reads

$$d\boldsymbol{\varepsilon}^p \in d\lambda\ \partial F(\boldsymbol{\sigma} \quad \text{or} \quad d\boldsymbol{\varepsilon}^p \in d\lambda\ \partial_{\boldsymbol{\eta}} F(\boldsymbol{\sigma}, \boldsymbol{\eta}) \qquad d\lambda \geq 0 \tag{6.55}$$

where $\partial F(\boldsymbol{\sigma})$ and $\partial_{\boldsymbol{\eta}} F(\boldsymbol{\sigma}, \boldsymbol{\eta})$ represent the subdifferential of $F(\boldsymbol{\sigma})$ and $\partial_{\boldsymbol{\eta}} F(\boldsymbol{\sigma}, \boldsymbol{\eta})$ respectively with respect to $\boldsymbol{\sigma}$.

The hypothesis of maximal plastic work implies both the normality of the flow rule and the convexity of the plastic criterion. The hypothesis of maximal plastic work and the hypothesis of both the normality of the flow rule and the convexity of the plastic criterion with respect to $\boldsymbol{\sigma}$ are equivalent.

A material is said to be standard if it satisfies the maximal plastic work hypothesis. The plastic criterion is then convex and the flow rule is normal.

6.5.2 The Associated Flow Rule

If a material satisfies the maximal plastic work hypothesis, then we say that the material is standard. The plasticity criterion is then convex and the flow rule is normal and we say that the flow rule is associated with the criterion of plasticity.

Consider an ideal plastic standard material, Eqs. (6.33) and (6.55) lead to write the flow rule as

$$d\boldsymbol{\varepsilon}^p \in d\lambda\ \partial F(\boldsymbol{\sigma}) \quad d\lambda \geq 0 \text{ where } \begin{array}{lll} d\lambda \geq 0 & \text{if } F = 0 & \text{and } dF = 0 \\ d\lambda = 0 & \text{if } F < 0 & \text{and } dF < 0 \end{array} \tag{6.56}$$

and for any regular point on the boundary of the elasticity domain in the case of plastic loading

$$d\boldsymbol{\varepsilon}^p = d\lambda \frac{\partial F}{\partial \sigma} \qquad d\lambda \geq 0 \qquad \text{if } F = dF = 0 \tag{6.57}$$

Consider a hardening standard material, Eqs. (6.41), (6.42) and (6.57) give for a hardening standard material

$$(d\boldsymbol{\varepsilon}^p, d\mathbf{m}) \in \frac{d_{\boldsymbol{\eta}}F}{H} \partial_\eta F(\boldsymbol{\sigma}, \boldsymbol{\eta}) \qquad \text{if } F = 0 \text{ and } d_{\boldsymbol{\eta}}F > 0$$
$$d\boldsymbol{\varepsilon}^p = d\mathbf{m} = 0 \qquad \text{if } F < 0 \quad \text{or if } F = 0 \text{ and } d_{\boldsymbol{\eta}}F \leq 0 \tag{6.58}$$

For softening (H < 0) we get

$$(d\boldsymbol{\varepsilon}^p, d\mathbf{m}) \in \frac{d_{\boldsymbol{\eta}}F(\boldsymbol{\sigma}, \boldsymbol{\eta})}{H} \partial_{\boldsymbol{\eta}} F(\boldsymbol{\sigma}, \boldsymbol{\eta}) \quad \text{or } d\boldsymbol{\varepsilon}^p = d\mathbf{m} = 0 \text{if } F = 0 \text{ and } d_{\boldsymbol{\eta}}F < 0$$
$$d\boldsymbol{\varepsilon}^p = d\mathbf{m} = 0 \qquad \text{if } F < 0 \qquad \text{or if } F = 0 \text{ and } d_{\boldsymbol{\eta}}F = 0 \tag{6.59}$$

Consider a regular point on the boundary of the elasticity domain and non-zero plastic increments $d\boldsymbol{\varepsilon}^p$. Then we have

$$d\boldsymbol{\varepsilon}^p = \frac{d_{\boldsymbol{\eta}}F}{H} \left(\frac{\partial F}{\partial \boldsymbol{\sigma}} \right) \tag{6.60}$$

6.5.3 Stability

The stability of a material can be explained in the example of a one dimensional simple tension test. The characteristics of stability for a hardening material model are illustrated in Fig. 6.15. The plastic strain increment shown is $d\varepsilon > 0$ and the corresponding strain increment is $d\sigma > 0$. For ideal plastic material $d\sigma = 0$ and then $d\sigma \cdot d\varepsilon = 0$ which is illustrated in Fig. 6.16.

Characteristics of unstability of a material are shown in Fig. 6.17. Metals possess the stability in the beginning of deformation, then the material becomes unstable, because $d\boldsymbol{\varepsilon} > 0$ and $d\boldsymbol{\sigma} < 0$.

The thermodynamic stability of a material will be considered first in a stress-free state when $\boldsymbol{\varepsilon}^p = 0$ i.e. in an elastic range. Consider a domain V bounded by a surface A in thermal equilibrium with the outside temperature T_o. The thermodynamic stability of a material in a stress-free reference state is analyzed in the material domain V, without a change in external criteria.

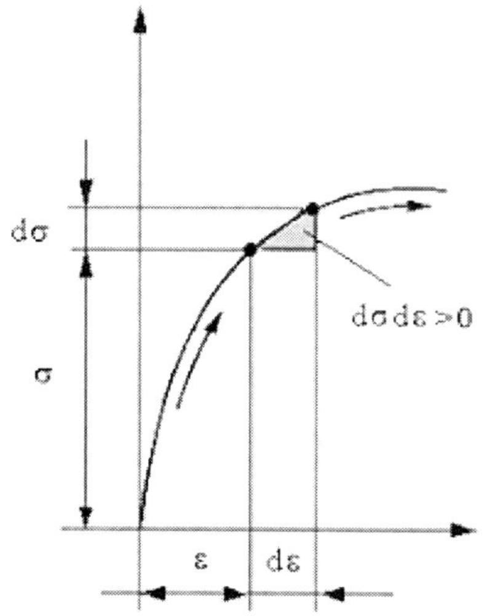

Fig. 6.15. Characteristics of material stability in a tension test

The thermodynamic state of a material is characterized by the strain **ε**, and the variation of temperature T, so the thermodynamic stability of a material refers to these variables. If we analyze the stability of the whole domain, a stability problem is expressed in terms of displacement. Body forces and inertia effects are not considered in the analysis. Denote by T_{oo} and $\boldsymbol{\varepsilon}_{oo}$ temperature and strain fields at time t = 0 respectively.

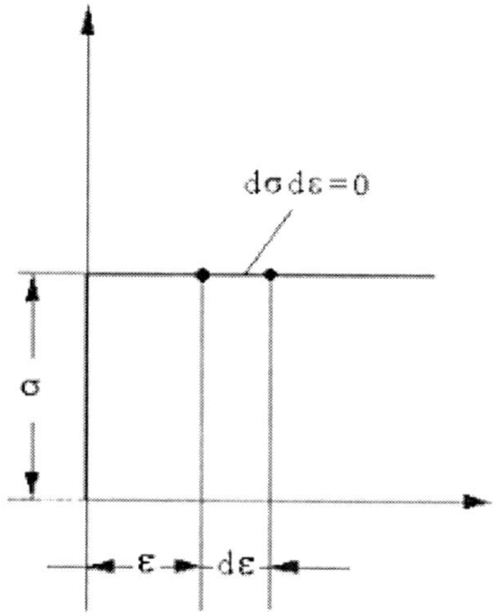

Fig. 6.16. Characteristic of stability for ideal plastic material

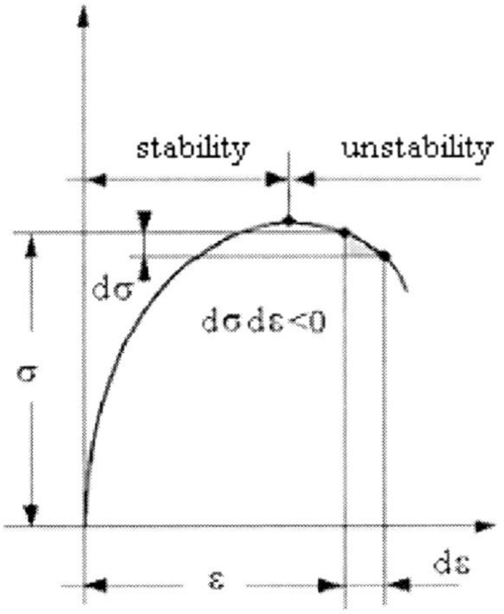

Fig. 6.17. Characteristics of stability and unstability of a material

The temperature field is different from T_o and the other fields are different from zero on account of initial disturbances. The definition of thermodynamic stability of the reference state is

$$\text{if } \|\boldsymbol{\varepsilon}_{oo}\| < \beta \quad \|\boldsymbol{\varepsilon}\| < \alpha \text{ for every } \alpha > 0 \text{ there exists } \beta \text{ such that}$$
$$\|T_{oo} - T_o\| < \beta \text{, then for every } t > 0 \quad \|T - T_o\| < \alpha$$

where $\|\cdot\|$ represents the functional norm defined for every $\mathbf{g}$ by (6.61)

$$\|\mathbf{g}\| = \left(\frac{1}{2}\int_V \mathbf{g}\cdot\mathbf{g}\,dV\right)^{\frac{1}{2}}$$

where $\mathbf{g}$ is a one-component vector when considering the temperature T. It is a vector of six components when considering the strain tensor.

In Eq. (6.61) we consider small-uncontrolled initial disturbances with regard to the reference state. It will always remain close to that state when left to itself. The distance associated with the norm $\|\cdot\|$ quantifies the vicinity of states.

A stability criterion with regard to any variable consists in establishing a criterion, which ensures the existence of a Lyapunov functional of this variable. A Lyapunov functional is a non-increasing function of time, which is the integral over domain V of a function strictly convex with respect to the analyzed variables. A Lyapunov functional of the variables $\boldsymbol{\varepsilon}$ and T are described from the first two laws of thermodynamics.

The criterion of the Lyapunov functional to be a non-increasing function of time can be expressed through the non-negativeness of the opposite of its time derivative.

If the external temperature is held at constant values T_o we have

$$-\left(T_o - T\right)\frac{\mathbf{Q}}{T}\cdot\mathbf{N} \geq 0 \tag{6.62}$$

when $\mathbf{Q}$ is the flow vector in the Lagrangian approach. The expression (6.62) says that the internal entropy production associated with heat exchange that occurs at the boundary surface A is non-negative.

The first law of thermodynamics (4.5) in the absence of volume heat sources R, body forces $\mathbf{P}$ and surface forces $\mathbf{t}$, and the neglecting inertia effects is

$$\frac{D}{Dt}\int_{V_r} E\,dV_r = \int_{A_r}\left(-\mathbf{Q}\cdot\mathbf{N}\right)dA_r \tag{6.63}$$

where D/Dt stands for the material derivative of the involved integral and E is the internal energy density in the Lagrangian approach. The expression (6.63) can be rewritten as

$$\frac{d}{dt}\int_{V_r} E\,dV_r = -\int_{A}\mathbf{Q}\cdot\mathbf{N}\,dA \tag{6.64}$$

The second law of thermodynamics says that the internal entropy production is non-negative. In the absence of the volume heat sources R, this non-negativeness is

$$\frac{D}{Dt}\int_{V_r} S\,dV_r \geq -\int_{A}\frac{\mathbf{Q}\cdot\mathbf{N}}{T}\,dA \tag{6.65}$$

which can be rewritten in the form

$$\frac{d}{dt}\int_{V_r} S\,dV_r \geq -\int_{A}\mathrm{Div}\left(\frac{\mathbf{Q}}{T}\right)dA \tag{6.66}$$

By multiplying the above by $-\,T_o$, and adding the resulting inequality to Eq. (6.64), it follows that

$$\frac{d}{dt}\int_{V_r}\left[E - T_o S\right]dV_r \leq -\int_{A}\mathbf{Q}\cdot\mathbf{N}\frac{T - T_o}{T}\,dA \tag{6.67}$$

By the definition $\Psi = E - TS$ of the volume density of free energy Ψ, Eq. (6.67) can be rewritten in the form

$$\frac{d}{dt}\int_{V_r}\left[\Psi + \left(T - T_o\right)S\right]dV_r \leq -\int_{A}\mathbf{Q}\cdot\mathbf{N}\frac{T - T_o}{T}\,dA \tag{6.68}$$

By (6.62) Eq. (6.67) gives

$$\frac{d}{dt}\int_{V_r}\left[\Psi+\left(T-T_o\right)S\right]dV_r \leq 0 \tag{6.69}$$

Let J (t) be the integral that appears in the left-hand side of Eq. (6.68). By (6.68) the function J (t) is a non-increasing function of time, and thus has its initial value as an upper bound

$$J(t)=\int_{V_r}\left[\Psi+\left(T-T_o\right)S\right]dV_r \leq J(0) \tag{6.70}$$

Assume that $-\Psi = -\Psi(T, \boldsymbol{\varepsilon})$ is a strictly convex function of temperature T. This function is continuously differentiable with respect to T on account of expression (6.10) of Ψ for $\boldsymbol{\varepsilon}^p = 0$

$$\Psi = \boldsymbol{\sigma}^o\boldsymbol{\varepsilon} - S_o\theta + \frac{1}{2}\boldsymbol{\varepsilon}\,\mathbf{C}\,\boldsymbol{\varepsilon} - \theta\,\mathbf{A}\,\boldsymbol{\varepsilon} - \frac{1}{2}b\theta^2 \tag{6.71}$$

Then substitute $-\Psi$ for T and T_o respectively to the property of the strict convexity of F defined by the expression

$$F(\mathbf{x}) - F(\mathbf{y}) > \frac{\partial F(\mathbf{y})}{\partial \mathbf{y}}\cdot(\mathbf{x}-\mathbf{y}) \qquad \text{for every } \mathbf{x} \text{ and } \mathbf{y} \neq \mathbf{x} \tag{6.72}$$

we have

$$\left[-\Psi(T_o,\boldsymbol{\varepsilon})\right]-\left[-\Psi(T,\boldsymbol{\varepsilon})\right] > \frac{\partial\left[-\Psi(T,\boldsymbol{\varepsilon})\right]}{\partial T}\left(T_o - T\right) \tag{6.73}$$

Since $-\Psi = -\Psi(T, \boldsymbol{\varepsilon})$ is also continuously differentiable twice in a row with regard to T the strict convexity of $-\Psi$ with respect to T is ensured if

$$-\frac{\partial^2\Psi}{\partial T^2} > 0 \tag{6.74}$$

By the expression (6.10) of Ψ, the criterion of the strict convexity criterion (6.74) gives

$$C_\Delta = T_o b > 0 \tag{6.75}$$

The state equations (6.9) and Eq. (6.73) gives

$$\Psi(T, \boldsymbol{\varepsilon}) + (T - T_o)\,S > \Psi(T_o, \boldsymbol{\varepsilon}) \tag{6.76}$$

The expression (6.10) of Ψ, when $T - T_o$ and $\boldsymbol{\varepsilon}^o = 0$ with (6.69) and (6.75) give

$$\frac{1}{2}\int_{V_r}\left[\boldsymbol{\varepsilon}\,\mathbf{C}_o\,\boldsymbol{\varepsilon}\right]dV_r < J(0) \tag{6.77}$$

Assume the following

$$\boldsymbol{\varepsilon}\,\mathbf{C}_o\,\boldsymbol{\varepsilon} > 0 \tag{6.78}$$

for every $\boldsymbol{\varepsilon} \neq 0$. Then the free energy Ψ is a strictly convex function with respect to strain tensor $\boldsymbol{\varepsilon}$. The relations (6.76) and (6.77) are assumed to be satisfied. The function J (t) defined by (6.70) can play the role of a Lyapunov functional. The local stability of the material remains $\mathbf{C}_o$ to be constant within the domain V. By (6.77) and (6.78)

$$\|\,\boldsymbol{\varepsilon}\,\| \leq \left[J(0)/\lambda_{inf}\right]^{\frac{1}{2}} \tag{6.79}$$

where λ_{inf} is the smallest of the eigenvalues associated with the positively defined symmetric quadratic form $\mathbf{C}_o$.

The thermodynamic stability of the elastic material is ensured because the integral J (0) depends continuously on the initial criterions $\boldsymbol{\varepsilon}_{oo}$, and $T_{oo} - T_o$. Thus sufficient criterion of thermodynamic stability of the elastic material is given by criteria (6.75) and (6.78).

The heat capacity C_Δ in iso-deformation is positive which indicates that the temperature of the elementary system increases with a positive external heat supply in an undrained iso-deformation experiment. The positivity of C_Δ ensures the increase of entropy with the temperature

$$\frac{\partial S}{\partial T} > 0 \tag{6.80}$$

which is a criterion of thermal stability.

The theorem of the virtual work rate (2.32) reducing to the surface force $\mathbf{t}$ gives

$$\int_V \boldsymbol{\sigma}\,\dot{\boldsymbol{\varepsilon}}\,dV = \int_A \mathbf{V}\cdot\mathbf{t}\,dA \tag{6.81}$$

Since it holds for $\boldsymbol{\sigma}$ and $\boldsymbol{\sigma}^o$, the substraction of the two corresponding equalities gives

$$\int_{V_r} \left(\boldsymbol{\sigma} - \boldsymbol{\sigma}^o\right)\dot{\boldsymbol{\varepsilon}}\,dV = 0 \tag{6.82}$$

The expression (6.82) may be used in any stability analysis.

In the case of plastic behaviour

$$\int_V \left[\left(\boldsymbol{\sigma} - \boldsymbol{\sigma}^o\right)\dot{\boldsymbol{\varepsilon}}^e\right]dV \leq -\int_V \left[\left(\boldsymbol{\sigma} - \boldsymbol{\sigma}^o\right)\dot{\boldsymbol{\varepsilon}}^p\right]dV \tag{6.83}$$

Consider the elasticity domain E_D^o of the state for the stability analysis. If the hardening is non-negative, for any evolution from the state which the stability is analyzed corresponding to a plastic loading, the present loading point

$(\boldsymbol{\sigma}^o + \Delta\boldsymbol{\sigma}) = (\boldsymbol{\sigma})$ is on the boundary of the present domain of elasticity E_D and is not interior to E_D. Assume that the maximal plastic work hypothesis is verified. Thus, the right-hand member of Eq. (6.83) is negative or zero. If criterion (6.77) is satisfied the potential W defined as

$$W = \boldsymbol{\sigma}^o\left(\boldsymbol{\varepsilon} - \boldsymbol{\varepsilon}^p\right) + \frac{1}{2}\left(\boldsymbol{\varepsilon} - \boldsymbol{\varepsilon}^p\right)\mathbf{C}_o\left(\boldsymbol{\varepsilon} - \boldsymbol{\varepsilon}^p\right) \tag{6.84}$$

corresponds to a positively defined quadratic form. Let $W^*(\boldsymbol{\sigma})$ be the Legendre-Fenchel transform of W given as

$$W^* = \boldsymbol{\sigma}\,\boldsymbol{\varepsilon}^e - W \tag{6.85}$$

The potential W* can be written as

$$W^* = \frac{1}{2}\left(\boldsymbol{\sigma} - \boldsymbol{\sigma}^o\right)\mathbf{C}_o^{-1}\left(\boldsymbol{\sigma} - \boldsymbol{\sigma}^o\right) \tag{6.86}$$

The potential W* is also a positively defined quadratic form. By the properties of the Legendre-Fenchel transform

$$\boldsymbol{\varepsilon} - \boldsymbol{\varepsilon}^p = \boldsymbol{\varepsilon}^e = \frac{\partial W^*}{\partial \boldsymbol{\sigma}} \tag{6.87}$$

By (6.86) and (6.82) we get

$$\frac{d}{dt}\int_V W^*\, dV \le -\int_V \left[\left(\boldsymbol{\sigma} - \boldsymbol{\sigma}_o\right)\frac{d\boldsymbol{\varepsilon}^p}{dt}\right] dV \tag{6.88}$$

The integral of W* over volume V plays the role of a Lyapunov functional with respect to stress $\boldsymbol{\sigma}$.

The hypothesis of the maximal plastic work with the strict positiveness of hardening ensures the stability of the present thermodynamic equilibrium of plastic material.

The sufficient criterion for isothermal stability of plastic materials with respect to variables $(\boldsymbol{\sigma})$ is given as

$$\boldsymbol{\varepsilon}\ \mathbf{C}_o\ \boldsymbol{\varepsilon} > 0 \tag{6.89}$$

for every $\boldsymbol{\varepsilon} \neq 0$ and $H > 0$ as a validity of the hypothesis of maximal plastic work. Criterion (6.89) concerns isothermal stability. With the relation (6.75) of positivity volume heat capacity C_Δ it can be shown that the stability with respect to the temperature is also ensured.

6.5.4 The Drucker Stability Postulate

The Drucker stability postulate can be formulated as follows. Any external loading increment added to the existing external loading has to develop a non-negative work during its application

$$d\boldsymbol{\sigma}\, d\boldsymbol{\varepsilon}^e \geq 0 \tag{6.90}$$

$$d\boldsymbol{\sigma}\, d\boldsymbol{\varepsilon}^p \geq 0 \tag{6.91}$$

Consider an ideal plastic material. If the loading point is at a singular point of the yield locus all directions of the cone of outward normals will be admitted by the Drucker postulate for the plastic increment $d\boldsymbol{\varepsilon}^p$ only if the loading point stays at the singular point or on the singular line. If the loading increment $d\boldsymbol{\sigma}$ is such that the loading point leaves the singular point when remaining on the yield locus, the Drucker postulate requires the plastic increment $d\boldsymbol{\varepsilon}$ to be normal to the loading increment (Fig. 6.18). The hypothesis of maximal plastic work requires the plastic increment $d\boldsymbol{\varepsilon}^p$ belong to the cone of outward normal at the singular point or on the singular line which the loading point leaves. The Drucker stability postulate is more restrictive than the maximal plastic work hypothesis.

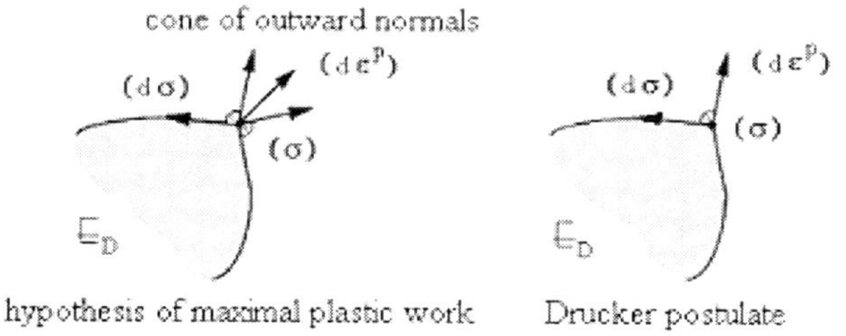

Fig. 6.18. Ideal plastic material in the Drucker stability postulate and the hypothesis of maximal plastic work

The Drucker postulate may constitute one of the sufficient criteria for thermodynamic stability of plastic materials, in replacing the hypothesis of maximal plastic work in criteria (6.89) sufficient for stability.

6.6 The Non-associated Flow Rule

For many materials the hypothesis of normality of dissipative mechanisms is not associated with the flow rule. The flow rule of these materials is presented then by the non-associated potentials, which are different from the loading function F. Then we say that plastic materials are non-standard (Fig. 6.19).

In a general case the loading function F and the non-associated potentials are convex. Introduce a function G different from the loading function F, but with the same arguments. The function G for the ideal plastic material is a non-associated potential, if

$$\mathbf{d\varepsilon}^{P} \in d\lambda\ \partial G(\boldsymbol{\sigma})\ \ d\lambda \geq 0 \quad \text{where} \begin{cases} d\lambda \geq 0 & \text{if} \quad F = 0 \quad \text{and} \quad dF = 0 \\ d\lambda = 0 & \text{if} \quad F < 0 \quad \text{or} \quad dF < 0 \end{cases} \tag{6.92}$$

In Eq. (8.25) $\partial G(\boldsymbol{\sigma})$ is the subdifferential of the non-associated potential $G(\boldsymbol{\sigma})$ with re spect to $\boldsymbol{\sigma}$. The flow rule in the case of plastic loading if G is without singular points is

$$\mathbf{d\varepsilon}^{p} = d\lambda \frac{\partial G}{\partial \boldsymbol{\sigma}} \quad d\lambda \geq 0 \quad \text{if } F = dF = 0 \tag{6.93}$$

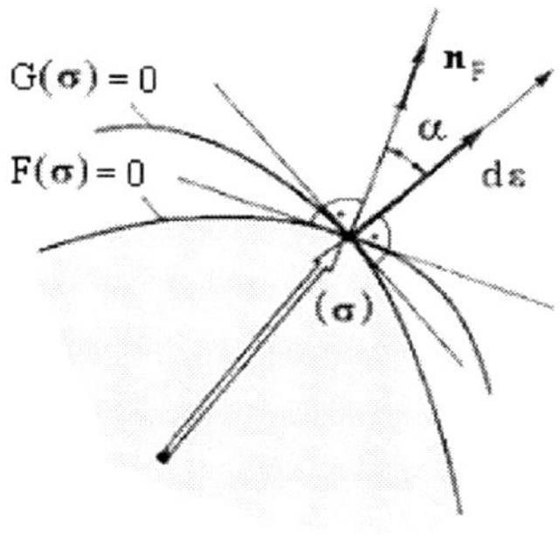

Fig. 6.19. Geometrical interpretation of the non-associated flow rule

The flow rule for a hardening material (H > 0) is

$$\mathbf{d\varepsilon}^{p} \in \frac{d_{\eta}F}{H} \partial_{\eta} G(\boldsymbol{\sigma}, \boldsymbol{\eta}) \quad \text{if } F = 0 \quad \text{and} \quad d_{\eta}F > 0$$

$$\mathbf{d\varepsilon}^{p} = \mathbf{dm} = 0 \quad \text{if } F < 0 \quad \text{or if } F = 0 \quad \text{and} \quad d_{\eta}F < 0$$

and for softening (H < 0) (6.94)

$$\mathbf{d\varepsilon}^{p} \in \frac{d_{\eta}F}{H} \partial_{\eta} G(\boldsymbol{\sigma}, \boldsymbol{\eta}) \quad \text{if } F = 0 \quad \text{and} \quad d_{\eta}F < 0$$

$$\mathbf{d\varepsilon}^{p} = \mathbf{dm} = 0 \quad \text{if } F < 0 \quad \text{or if } F = d_{\eta}F = 0$$

In the above equations $\partial_{\eta} G$ is the subdifferential of a non-associated potential $G(\boldsymbol{\sigma}, \boldsymbol{\eta})$ with respect to $\boldsymbol{\sigma}$ at a constant hardening force $\boldsymbol{\eta}$. If G does not present singular points, in the case of plastic loading the flow rule is

$$\mathbf{d\varepsilon}^{p} = \frac{d_{\eta}F}{H} \frac{\partial G}{\partial \boldsymbol{\sigma}} \tag{6.95}$$

The non-associated potentials can be extended to hardening variables. Consider the subdifferential $\partial G(\boldsymbol{\sigma}, \boldsymbol{\eta})$ of the non-associated potential $G(\boldsymbol{\sigma}, \boldsymbol{\eta})$ with respect to the extended forces ($\boldsymbol{\sigma}$, $\boldsymbol{\eta}$). If G is without singular points, in the case of plastic loading the extended flow rule is

$$\left(d\boldsymbol{\varepsilon}^{p}, d\mathbf{m}\right) = \frac{d_{\eta}F}{H}\left(\frac{\partial G}{\partial \boldsymbol{\sigma}}, \frac{\partial G}{\partial \boldsymbol{\eta}}\right) \tag{6.96}$$

The potential G expressing the evolution for hardening variables may differ from the one used in the flow rule. Denote it by G*

$$\left(d\boldsymbol{\varepsilon}^{p}, d\mathbf{m}\right) = \frac{d_{\eta}F}{H}\left(\frac{\partial G^{*}}{\partial \boldsymbol{\sigma}}, \frac{\partial G}{\partial \boldsymbol{\eta}}\right) \tag{6.97}$$

For standard material G* = F and by Eq. (6.97)

$$\left(d\boldsymbol{\varepsilon}^{p}, d\mathbf{m}\right) = \frac{d_{\eta}F}{H}\left(\frac{\partial F}{\partial \boldsymbol{\sigma}}, \frac{\partial G}{\partial \boldsymbol{\eta}}\right) \tag{6.98}$$

The above developed hardening laws indicate that. The direction $\mathcal{G}_{\mathbf{m}}$ is identified with $\partial G/\partial \boldsymbol{\eta}$.

The hardening modulus H is expressed in the form

$$H = \frac{\partial F}{\partial \boldsymbol{\eta}} \cdot \frac{\partial^{2} U}{\partial \mathbf{m}^{2}} \cdot \frac{\partial G}{\partial \boldsymbol{\eta}} = \frac{\partial F}{\partial \boldsymbol{\eta}} \cdot \mathrm{H} \cdot \frac{\partial G}{\partial \boldsymbol{\eta}} \tag{6.99}$$

Consider a potential G without singular points for an ideal plastic material. According to the flow rule (6.92) the non-negativeness (6.24) of the intrinsic dissipation $\daleth_1$ leads to the following restriction upon the potential

$$\boldsymbol{\sigma}\frac{\partial G}{\partial \boldsymbol{\sigma}} \geq 0 \quad \text{if } F = 0 \tag{6.100}$$

For a potential G without singular points and a hardening material following the flow rule (6.94) at a given hardening state, due to relation (6.36) the condition for the potential is

$$\boldsymbol{\sigma}\frac{\partial G}{\partial \boldsymbol{\sigma}} + \boldsymbol{\eta} \cdot \mathcal{G}_{\mathrm{m}} \geq 0 \quad \text{if } F = 0 \tag{6.101}$$

Consider a potential G without singular points and a hardening material with the flow rule (6.96) and therefore $\mathcal{G}_{\mathbf{m}} = \partial G/\partial \boldsymbol{\eta}$. The condition upon G is

$$\boldsymbol{\sigma}\frac{\partial G}{\partial \boldsymbol{\sigma}} + \boldsymbol{\eta} \cdot \frac{\partial G}{\partial \boldsymbol{\eta}} \geq 0 \quad \text{if } F = 0 \tag{6.102}$$

Consider the non-associated potential G used to express the hardening law which is different from the non-associated potential G* used to express the flow rule. The condition that concerns G and G* is

$$\sigma \frac{\partial G^*}{\partial \sigma} + \eta \cdot \frac{\partial G}{\partial \eta} \geq 0 \quad \text{if } F = 0 \tag{6.103}$$

and if F = G i.e. for standard material

$$\sigma \frac{\partial F}{\partial \sigma} + \eta \cdot \frac{\partial G}{\partial \eta} \geq 0 \quad \text{if } F = 0 \tag{6.104}$$

For the non-standard materials with the flow rule expressed by non-associated potentials inherent modeling problems are involved. The hypothesis of maximal plastic work is not followed and such materials do not satisfy the sufficient conditions ensuring thermodynamic stability, even if they exhibit hardening. The loading increment $d\boldsymbol{\sigma}$ and the response increment $d\boldsymbol{\varepsilon}$ are linked by the relation which is not necessarily biunivocal.

6.7 Incremental Formulation

The relation between the loading increments $d\boldsymbol{\sigma}$ and the increments of the external variables $d\boldsymbol{\varepsilon}$ is descried by the constitutive equations of plasticity in incremental form. In this point the discussion of the equation is restricted to isothermal situations and to criteria and potentials, which do not include singular points. The expression (6.21) in isothermal case has the form

$$d\boldsymbol{\varepsilon} = \mathbf{C}^{-1}\, d\boldsymbol{\sigma} + d\boldsymbol{\varepsilon}^{p} \tag{6.105}$$

where $\mathbf{C}^{-1}$ is the inverse of the tensor $\mathbf{C}$ of the moduli of elasticity. In the case of a standard material with (positive) hardening (H > 0) and a loading function F without singular points, Eqs. (6.105) and (6.58) give

$$d\boldsymbol{\varepsilon} = \mathbf{C}^{-1}\, d\boldsymbol{\sigma} + \frac{1}{H} Y(F) \left(\frac{\partial F}{\partial \boldsymbol{\sigma}}\, d\boldsymbol{\sigma} \right) \frac{\partial F}{\partial \boldsymbol{\sigma}} \tag{6.106}$$

where Y(x) is the Heaviside function defined by

$$Y(x) = 0 \quad \text{if } x < 0 \quad \text{and } Y(x) = 1 \quad \text{if } x \geq 0 \tag{6.107}$$

Consider the following incremental potential W*

$$W^* = \frac{1}{2} d\boldsymbol{\sigma}\, \mathbf{C}^{-1}\, d\boldsymbol{\sigma} + \frac{1}{2H} Y(F) \left(\frac{\partial F}{\partial \boldsymbol{\sigma}}\, d\boldsymbol{\sigma} \right)^{2} \tag{6.108}$$

By Eq. (6.106) we have

$$d\boldsymbol{\varepsilon} = \frac{\partial W^*}{\partial (d\boldsymbol{\sigma})} \tag{6.109}$$

The incremental potential W* consists of a positively defined quadratic form plus a non-negative form of the argument $d\boldsymbol{\sigma}$. From the above the potential W* is a strictly convex function of $d\boldsymbol{\sigma}$.

Consider the Legendre-Fenchel transform W of W*

$$W = d\boldsymbol{\sigma}\, d\boldsymbol{\varepsilon} - W^* \tag{6.110}$$

where $d\boldsymbol{\sigma}$ and $d\boldsymbol{\varepsilon}$ are given by Eq. (6.109).

By the properties of the Legendre-Fenchel transform,

$$d\boldsymbol{\sigma} = \frac{\partial W}{\partial(d\boldsymbol{\varepsilon})} \tag{6.111}$$

Assume that the loading increment corresponds to a plastic loading with non-zero plastic increments $d\boldsymbol{\varepsilon}^p$. By Eq. (6.105) we have

$$d\boldsymbol{\varepsilon} = \mathbf{C}^{-1}\, d\boldsymbol{\sigma} + \frac{1}{H}\left(\frac{\partial F}{\partial \boldsymbol{\sigma}}\, d\boldsymbol{\sigma}\right)\frac{\partial F}{\partial \boldsymbol{\sigma}} \tag{6.112}$$

or

$$d\boldsymbol{\sigma} = \mathbf{C}\, d\boldsymbol{\varepsilon} - \frac{1}{H}\left(\frac{\partial F}{\partial \boldsymbol{\sigma}}\, d\boldsymbol{\sigma}\right)\left(\mathbf{C}\frac{\partial F}{\partial \boldsymbol{\sigma}}\right) \tag{6.113}$$

The contracted tensor product of the above with tensor $\partial F/d\boldsymbol{\sigma}$ gives

$$\frac{\partial F}{\partial \boldsymbol{\sigma}}\, d\boldsymbol{\sigma} = \frac{\left(\frac{\partial F}{\partial \boldsymbol{\sigma}}\mathbf{C}\right) d\boldsymbol{\varepsilon}}{1 + \frac{1}{H}\left(\frac{\partial F}{\partial \boldsymbol{\sigma}}\mathbf{C}\frac{\partial F}{\partial \boldsymbol{\sigma}}\right)} \tag{6.114}$$

By the stability hypothesis for elastic material and the assumption of positivity of hardening the positivity of the numerator of the right-hand member results from the positivity of the left-hand side of Eq. (6.114). By substituting (6.114) into (6.113), the relation (6.111) is written as

$$d\boldsymbol{\sigma} = \mathbf{C}\, d\boldsymbol{\varepsilon} - Y(F)\left(\mathbf{C}\frac{\partial F}{\partial \boldsymbol{\sigma}}\right)\frac{\left(\frac{\partial F}{\partial \boldsymbol{\sigma}}\mathbf{C}\, d\boldsymbol{\varepsilon}\right)}{H + \frac{\partial F}{\partial \boldsymbol{\sigma}}\mathbf{C}\frac{\partial F}{\partial \boldsymbol{\sigma}}} \tag{6.115}$$

The potential W from which increments $d\boldsymbol{\sigma}$ are derived is of the form

$$W = \frac{1}{2} d\boldsymbol{\varepsilon}\, \mathbf{C}\, d\boldsymbol{\varepsilon} - \frac{1}{2} Y(F) \frac{\left(\frac{\partial F}{\partial \boldsymbol{\sigma}} \mathbf{C}\, d\boldsymbol{\varepsilon} \right)^2}{H + \frac{\partial F}{\partial \boldsymbol{\sigma}} \mathbf{C} \frac{\partial F}{\partial \boldsymbol{\sigma}}} \tag{6.116}$$

By Eq. (6.11) and the relation (6.115) we get

$$d\boldsymbol{\varepsilon}^p = Y(F) \frac{\left(\frac{\partial F}{\partial \boldsymbol{\sigma}} \mathbf{C}\, d\boldsymbol{\varepsilon} \right)}{H + \frac{\partial F}{\partial \boldsymbol{\sigma}} \mathbf{C} \frac{\partial F}{\partial \boldsymbol{\sigma}}} \frac{\partial F}{\partial \boldsymbol{\sigma}} \tag{6.117}$$

The plastic multiplier $d\lambda$ is finally identified as

$$d\lambda = Y(F) \frac{\left(\frac{\partial F}{\partial \boldsymbol{\sigma}} \mathbf{C}\, d\boldsymbol{\varepsilon} \right)}{H + \frac{\partial F}{\partial \boldsymbol{\sigma}} \mathbf{C} \frac{\partial F}{\partial \boldsymbol{\sigma}}} \tag{6.118}$$

The relation (6.105) can be written in the form

$$d\boldsymbol{\sigma} = \mathbf{C}\,(d\boldsymbol{\varepsilon} - d\boldsymbol{\varepsilon}^p) \tag{6.119}$$

For plastic loading and standard material, the flow rule is

$$d\boldsymbol{\varepsilon}^p = \frac{1}{H} \left(\frac{\partial F}{\partial \boldsymbol{\sigma}} \cdot d\boldsymbol{\sigma} \right) \frac{\partial F}{\partial \boldsymbol{\sigma}} \tag{6.120}$$

The elastic material satisfies the sufficient condition of stability. Then the quadratic form associated with $\mathbf{C}^{-1}$ is positively defined.

The scalar product $<d\boldsymbol{\sigma}, d\boldsymbol{\sigma}^*>$ of any couple of vectors $(d\boldsymbol{\sigma}, d\boldsymbol{\sigma}^*)$ and the norm $\| d\boldsymbol{\sigma} \|$ of vector $d\boldsymbol{\sigma}$ is

$$<d\boldsymbol{\sigma}, d\boldsymbol{\sigma}^*> = \frac{1}{2} d\boldsymbol{\sigma} \cdot \mathbf{C}^{-1} \cdot d\boldsymbol{\sigma}^* \tag{6.121}$$

$$\| d\boldsymbol{\sigma} \| = \left(\frac{1}{2} d\boldsymbol{\sigma} \cdot \mathbf{C}^{-1} \cdot d\boldsymbol{\sigma} \right)^{\frac{1}{2}} \tag{6.122}$$

The equality

$$\| d\boldsymbol{\sigma} \| \, \| d\boldsymbol{\sigma}^* \| \geq < d\boldsymbol{\sigma}, d\boldsymbol{\sigma}^* > \tag{6.123}$$

leads to the expression for incremental potential W

$$W = \left\| \mathbf{C} \cdot d\boldsymbol{\varepsilon} \right\|^p - Y(F) \frac{\left(\left\langle \mathbf{C} \cdot \frac{\partial F}{\partial \boldsymbol{\sigma}}, \mathbf{C} \cdot d\boldsymbol{\varepsilon} \right\rangle \right)^2}{\frac{H}{2} + \left\| \mathbf{C} \cdot \frac{\partial F}{\partial \boldsymbol{\sigma}} \right\|^2} \tag{6.124}$$

The expression (6.123) is strictly positive, so the quadratic form W defined by (6.124) is strictly convex in the case of strict hardening ($H > 0$). W is the Legendre-Fenchel transform of the strictly convex function W* so this positivity is guaranteed by this formulation. The strict convexity of potentials W* and W ensures the inversibility of relation (6.115).

Consider the case when $H \to 0$. Then the quadratic form W is non-negative and we do not guarantee the biunivocity of the correspondence between the increments $d\boldsymbol{\sigma}$ and $d\boldsymbol{\varepsilon}$.

If $H = 0$ Eq. (6.117) gives the relation

$$2W = d\boldsymbol{\varepsilon} \cdot \mathbf{C} \cdot d\boldsymbol{\varepsilon} - d\boldsymbol{\varepsilon}^p \cdot \mathbf{C} \cdot d\boldsymbol{\varepsilon}^p \tag{6.125}$$

The following relation in the ideal plastic case holds

$$d\boldsymbol{\sigma} \cdot d\boldsymbol{\varepsilon}^p = 0 \tag{6.126}$$

because $d\boldsymbol{\varepsilon}^p = 0$ or $d\boldsymbol{\varepsilon}^p$ is normal to the boundary of the elasticity domain defined by $F(\boldsymbol{\sigma}) = 0$ and vector $d\boldsymbol{\sigma}$ is tangential to the latter.

The relation $2W = d\boldsymbol{\sigma} \cdot d\boldsymbol{\varepsilon}$ can be derived in the standard ideal plastic case using the state equation (6.119).

Let us go to a geometrical interpretation of the relations (6.115) and (6.117). Let $\mathbf{n}$ be the outward unit normal to the boundary of the elasticity domain E_D at the point ($\boldsymbol{\sigma}$), according to the scalar product (6.120)

$$\mathbf{n} = \frac{\mathbf{C} \cdot \frac{\partial F}{\partial \boldsymbol{\sigma}}}{\sqrt{\frac{1}{2} \frac{\partial F}{\partial \boldsymbol{\sigma}} \cdot \mathbf{C} \cdot \frac{\partial F}{\partial \boldsymbol{\sigma}}}} \tag{6.127}$$

The vector $d\boldsymbol{\sigma}^*$ tangent to the boundary at the point ($\boldsymbol{\sigma}$) satisfies

$$\frac{\partial F}{\partial \boldsymbol{\sigma}} \cdot d\boldsymbol{\sigma}^* = 0 \tag{6.128}$$

The vector $\mathbf{n}$ defined by (6.127) is such that $\langle \mathbf{n}, d\boldsymbol{\sigma}^* \rangle = 0$ and $\left\| \mathbf{n} \right\| = 1$ for any vector $d\boldsymbol{\sigma}^*$ satisfying Eq. (6.128).

In the case of ideal plastic material ($H = 0$), and plastic loading Eqs. (6.115) and (6.117) can be written in the form

$$\mathbf{C} \cdot d\boldsymbol{\varepsilon} = d\boldsymbol{\sigma} + \mathbf{C} \cdot d\boldsymbol{\varepsilon}^p \tag{6.129}$$

$$\mathbf{C} \cdot d\boldsymbol{\varepsilon}^p = <\mathbf{n},\ \mathbf{C} \cdot d\boldsymbol{\varepsilon}> \mathbf{n} \qquad (6.130)$$

In plastic loading, $\mathbf{C} \cdot d\boldsymbol{\varepsilon}^p$ given in Eqs. (6.129) and (6.130) represents the orthogonal projection of $\mathbf{C} \cdot d\boldsymbol{\varepsilon}$ onto $\mathbf{n}$ i.e. the normal to the boundary of elasticity domain E_D at the point $(\boldsymbol{\sigma})$ as illustrated in Fig. 6.20

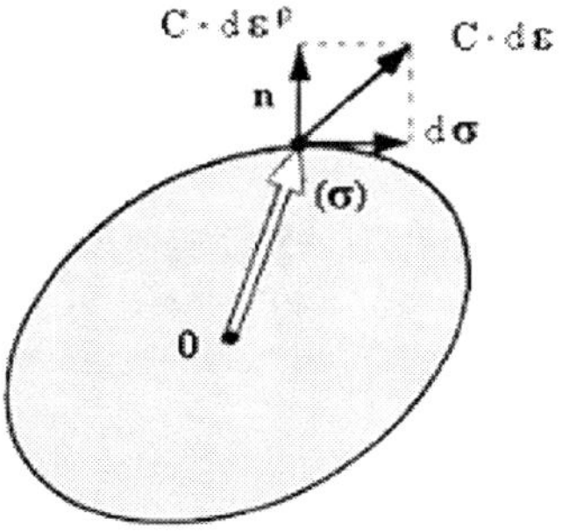

Fig. 6.20. Incremental formulation of plasticity for a standard ideal plastic material in the space $\{\boldsymbol{\sigma}\}$

Consider a non-standard material defined by a non-associated potential G. The state equations (6.119) become unchanged. If the plastic loading occurs the flow rule (6.120) must be replaced by

$$d\boldsymbol{\varepsilon}^p = \frac{1}{H}\left(\frac{\partial F}{\partial \boldsymbol{\sigma}} \cdot d\boldsymbol{\sigma}\right)\frac{\partial G}{\partial \boldsymbol{\sigma}} \qquad (6.131)$$

The expressions (6.119) and (6.131) give

$$\frac{\partial F}{\partial \boldsymbol{\sigma}} \cdot d\boldsymbol{\sigma} = \frac{\dfrac{\partial F}{\partial \boldsymbol{\sigma}} \cdot \mathbf{C} \cdot d\boldsymbol{\varepsilon}}{1 + \dfrac{1}{H}\dfrac{\partial F}{\partial \boldsymbol{\sigma}} \cdot \mathbf{C} \cdot \dfrac{\partial G}{\partial \boldsymbol{\sigma}}} \qquad (6.132)$$

By Eqs. (6.119), (6.131) and (6.132) we have

$$d\boldsymbol{\sigma} = \mathbf{C} \cdot d\boldsymbol{\varepsilon} - Y(F)\frac{\dfrac{\partial F}{\partial \boldsymbol{\sigma}} \cdot \mathbf{C} \cdot d\boldsymbol{\varepsilon}}{H + \dfrac{\partial F}{\partial \boldsymbol{\sigma}} \cdot \mathbf{C} \cdot \dfrac{\partial G}{\partial \boldsymbol{\sigma}}} \cdot \mathbf{C} \cdot \frac{\partial G}{\partial \boldsymbol{\sigma}} \qquad (6.133)$$

with

$$d\boldsymbol{\varepsilon}^{p} = Y(F)\frac{\dfrac{\partial F}{\partial \boldsymbol{\sigma}}\cdot \mathbf{C}\cdot d\boldsymbol{\varepsilon}}{H + \dfrac{\partial F}{\partial \boldsymbol{\sigma}}\cdot \mathbf{C}\cdot \dfrac{\partial G}{\partial \boldsymbol{\sigma}}}\frac{\partial G}{\partial \boldsymbol{\sigma}} \tag{6.134}$$

$$d\lambda = Y(F)\frac{\dfrac{\partial F}{\partial \boldsymbol{\sigma}}\cdot \mathbf{C}\cdot d\boldsymbol{\varepsilon}}{H + \dfrac{\partial F}{\partial \boldsymbol{\sigma}}\cdot \mathbf{C}\cdot \dfrac{\partial G}{\partial \boldsymbol{\sigma}}} \tag{6.135}$$

If the plastic loading actually occurs in the case of softening the above hold irrespective of the hardening sign. In the case of ideal plastic material we get the relation by taking $H = 0$ in the above equations. For such a material, in the case of plastic loading Eqs. (6.133) and (6.134) give

$$\mathbf{C}\cdot d\boldsymbol{\varepsilon} = d\boldsymbol{\sigma} + \mathbf{C}\cdot d\boldsymbol{\varepsilon}^{p} \tag{6.136}$$

$$\mathbf{C}\cdot d\boldsymbol{\varepsilon}^{p} = \frac{<\mathbf{n}, \mathbf{C}\cdot d\boldsymbol{\varepsilon}>}{<\mathbf{n}, \mathbf{u}_h>}\mathbf{u}_h \tag{6.137}$$

where $\mathbf{u}_h$ is the unit vector of the direction of the extended vector $\mathbf{C}\cdot\boldsymbol{\varepsilon}^{p}$ in the loading space $\{\boldsymbol{\sigma}\}$. We have

$$\mathbf{u}_h = \frac{\mathbf{C}\cdot \dfrac{\partial G}{\partial \boldsymbol{\sigma}}}{\sqrt{\dfrac{1}{2}\dfrac{\partial G}{\partial \boldsymbol{\sigma}}\cdot \mathbf{C}\cdot \dfrac{\partial G}{\partial \boldsymbol{\sigma}}}} \tag{6.138}$$

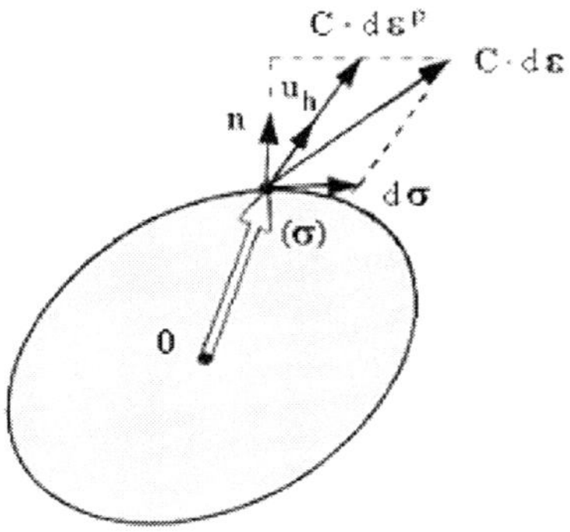

Fig. 6.21. Incremental formulation of plasticity for a non-standard ideal plastic material in space $\{\boldsymbol{\sigma}\}$

In the case of a non-standard ideal plastic material Eq. (6.137) says that according to the scalar product < >, the vectors $\mathbf{C} \cdot d\boldsymbol{\varepsilon}$ and $\mathbf{C} \cdot d\boldsymbol{\varepsilon}^p$ have the same orthogonal projections onto $\mathbf{n}$, which is the outward unit normal to the boundary of the elasticity domain at the present loading point ($\boldsymbol{\sigma}$). The geometrical interpretation of the incremental formulation in loading space $\{\boldsymbol{\sigma}\}$ for a non-standard ideal plastic material is given in Fig. 6.21. The expression (6.133) in the case of plastic loading can be rewritten in the form

$$d\boldsymbol{\sigma} = \mathbf{C}^T \; d\boldsymbol{\varepsilon} \tag{6.139}$$

where

$$\mathbf{C}^T = \mathbf{C} - \frac{\mathbf{C}\dfrac{\partial G}{\partial \boldsymbol{\sigma}} \otimes \dfrac{\partial G}{\partial \boldsymbol{\sigma}}\mathbf{C}}{H + \dfrac{\partial F}{\partial \boldsymbol{\sigma}}\mathbf{C}\dfrac{\partial G}{\partial \boldsymbol{\sigma}}} \tag{6.140}$$

The tensor $\mathbf{C}^T$ is often called the tensor of tangent moduli. The tensor $\mathbf{C}^T$ does not possess the property of symmetry $(\cdot)_{ijkl} = (\cdot)_{ikjl}$ verified by the corresponding elastic moduli $\mathbf{C}$. If the tensor $\mathbf{C}$ satisfies the symmetry relative to an isotropic elastic behaviour, the tensor $\mathbf{C}^T$ does not necessarily possess these symmetries, by the anisotropy induced by plasticity.

6.8 Incremental Formulation for Thermal Hardening

Consider Eq. (6.11) with its incremental form

$$d\boldsymbol{\sigma} = \mathbf{C}\,(d\boldsymbol{\varepsilon} - d\boldsymbol{\varepsilon}^p) - \mathbf{A}d\theta \tag{6.141}$$

By Eq. (6.41) and (6.52) we get

$$d\boldsymbol{\sigma} = \mathbf{C}\,d\boldsymbol{\varepsilon} - \mathbf{A}d\theta - \frac{1}{H}\left(\frac{\partial F}{\partial \boldsymbol{\sigma}} + \frac{\partial F}{\partial \eta}\frac{\partial S^f}{\partial \mathbf{m}}d\theta\right)\mathbf{C}_o\frac{\partial F}{\partial \boldsymbol{\sigma}} \tag{6.142}$$

By contracting Eq. (6.141) with the tensor $\partial F/\partial \boldsymbol{\sigma}$ and adding the result to the quantity $\dfrac{\partial F}{\partial \eta}\dfrac{\partial S^f}{\partial \mathbf{m}}d\theta$

$$\frac{\partial F}{\partial \boldsymbol{\sigma}} + \frac{\partial F}{\partial \eta}\frac{\partial S^f}{\partial \mathbf{m}}d\theta = \frac{\dfrac{\partial F}{\partial \boldsymbol{\sigma}}\mathbf{C}\,d\boldsymbol{\varepsilon} + \left(\dfrac{\partial F}{\partial \eta}\dfrac{\partial S^*}{\partial \mathbf{m}} - \dfrac{\partial F}{\partial \boldsymbol{\sigma}}\mathbf{A}\right)d\theta}{1 + \dfrac{1}{H}\dfrac{\partial F}{\partial \boldsymbol{\sigma}}\mathbf{C}\dfrac{\partial F}{\partial \boldsymbol{\sigma}}} \tag{6.143}$$

$$d\boldsymbol{\sigma} = \mathbf{C}\, d\boldsymbol{\varepsilon} - \mathbf{A}\, d\theta - Y(F)\frac{\left[\frac{\partial F}{\partial \boldsymbol{\sigma}}\mathbf{C}\, d\boldsymbol{\varepsilon} + \left(\frac{\partial F}{\partial \eta}\frac{\partial S^*}{\partial \mathbf{m}} - \mathbf{A}\frac{\partial F}{\partial \boldsymbol{\sigma}}\right) d\theta\right]\mathbf{C}\frac{\partial F}{\partial \boldsymbol{\sigma}}}{H + \frac{\partial F}{\partial \boldsymbol{\sigma}}\mathbf{C}\frac{\partial F}{\partial \boldsymbol{\sigma}}} \tag{6.144}$$

By (6.144) we get

$$d\boldsymbol{\sigma} = \frac{\partial W^*}{\partial(d\boldsymbol{\varepsilon})} \tag{6.145}$$

where $W^*(d\boldsymbol{\varepsilon})$ is the incremental potential defined by

$$W^* = \frac{1}{2} d\boldsymbol{\varepsilon}\, \mathbf{C}\, d\boldsymbol{\varepsilon} - d\theta\, \mathbf{A}\, d\boldsymbol{\varepsilon} - \frac{1}{2} Y(F)\frac{\left[\frac{\partial F}{\partial \boldsymbol{\sigma}}\mathbf{C}\, d\boldsymbol{\varepsilon} + \left(\frac{\partial F}{\partial \eta}\frac{\partial S^*}{\partial \mathbf{m}} - \mathbf{A}\frac{\partial F}{\partial \boldsymbol{\sigma}}\right) d\theta\right]^2}{H + \frac{\partial F}{\partial \boldsymbol{\sigma}}\mathbf{C}\frac{\partial F}{\partial \boldsymbol{\sigma}}} \tag{6.146}$$

6.9 Models of Plasticity

6.9.1 The Isotropic Model

The loading function F, the flow rule and the non-associated potential G have to be specified to solve any problems of plasticity.

In isotropic material the loading function F involves the principal components of symmetric stress tensor $\boldsymbol{\sigma}$, i.e. the three principal stresses σ_1, σ_2 and σ_3. The principal stresses can be expressed in terms of the three first invariants of the stress tensor. We denote the first invariant of the stress tensor by $J_{1\sigma}$, the second invariant of the stress deviator tensor $\mathbf{s} = \boldsymbol{\sigma} - (\mathrm{tr}\, \boldsymbol{\sigma}/3)\mathbf{1}$ by J_{2s} and the third invariant of the stress deviator tensor by J_{3s}

$$J_{1\sigma}(\boldsymbol{\sigma}) = \mathrm{tr}\, \boldsymbol{\sigma} = 3p \tag{6.147}$$

$$h^2 = J_{2s}(\boldsymbol{\sigma}) = \frac{1}{2} s_{ij} s_{ji} = \frac{1}{2}\mathrm{tr}(\mathbf{s}\cdot\mathbf{s}) \tag{6.148}$$

$$J_{3s}(\boldsymbol{\sigma}) = \frac{1}{3} s_{ij} s_{jk} s_{kl} \tag{6.149}$$

By (6.148) we have

$$h^2 = \frac{1}{6}\left[(\sigma_1 - \sigma_2)^2 + (\sigma_1 - \sigma_3)^2 + (\sigma_2 - \sigma_3)^2\right] \tag{6.150}$$

The loading function for isotropic models of plasticity can be represented by

$$F = F(h, p) \tag{6.151}$$

In the space $\{\sigma_i\}$ with **0** as the origin, let the point **1** (σ_1, σ_2, σ_3) represent the stress state $\boldsymbol{\sigma}$. Let the point **2** is its orthogonal projection with regard to the Euclidean product onto trisector (**Δ**) defined by the unit vector with $\left(1/\sqrt{3}, 1/\sqrt{3}, 1/\sqrt{3}\right)$ as cosine directions. The distances $\overline{\mathbf{02}}$ and $\overline{\mathbf{12}}$ can by expressed as

$$\overline{\mathbf{02}} = \sqrt{3}\sigma \qquad \overline{\mathbf{12}}^2 = \overline{\mathbf{01}}^2 - \overline{\mathbf{02}}^2 = 2h^2 \tag{6.152}$$

The relation (6.146) shows that in the loading point space $\{\sigma_i\}$ the yield surface defined by $F = 0$ is the axisymmetric surface around the trisector (Δ) as illustrated in Fig. 6.22.

In the case of isotropic hardening materials, the loading function F is expressed in the form

$$F = F(h, p, \boldsymbol{\eta}) \tag{6.153}$$

where $\boldsymbol{\eta}$ is the hardening force describing the evolution of the yield surface in loading point space $\{\sigma_i\}$. In isotropic hardening the yield surface is derived through a homothety of center **0** in the loading point space $\{\sigma_i\}$. Then the hardening force $\boldsymbol{\eta}$ reduces to a scalar variable η which defines this homothety.

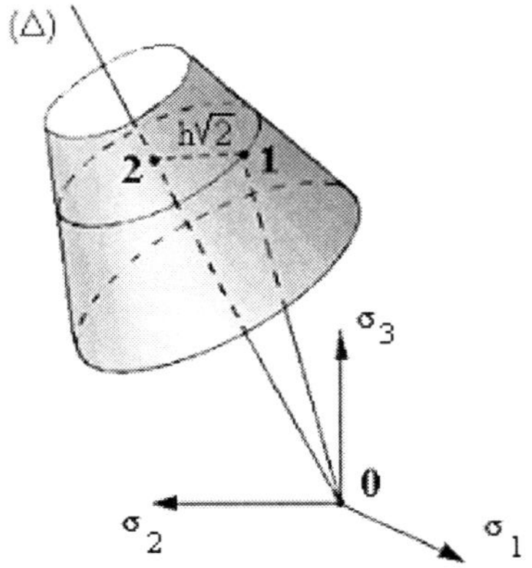

Fig. 6.22. Isotropic criteria of plasticity in the space $\{\sigma_i\}$

The expression (6.153) can be written as

$$F = F\,(h, p, \eta) \tag{6.154}$$

The loading function given by (6.154) can be expressed as a homogeneous polynome of degree n with regard to h and η

$$F = F\,(h, p, \eta) = \eta^n F\,(h/\eta, p/\eta, 1) \tag{6.155}$$

where by convection η is specified as the ratio of the homothety that transforms the yield surface defined by $\eta = 1$ into the present yield surface. In kinematic hardening, the yield surfaces are defined from each other through a translation in the loading point space $\{\sigma_i\}$. The hardening force $\boldsymbol{\eta}$ reduces to a second-order symmetric tensor $\boldsymbol{\eta}$ that defines this translation

$$F = F\,[J_{2s}(\boldsymbol{\sigma} + \boldsymbol{\eta}), J_{1\sigma}(\boldsymbol{\sigma} + \boldsymbol{\eta})] \tag{6.156}$$

In space $\{\sigma_i\}$ vector $(\boldsymbol{\eta})$ represents the vector of translation that transforms the yield surface defined by $(\boldsymbol{\eta}) = (0)$ into the present yield surface.

The non-associated potential analogous to Eqs. (6.152), (6.153) and (6.156) can be expressed as

$$G = G\,(h, p) \tag{6.157}$$

$$G = \eta^n G\,(h/\eta, p/\eta, 1) \tag{6.158}$$

$$G = G\,[J_{2s}(\boldsymbol{\sigma} + \boldsymbol{\eta}), J_{1\sigma}(\boldsymbol{\sigma} + \boldsymbol{\eta})] \tag{6.159}$$

For any function $G\,(\mathbf{s}, \boldsymbol{\sigma})$ the following relation holds

$$\frac{\partial G(\mathbf{s},\boldsymbol{\sigma})}{\partial \boldsymbol{\sigma}} = \frac{\partial G}{\partial \mathbf{s}} - \frac{1}{3}\left(\mathrm{tr}\frac{\partial G}{\partial \mathbf{s}}\right)\mathbf{1} + \frac{1}{3}\frac{\partial G}{\partial \mathbf{s}}\mathbf{1} \tag{6.160}$$

Let $\mathbf{e}^p$ be the plastic deviator strain tensor and $\boldsymbol{\gamma}^p = 2\mathbf{e}^p$ the associated plastic distortion tensor

$$\mathbf{e}^p = \boldsymbol{\varepsilon}^p - (\mathrm{tr}\,\boldsymbol{\varepsilon}^p/3)\mathbf{1} \tag{6.161}$$

$$\boldsymbol{\gamma}^p = 2\mathbf{e}^p \tag{6.162}$$

Define the plastic dilatancy factor β by

$$d\varepsilon^p_{ii} = \beta\left|d\gamma^p\right| = \beta\sqrt{\frac{1}{2}d\gamma^p_{ij}d\gamma^p_{ji}} \tag{6.163}$$

or for any evolution and for an isotropic behaviour

$$d\varepsilon^p_{ii} = \beta\, d\gamma^p_{eq} \qquad \text{where} \quad d\gamma^p_{eq} = \sqrt{\frac{1}{2}d\gamma^p_{ij}d\gamma^p_{ji}} \tag{6.164}$$

where γ^{p}_{eq} is the equivalent plastic distortion which is an increasing function of time and β is the plastic dilatancy factor.

The dilatancy factor β is not necessarily constant and may depend on the present state of the material. The factor β can be positive or negative, depending on whether the material exhibits plastic volume dilatation or plastic volume contraction, respectively.

The plastic increment for the flow rule defined by a non-associated potential G is given as

$$d\boldsymbol{\varepsilon}^{p} = d\lambda \frac{\partial G}{\partial \sigma} \qquad d\lambda \geq 0 \tag{6.165}$$

In an isotropic behaviour the potential G may be expressed as one of the form (6.157). By (6.160) from Eq. (6.165) we get

$$d\boldsymbol{\varepsilon}^{p} = d\lambda \left(\frac{\partial G}{\partial h} \frac{\mathbf{s}}{2h} + \frac{1}{3} \frac{\partial G}{\partial p} \mathbf{1} \right) \tag{6.166}$$

By the relation (6.166) the stress tensor $\boldsymbol{\sigma}$ and the plastic strain increment $d\boldsymbol{\varepsilon}^{p}$ have the same principal directions.

By definitions (6.161), (6.162) and (6.163) the flow rule (6.165) and the relation (6.166)

$$d\varepsilon^{p}_{ii} = d\lambda \frac{\partial G}{\partial \boldsymbol{\sigma}} \qquad d\lambda \geq 0 \tag{6.167}$$

$$d\,\mathbf{e}^{p} = d\lambda \frac{\partial G}{\partial \mathbf{s}} = d\lambda \frac{\partial G}{\partial h} \frac{s}{\partial h} \qquad d\gamma^{p}_{eq} = d\lambda \frac{\partial G}{\partial h} \qquad d\lambda \geq 0 \tag{6.168}$$

The relation (6.168) assumes that $\frac{\partial G}{\partial h}$ is non-negative. If G = F in the above equations, the flow rule is associated. When G = F in Eq. (6.168) the flow rule is to satisfy deviatoric normality. An associated flow rule satisfies deviatoric normality for an isotropic material if the loading function F is independent of $J_{3\sigma}$ and then $\mathrm{tr}\left(\frac{\partial F}{\partial s}\right) = 0$.

Thermodynamic restriction to the plasticity models is based on the second principle of thermodynamics that requires the intrinsic dissipation Π_1 to be non-negative. The intrinsic dissipation Π_1 and the plastic dissipation Π^{p} for ideal plastic material are equal

$$\Pi_1 = \Pi^{p} = \boldsymbol{\sigma} \frac{d\boldsymbol{\varepsilon}^{p}}{dt} \geq 0 \tag{6.169}$$

The expression (6.169) is guaranteed for a standard material. If the flow is defined by a non-associated potential G given by (6.157), based on Eqs. (6.166) - (6.168), expression (6.169) can be rewritten in the form

$$\Pi_1 = \Pi^p = \boldsymbol{\sigma}\frac{d\boldsymbol{\varepsilon}^p}{dt} + h\frac{d\gamma^p_{eq}}{dt} \geq 0 \tag{6.170}$$

$$\boldsymbol{\sigma}\frac{\partial G}{\partial \boldsymbol{\sigma}} + h\frac{\partial G}{\partial h} \geq 0 \quad \text{if } F = 0 \tag{6.171}$$

The non-negativeness of the intrinsic dissipation for hardening material is

$$\Pi_1 = \boldsymbol{\sigma}\frac{d\boldsymbol{\varepsilon}^p}{dt} + \boldsymbol{\eta}\frac{d\mathbf{m}}{dt} \geq 0 \tag{6.172}$$

If the flow rule is defined by a non-associated potential G given by (6.157), and using (6.166) - (6.167), expression (6.172) can be rewritten as

$$\Pi_1 = \boldsymbol{\sigma}\frac{d\boldsymbol{\varepsilon}^p}{dt} + h\frac{d\gamma^p_{eq}}{dt} + \boldsymbol{\eta}\cdot\frac{d\mathbf{m}}{dt} \geq 0 \tag{6.173}$$

By (6.173) in the isotropic case Eqs. (6.102) – (6.104) have the form

$$\boldsymbol{\sigma}\frac{\partial G}{\partial \boldsymbol{\sigma}} + h\frac{\partial G}{\partial h} + \mathbf{m}\cdot\frac{\partial G}{\partial \mathbf{m}} \geq 0 \qquad \text{if } F = 0 \tag{6.174}$$

$$\boldsymbol{\sigma}\frac{\partial G^*}{\partial \boldsymbol{\sigma}} + h\frac{\partial G^*}{\partial h} + \boldsymbol{\eta}\cdot\frac{\partial G}{\partial \boldsymbol{\eta}} \geq 0 \qquad \text{if } F = 0 \tag{6.175}$$

$$\boldsymbol{\sigma}\frac{\partial F}{\partial \boldsymbol{\sigma}} + h\frac{\partial F}{\partial h} + \mathbf{m}\cdot\frac{\partial \sigma}{\partial \mathbf{m}} \geq 0 \qquad \text{if } F = 0 \tag{6.176}$$

6.9.2 Loading Functions

Assume the convex loading function for the isotropic plastic material

$$F(h, p) = h + \alpha p - q \tag{6.177}$$

where α and q are material characteristics. The constant q is necessarily non-negative to ensure that the zero loading point satisfies $F(0,0) \leq 0$. The coefficient α is non-negative to describe an infinite tensile stress. The yield surface given by (6.177) is an axisymmetric surface around the trisector in principal stress space $\{\sigma_i\}$. If $\alpha = 0$ the loading function reduces to the Huber-Mises loading function.

The form of the Huber-Mises loading function is

$$F=\frac{1}{\sqrt{3}}\sqrt{\sigma_{11}^2+\sigma_{22}^2+\sigma_{33}^2-\sigma_{11}\sigma_{22}-\sigma_{22}-\sigma_{33}-\sigma_{33}\sigma_{11}+3\left(\sigma_{12}^2+\sigma_{23}^2+\sigma_{31}^2\right)}-q \tag{6.178}$$

or for principal directions

$$F=\frac{1}{\sqrt{3}}\sqrt{\sigma_1^2+\sigma_2^2+\sigma_3^2-\sigma_1\sigma_2-\sigma_2\sigma_3-\sigma_3\sigma_1}-q \tag{6.179}$$

Another form of the Huber-Mises loading function is

$$F=\frac{1}{\sqrt{6}}\sqrt{\left(\sigma_{11}-\sigma_{22}\right)^2+\left(\sigma_{22}-\sigma_{33}\right)^2+\left(\sigma_{33}-\sigma_{11}\right)^2+6\left(\sigma_{12}^2+\sigma_{23}^2+\sigma_{31}^2\right)}-q \tag{6.180}$$

or

$$F=\frac{1}{\sqrt{6}}\sqrt{\left(\sigma_1-\sigma_2\right)^2+\left(\sigma_2-\sigma_3\right)^2+\left(\sigma_3-\sigma_1\right)^2}-q \tag{6.181}$$

The Huber-Mises loading function can be transformed to the equivalent forms if we introduce material parameter $q={}^{1}\!/\!_{\sqrt{3}}\,\sigma_o$, where σ_o is the yield point of the material in uniaxial tension. Then the Huber-Mises loading function is expressed in the frequently met form:

$$F=\frac{1}{\sqrt{3}}\sqrt{\sigma_{11}^2+\sigma_{22}^2+\sigma_{33}^2-\sigma_{11}\sigma_{22}-\sigma_{22}-\sigma_{33}-\sigma_{33}\sigma_{11}+3\left(\sigma_{12}^2+\sigma_{23}^2+\sigma_{31}^2\right)}-\frac{1}{\sqrt{3}}\sigma_o \tag{6.182}$$

$$F=\frac{1}{\sqrt{3}}\sqrt{\sigma_1^2+\sigma_2^2+\sigma_3^2-\sigma_1\sigma_2-\sigma_2\sigma_3-\sigma_3\sigma_1}-\frac{1}{\sqrt{3}}\sigma_o \tag{6.183}$$

$$F=\frac{1}{\sqrt{6}}\sqrt{\left(\sigma_{11}-\sigma_{22}\right)^2+\left(\sigma_{22}-\sigma_{33}\right)^2+\left(\sigma_{33}-\sigma_{11}\right)^2+6\left(\sigma_{12}^2+\sigma_{23}^2+\sigma_{31}^2\right)}-\frac{1}{\sqrt{3}}\sigma_o \tag{6.184}$$

or

$$F=\frac{1}{\sqrt{6}}\sqrt{\left(\sigma_1-\sigma_2\right)^2+\left(\sigma_2-\sigma_3\right)^2+\left(\sigma_3-\sigma_1\right)^2}-\frac{1}{\sqrt{3}}\sigma_o \tag{6.185}$$

In order to present its geometrical interpretation, the Huber-Mises criterion is rewritten using principal stress deviator components as

$$F=\frac{1}{\sqrt{2}}\sqrt{s_1^2+s_2^2+s_3^2}-q=0 \tag{6.186}$$

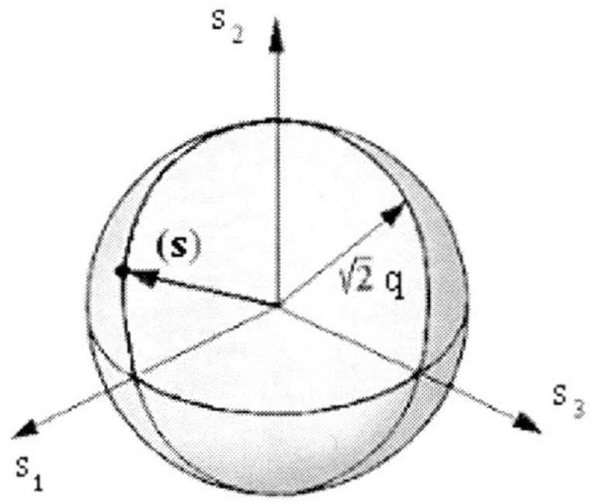

Fig. 6.23. The Huber-Mises yield locus in the space $\{s_i\}$ of principal stress deviators

In the space $\{s_i\}$ the expression (6.185) represents spherical surface of the radius $h\sqrt{2}$. The points inside the spherical surface represent the elastic state. If the material is in a plastic range then the point (**s**) is on the surface of the sphere. In the space $\{\sigma_i\}$ of principal stresses the Huber-Mises yield criterion represents a circular cylinder with an axis of unit vector with $\left(1/\sqrt{3},\ 1/\sqrt{3},\ 1/\sqrt{3}\right)$ as the cosine directors.

In the space $\{\sigma_i\}$ of principal stresses the stress tensor and its isotropic or deviatoric part are described by three components so in this space can be treated as vectors

$$\boldsymbol{\sigma} = (\sigma_1, \sigma_2, \sigma_3) \qquad \mathbf{p} = (p, p, p) \tag{6.187}$$

$$\mathbf{s} = (\sigma_1 - p, \sigma_2 - p, \sigma_3 - p) \tag{6.188}$$

where

$$\mathbf{p} = (\mathrm{tr}\,\boldsymbol{\sigma}/3)\,\mathbf{1} \tag{6.189}$$

The geometrical interpretation of an isotropic part of stress tensor is the trisector defined by the unit vector with $\left(1/\sqrt{3},\ 1/\sqrt{3},\ 1/\sqrt{3}\right)$ as cosine directors. Since $\boldsymbol{\sigma} = \mathbf{s} + \mathbf{p}$, the deviatoric stress represents deviation of the stress $\boldsymbol{\sigma}$ from the axis of the cylinder, which is presented in Fig. 6.24. A deviation of stress from the axis of the cylinder symmetry is the measure of material effort. This distance is

$$|\mathbf{s}| = \sqrt{s_i s_i} = \sqrt{s_1^2 + s_2^2 + s_3^2} \tag{6.190}$$

and is equal to the radius of the Huber-Mises cylinder.

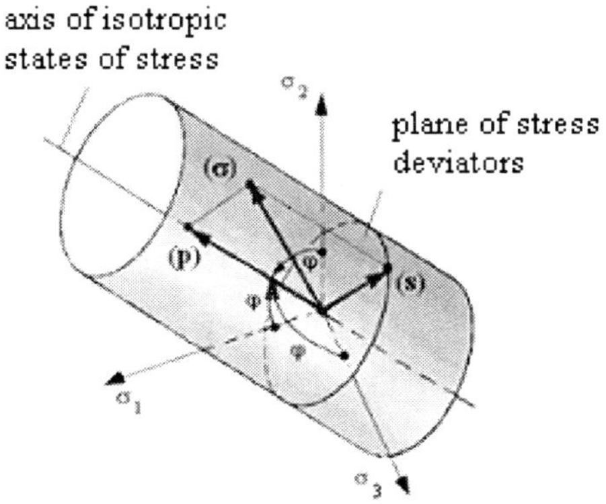

Fig. 6.24. The Huber-Mises yield locus in the space $\{\sigma_i\}$ of principal stresses

In the case of a plane state of strain the Huber-Mises yield criterion represents in the space $\{\sigma_{11}, \sigma_{22}, \sigma_{12}\}$ an elliptic cylinder with the axis on the plane $\{\sigma_{11}, \sigma_{22}\}$ defined by a unit vector with $\left(1/\sqrt{2},\ 1/\sqrt{2}\right)$ as cosine directors (Fig. 6.25).

In the case of a plane state of stress the Huber-Mises yield criterion in the space $\{\sigma_1, \sigma_2\}$ is represented by an ellipse being the trace of the cross section of the Huber-Mises cylinder by the plane $\sigma_3 = 0$ (Fig. 6.26).

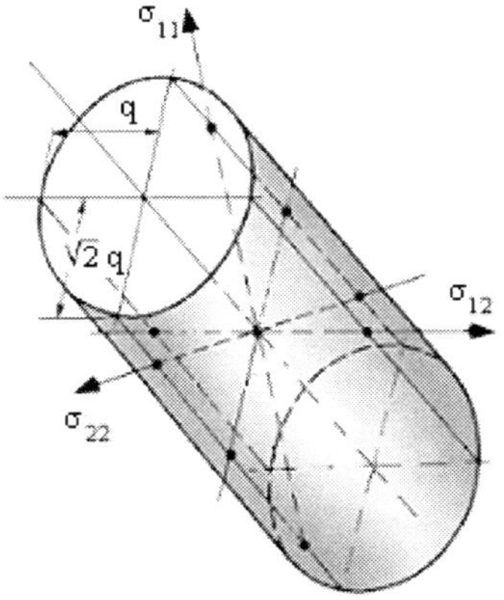

Fig. 6.25. The Huber-Mises yield locus for the plane state of strain in the space $\{\sigma_1 \times \sigma_{21} \times \sigma_{12}\}$

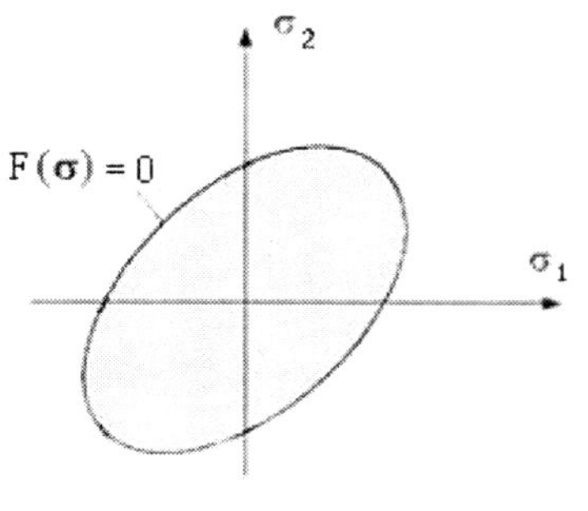

Fig. 6.26. The Huber-Mises yield locus for the plane state of stress in the space $\{\sigma_1 \times \sigma_2\}$

Based on Eq. (6.185) the Huber-Mises yield criterion can be written as

$$\left|\sigma_1 - \sigma_2\right|^n + \left|\sigma_2 - \sigma_3\right|^n + \left|\sigma_3 - \sigma_1\right|^n = 2\sigma_o^n \tag{6.191}$$

where n = 2. If $n \to \infty$ in Eq. (6.191) the yield criterion became the so-called the Treska yield criterion. According to the Treska criterion the loading function reads

$$F = \mathrm{Sup}_{i,j=1,2,3}\left(\sigma_i - \sigma_j\right) \tag{6.192}$$

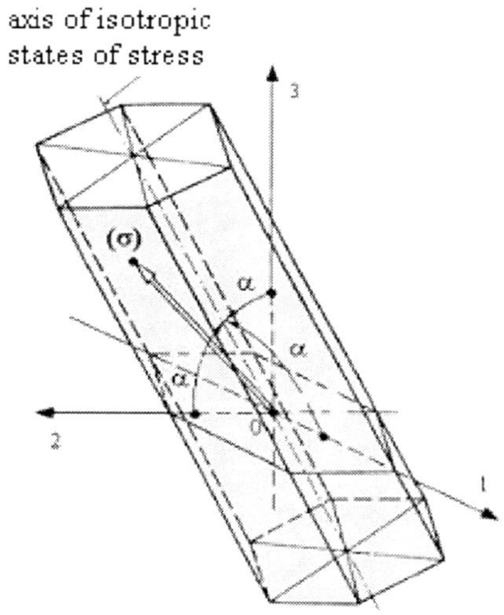

Fig. 6.27. The Treska yield criterion in the space $\{\sigma_i\}$

The Treska yield criterion can be written as

$$\left[(\sigma_1-\sigma_2)^2-\sigma_o^2\right]\left[(\sigma_2-\sigma_3)^2-\sigma_o^2\right]\left[(\sigma_3-\sigma_1)^2-\sigma_o^2\right]=0 \qquad (6.193)$$

The geometrical interpretation of the Treska yield criterion is given in Fig. 6.27. The Treska yield criterion for a plane state of stress is

$$(\sigma_{11}-\sigma_{22})^2+4\sigma_{12}^2=4q^2 \qquad (6.194)$$

It has the identical form as the Huber-Mises yield criterion (6.181) if we put $\sigma_{13}=\sigma_{23}=0$ and $\sigma_3=\frac{1}{2}(\sigma_{11}+\sigma_{22})$. The difference is when we change q onto σ_o. For the Treska yield criterion

$$\sigma_o=2q \qquad (6.195)$$

and for the Huber-Mises criterion

$$\sigma_o=\sqrt{3}q \qquad (6.196)$$

The Treska yield criterion represents a prism inscribed in a Huber-Mises cylinder. Any plane orthogonal to the trisector, i.e. any deviatoric plane defined by σ = const intersects with the loading surface along a regular hexagon. A comparison of the Huber-Mises and the Treska yield criteria in the space $\{\sigma_i\}$ is given in Fig. 6.28. and on the plane of deviators in Fig. 6.29. On the plane $\sigma_3=0$ representing a plane state of stress the Huber-Mises and the Treska yield criteria are presented in Fig. 6.30.

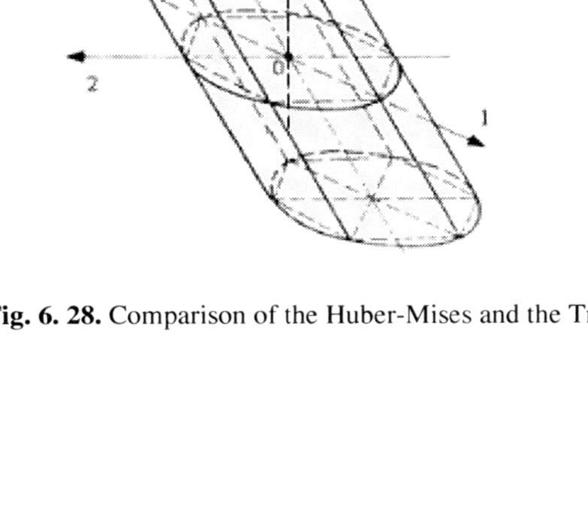

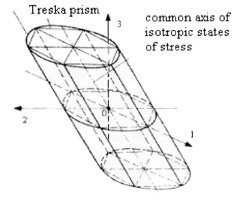

Fig. 6. 28. Comparison of the Huber-Mises and the Treska yield criteria in the space $\{\sigma_i\}$

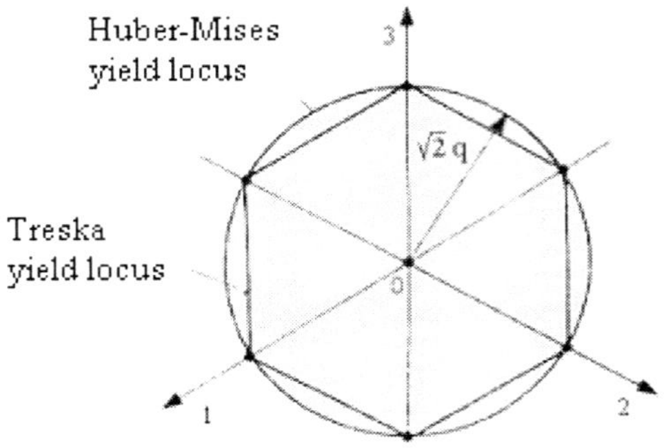

Fig. 6. 29. Comparison of the Huber-Mises and the Treska yield criteria on a plane of deviators; plane normal to the cylinder and prism axis

The previous loading function (6.177) for an isotropic material that exhibits isotropic hardening is extended in the form

$$F = h + \alpha p - \eta\, q \tag{6.197}$$

which agrees with Eq. (6.155) when n = 1. The yield surface that corresponds to the loading function (6.197) is an axisymmetric cone around the trisector in principal stress space $\{\sigma_i\}$. In this space, the three coordinates of the vertex of the

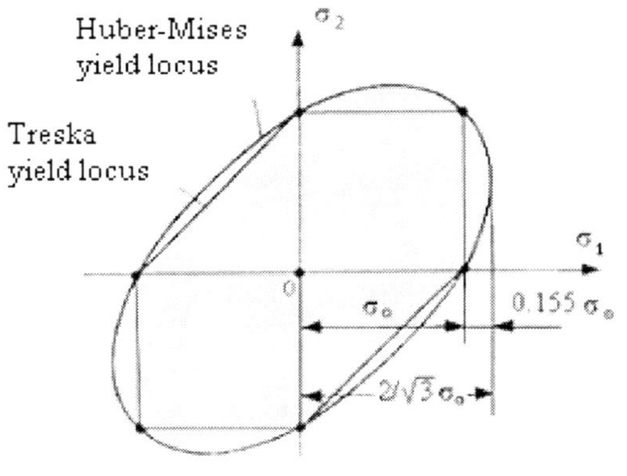

Fig. 6.30. Comparison of the Huber-Mises and the Treska yield criteria on a plane $\{\sigma_1, \sigma_2\}$

cone are equal to $q\eta/\alpha$. In the loading point space $\{\sigma_i\}$ the singular points are located on a line given by the equation $\sigma_1 = \sigma_2 = \sigma_3 = q\eta/\alpha$. The present yield surface is the transform of the surface defined by $\eta = 1$, through the homothety centred at the origin with η as the ratio. For an isotropic material exhibiting kinematic hardening

$$F = \sqrt{\frac{1}{2}\left(s_{ij} + \beta_{ij}\right)\left(s_{ji} + \beta_{ji}\right)} + \alpha\left(p + \eta_{ii}\right) - q \tag{6.198}$$

where the symmetric tensor $\boldsymbol{\beta}$ is the deviator of the tensor

$$\boldsymbol{\beta} = \boldsymbol{\eta} - 1/3\,(\mathrm{tr}\,\boldsymbol{\eta})\mathbf{1} \tag{6.199}$$

In the loading point space $\{\sigma_i\}$ the present yield surface is the transform of the surface defined by $\boldsymbol{\eta} = 0$ through the translation defined by the vector with $\boldsymbol{\eta}$ and the eigenvalues η_1, η_2, η_3 of the tensor $\boldsymbol{\eta}$ as components. In the loading point space $\{\sigma_i\}$ the singular points are located on the line given by the equation $\sigma_1 + \eta_1 = \sigma_2 + \eta_2 = \sigma_3 + \eta_3 = q/\alpha$. The expressions (6.197) and (6.198) may be combined to obtain

$$F = \sqrt{\frac{1}{2}\left(s_{ij} + \beta_{ij}\right)\left(s_{ji} + \beta_{ji}\right)} + \alpha\left(p + \eta_{ii}\right) - q\eta' \tag{6.200}$$

In the loading space $\{\sigma_i\}$ the present yield surface is the transform of the surface defined by $\boldsymbol{\eta} = 0$ and $\eta' = 1$ through the translation defined by the vector with (η_1, η_2, η_3) as components followed by the homothety of ratio η' and centred at the point (η_1, η_2, η_3), the translated point of origin.

In the loading point space $\{\sigma_i\}$ the singular points are located on the line given by the equation

$$\sigma_1 + \eta_1 = \sigma_2 + \eta_2 = \sigma_3 + \eta_3 = q\eta'/\alpha \tag{6.201}$$

6.9.3 The Flow Rule

The flow rule considering the non-associated potential G in the case of ideal plasticity gives

$$d\boldsymbol{\varepsilon}^p = d\lambda \frac{\partial G}{\partial \boldsymbol{\sigma}} \qquad d\lambda \geq 0 \tag{6.202}$$

In the case of hardening material

$$d\boldsymbol{\varepsilon}^p = d\lambda \left(\frac{\partial G}{\partial \boldsymbol{\sigma}}\right) \qquad d\lambda = \frac{d_{\eta}F}{H} \geq 0 \tag{6.203}$$

For ideal plastic material with the loading function defined by Eq. (6.177) consider the following expression for the non-associated potential

$$G = h + \delta p - q \tag{6.204}$$

where δ is the material parameter. Eq. (6.204) is of the form (6.157). For an ideal plastic material using Eqs. (6.166), (6.202) and (6.204), the flow rule is

$$d\boldsymbol{\varepsilon}^p = d\lambda\left(\frac{\mathbf{s}}{2h} + \frac{\delta}{3}\mathbf{1}\right) \qquad d\lambda \geq 0 \qquad \text{if } F = dF = 0 \tag{6.205}$$

For isotropic hardening material with the loading function given by Eq. (6.197), the multiplier $d\lambda$ is of the form

$$d\lambda = \frac{d_\eta F}{H} = \frac{1}{H}\left[\frac{\mathbf{s}}{2h} d\,\mathbf{s} + \alpha\, dp\right] \tag{6.206}$$

Let

$$G = h + \delta p - \eta q \tag{6.207}$$

be the corresponding non-associated potential, which is of the form (6.158) for $n = 1$. Using Eqs. (6.166), (6.203) and (6.204), the flow rule explicitly reads

$$d\boldsymbol{\varepsilon}^p = d\lambda\left(\frac{\mathbf{s}}{2h} + \frac{\delta}{3}\mathbf{1}\right) \quad d\lambda = \frac{d_\eta F}{H} \geq 0 \qquad \text{if } F = dF = 0 \tag{6.208}$$

where the plastic multiplier $d\lambda$ is expressed by Eq. (6.206). The flow rules (6.203) and (6.208) satisfies deviatoric normality.

In the case of kinematic hardening with the loading function given by Eq. (6.198) the plastic multiplier will be

$$d\lambda = \frac{d_\eta F}{H} = \frac{1}{H}\left[\frac{\mathbf{s}+\boldsymbol{\beta}}{2\sqrt{\mathrm{tr}(\mathbf{s}+\boldsymbol{\beta})\cdot(\mathbf{s}+\boldsymbol{\beta})}}(d\,\mathbf{s} + d\boldsymbol{\beta}) + \alpha(dp)\right] \tag{6.209}$$

Let

$$G = \sqrt{\frac{1}{2}(s_{ij} + \beta_{ij})\cdot(s_{ji} + \beta_{ji})} + \delta[p + \eta_{ii}] - q \tag{6.210}$$

be the corresponding non-associated potential, which is of the form (6.159). Using Eqs. (6.166), (6.203) and (6.210), the flow rule is of the form

$$d\boldsymbol{\varepsilon}^p = d\lambda\left[\frac{\mathbf{s}+\boldsymbol{\beta}}{2\sqrt{\mathrm{tr}(\mathbf{s}+\boldsymbol{\beta})\cdot(\mathbf{s}+\boldsymbol{\beta})}} + \frac{\delta}{3}\mathbf{1}\right] \tag{6.211}$$

$$d\lambda = \frac{d_\eta F}{H} \geq 0 \qquad \text{if } F = dF = 0 \tag{6.212}$$

If $F = G$ i.e. if $\alpha = \delta$, the material is standard and verifies the hypothesis of maximal plastic work. In all the above considered cases $\partial G/\partial h = 1$. By Eq. (6.168) the plastic multiplier is

$$d\lambda = d\gamma^{p}_{eq} \tag{6.213}$$

where $d\gamma^{p}_{eq}$ is the equivalent plastic distortion defined by Eq. (6.164). For all considered cases, it is $\partial G/\partial p = \delta$. The coefficient δ can actually be identified as the plastic dilatancy coefficient defined in the general case by Eq. (6.164).

6.9.4 Hardening

Assume the following transformation $\eta \rightarrow \eta^*$ in the case of an isotropic hardening material model

$$\eta q = -\eta^* + q \tag{6.214}$$

where the hardening forces η and η^* satisfy the expressions

$$\eta = -\frac{\partial U}{\partial m} \qquad \eta^* = -\frac{\partial U^*}{\partial m^*} \tag{6.215}$$

and $U^*(m) = U(m(m^*))$ is the frozen energy. We attach a new potential G^* by substituting Eq. (6.214) into (6.207) of the non-associated potential G

$$G^*(h, p, \eta^*) = h + \delta p + \eta^* - q \tag{6.216}$$

We express the hardening law by

$$dm^* = d\lambda \frac{\partial G^*}{\partial \eta^*} \tag{6.217}$$

where $d\lambda$ is given by (6.206). If the isotropic hardening variable m^* is identified by (6.213), then

$$dm^* = d\gamma^{p}_{eq} \tag{6.218}$$

By definition (6.164) of the equivalent plastic distortion

$$m^* = \gamma^{p}_{ep} = \int_0^t \left(\frac{1}{2} \dot{\gamma}^{p}_{ij} \dot{\gamma}^{p}_{ji} \right)^{1/2} dt \tag{6.219}$$

where the dot denotes time derivation. In strain space, the hardening variable m^* represents the length of the path covered from the initial state to the present state by the representative point of the plastic distortion i.e. the end of vector of components γ^{p}_{ij} . In the case $\partial F/\partial m^* = \partial G/\partial \eta^* = 1$, Eq. (6.99) leads to one, expressing the hardening modulus H as

$$H = \frac{\partial^2 U^*}{\partial m^{*2}} \tag{6.220}$$

which holds in any case. In the case of kinematic hardening, the hardening state is defined by a tensor variable **m**. Using Eq. (6.210) we get the hardening law

$$\mathrm{d}\,\mathbf{m} = \mathrm{d}\lambda \frac{\partial G}{\partial \boldsymbol{\eta}} \tag{6.221}$$

where the plastic multiplier dλ is given by Eq. (6.209). Using Eqs. (6.210) and (6.221) the hardening law is expressed as

$$\mathrm{d}\,\mathbf{m} = \mathrm{d}\lambda \left[\frac{\mathbf{s}+\boldsymbol{\beta}}{2\sqrt{\mathrm{tr}(\mathbf{s}+\boldsymbol{\beta})\cdot(\mathbf{s}+\boldsymbol{\beta})}} + \frac{\delta}{3}\mathbf{1} \right] \tag{6.222}$$

By Eqs. (6.222) and (6.211) it follows that

$$\mathbf{m} = \boldsymbol{\varepsilon}^{\mathrm{p}} \tag{6.223}$$

The hardening variable **m** is identified as the plastic strain tensor.

Part IV

Powder Forming Plasticity

7 Description of Powder Material

7.1 Introduction

Powder forming involves compacting a metal powder into a green preform and then after modifying the metallurgical structure, forming the preform by repressing or upsetting in a closed-die. In powder forming the matrix material may be composed of a solid part and a disconnected occluded space. The notion of porosity is then introduced as the ratio of the volume of the porous space to the total volume of the material. It is denoted by ϕ. In this section the assumption that porous space is an empty space is made. This simplified assumption in powder forming can be questionable for porous materials with porous space containing fluid. Such a case is however the rarity in powder forming so has been omitted in the book.

7.2 Infinitesimal Transformation

In infinitesimal transformation, the trace tr $\boldsymbol{\varepsilon}$ of the linearized stress tensor $\boldsymbol{\varepsilon}$ represents the volume change per unit volume in the deformation. It is called volume dilatation

$$\frac{dV - dV_r}{dV_r} = \operatorname{div}\mathbf{u} = \operatorname{tr}\boldsymbol{\varepsilon} = \varepsilon_{ii} = \boldsymbol{\varepsilon} \tag{7.1}$$

The observable volume dilatation of the matrix material is due to both variation of the porous space and the volume dilatation of the matrix material. The latter is noted by $\boldsymbol{\varepsilon}_m$ where the subscript m stands for the material matrix. If dV_r^m and dV^m denote the volumes occupied respectively, by the matrix in the reference and current configurations, the volume dilatation of the matrix is

$$\boldsymbol{\varepsilon}_m = \frac{dV^m - dV_r^m}{dV^m} \tag{7.2}$$

The matrix dilatation $\boldsymbol{\varepsilon}_m$ can be derived as a function of the material dilatation, and of the values of porosity, in the reference and present configuration, denoted by, respectively ϕ_o and ϕ. With $dV_r^m = (1-\phi_o)\, dV_r$ and $dV^m = (1 - \phi)\, dV$, Eqs. (7.1) and (7.2) give

$$(1-\phi_o)\,\boldsymbol{\varepsilon}_m = (1-\phi)\mathrm{tr}\,\boldsymbol{\varepsilon} - (\phi - \phi_o) \tag{7.3}$$

7.3 Mass Conservation

Denote the overall mass density of the porous material viewed as a whole by ρ, and assume that the porous space is an empty space. Assume that the mass contained in the infinitesimal volume dV is ρdV. We have

$$\rho dV = \rho_m\, (1 - \phi)\, dV \tag{7.4}$$

where ρ_m is the present mass density of the matrix material, and ϕ is the present porosity. Since there is no overall mass creation, the global mass balance is

$$\frac{D}{Dt}\int_V \rho dV = 0 \tag{7.5}$$

Using expression (1.43) and (1.45) of the material derivative with g = r the overall integral mass balance is

$$\int_V \frac{\partial \rho}{\partial t} dV + \int_a (\rho\, \mathbf{V}) \cdot \mathbf{n} \cdot da = 0 \tag{7.6}$$

The divergence theorem applied to Eq. (7.6) gives

$$\frac{\partial \rho}{\partial t} + \mathrm{div}(\rho\, \mathbf{V}) = 0 \tag{7.7}$$

or equivalently

$$\frac{d\rho}{dt} + \rho\ \mathrm{div}\, \mathbf{V} = 0 \tag{7.8}$$

The matrix mass conservation implies

$$\frac{d}{dt}\int_V \rho^m (1-\phi)\, dV = 0 \tag{7.9}$$

or equivalently

$$\frac{D}{Dt}\int_V \rho^m (1-\phi)\, dV = 0 \tag{7.10}$$

By Eq. (7.6) with $g = \rho^m (1 - \phi)$ and Eq. (7.10) yields the integral matrix material mass balance is

$$\int_V \frac{\partial}{\partial t} \rho^m (1-\phi)\, dV + \int_a \rho^m (1-\phi)\, \mathbf{V} \cdot \mathbf{n}\, da = 0 \tag{7.11}$$

By (7.11) the matrix material continuity equation is derived in terms of its apparent mass density $\rho^m (1-\phi)$ in the equivalent forms

$$\frac{\partial}{\partial t}\left[\rho^m (1-\phi)\right] + \mathrm{div}\left[\rho^m (1-\phi)\mathbf{V}\right] = 0 \tag{7.12}$$

$$\frac{d}{dt}\left[\rho^m (1-\phi)\right] + \rho^m (1-\phi) \mathrm{div} V = 0 \tag{7.13}$$

The overall mass ρdV, which is contained in volume dV, may be written as

$$\rho dV = \rho_o \, dV_r \tag{7.14}$$

The mass density ρ_o reads

$$\rho_o = \rho_o^m (1-\phi) \tag{7.15}$$

The overall mass conservation may be written as

$$\frac{D}{Dt} \int_{V_r} \rho_o dV_r = 0 \tag{7.16}$$

By Eqs. (7.4) and (7.15) the following transport formula is obtained

$$J\rho^m (1-\phi) = \rho_o^m (1-\phi_o) \tag{7.17}$$

which expresses the matrix material mass conservation in Lagrange variables. The above equation can be obtained by writing the matrix material mass conservation as

$$\frac{d}{dt} \int_{V_r} J\rho^m (1-\phi)\, dV_r = 0 \tag{7.18}$$

which gives

$$\frac{d}{dt}\left[J\rho^m (1-\phi)\right] = 0 \tag{7.19}$$

7.4 Momentum Balance

The Euler theorem (2.1) – (2.4) for porous material is formulated as follows. The resultant of the elementary body and surface forces and the resultant of the corresponding elementary moments for any porous material domain V are respectively, equal to the resultant and the overall moment of vectors

$$\frac{\partial}{\partial t}\left[\rho^{m}(1-\phi)\right]\mathbf{V}\,dV \tag{7.20}$$

distributed within the volume V, and vectors

$$\left\{\left[\rho^{m}(1-\phi)\,\mathbf{V}\right]\mathbf{V}\cdot\mathbf{n}\right\}da \tag{7.21}$$

distributed on the surface a which encloses the domain V.

The dynamic theorem is formulated as follows. For any porous material domain V, the resultant of the elementary body and surface forces and the resultant of the corresponding elementary moments are, respectively equal to the resultant and the overall moment of the elementary dynamic forces

$$\int_{V}\left[\rho^{m}(1-\phi)\boldsymbol{a}\right]dV = \int_{V}\left[\rho^{m}(1-\phi)\right]\mathbf{P}\,dV + \int_{a}\mathbf{t}\,da \tag{7.22}$$

$$\int_{V}\mathbf{x}\times\left[\rho^{m}(1-\phi)\boldsymbol{a}\right]dV = \int_{V}\mathbf{x}\times\left[\rho^{m}(1-\phi)\right]\mathbf{P}\,dV + \int_{a}\mathbf{x}\times\mathbf{t}\,da \tag{7.23}$$

The equation of motion (2.17) can be written as

$$\int_{V}\left\{\mathrm{div}\boldsymbol{\sigma}+\left[\rho^{m}(1-\phi)\right](\mathbf{P}-\boldsymbol{a})\right\}dV = 0 \tag{7.24}$$

Since Eq. (7.24) holds for any subdomain V, the equation of motion is derived in the form

$$\mathrm{div}\boldsymbol{\sigma}+\left[\rho^{m}(1-\phi)\right](\mathbf{P}-\boldsymbol{a}) = 0 \tag{7.25}$$

The theorem of virtual work rate (2.32) is formulated as follows. For any material domain V, and any velocity field $\mathbf{V}^*$, actual or virtual, the sum of work rates of the external forces R_{EF} ($\mathbf{V}^*$), of inertia force R_{IN} ($\mathbf{V}^*$) and of internal forces R_{IF} is equal to zero, where

$$R_{IN}(\mathbf{V}^*) = -\int_{V}\mathbf{V}^*\cdot\left[\rho^{m}(1-\phi)\boldsymbol{a}\right]dV \tag{7.26}$$

and

$$-R_{IF}(\mathbf{V}^*) = \int_{V}\mathbf{V}^*\cdot\rho\,\mathbf{P}\,dV + \int_{a}\mathbf{V}^*\cdot\mathbf{t}\,da - \int_{V}\mathbf{V}^*\cdot\left[\rho^{m}(1-\phi)\boldsymbol{a}\right]dV \tag{7.27}$$

The kinetic energy theorem is formulated as:
If the velocity field **V*** is the actual matrix material velocity field **V** in the virtual work rate theorem (2.32), together with expression (2.44) of the material derivative of the overall corrected kinetic energy, then the kinetic energy theorem reads

$$\frac{DK}{Dt} + R_{SR}(\mathbf{V}) = R_{EF}(\mathbf{V}) \tag{7.28}$$

where the kinetic energy K is expressed as

$$K = \int_V \frac{1}{2} \rho^m (1-\phi)\, \mathbf{V}^2 \, dV \tag{7.29}$$

The material derivative of (7.29) reads

$$\frac{DK}{Dt} = \int_V \left[\rho^m (1-\phi) \mathbf{V} \cdot \boldsymbol{a}\right] dV \tag{7.30}$$

The material derivative of the kinetic energy corresponds to the opposite of the work rate of the inertia forces to which they are related. The first law of thermodynamics for porous material is identical to this given by Eq. (4.5) when the kinetic energy is expressed by Eq. (7.29). The second law of thermodynamics is formulated as in Point 4.5. All considerations from Section 2 can be related to porous material so they can be recalled.

7.5 Physical Laws

The equations describing powder forming can be solved if the number of equations should equal to the number of unknown parameters, which have to be determined. This description with the use of Lagrange variables is made in Table 7.1. and 7.2. The indicated functions of space and time have arguments **X** and t. The number of unknowns in the thermodynamical process is given in Table 7.1. It is equal to 21 + n, where n is the number of internal variables. The number of equations governing the process is 21 + n which is shown in Table 7.2. The completely description of the thermodynamical process needs to specify the boundary conditions describing the values of the variables at the limit of the considered material as given in point 4.9. If the expressions of the free energy Ψ are specified and the evolutions laws governing the internal variables m_I are given then the system of equations is complete.

Table 7.1. Unknown parameters in thermodynamical process of powder forming

Unknowns	Number of unknowns
Temperature T	1
Heat flow vector **Q**	3
Entropy S	1
Porosity ϕ	1
Displacement **u**	3
Strain tensor **E**	6
Stress tensor **S**	6
Internal variables m_I	n

Table 7.2. Governing equations in thermodynamical process of powder forming

Equation	Number of equations
Thermal equation $T\left[\frac{dS}{dt}\right] = R - \text{Div}\,\mathbf{Q} + \Pi_2$	1
Conduction law $\mathbf{q} = -\mathbf{K}\cdot\text{Grad}\,T$	3
State equation $\mathbf{S} = \frac{d\Psi}{d\mathbf{E}} \quad S = -\frac{d\Psi}{dT}$	7
Mass conservation $\frac{D}{Dt}\int_V \rho^m (1-\phi)\, dV = 0$	1
Strain-displacement relation $2\,\mathbf{E} = \text{Grad}\,\mathbf{u} + {}^T\text{Grad}\,\mathbf{u} + {}^T\text{Grad}\,\mathbf{u}\cdot\text{Grad}\,\mathbf{u}$	6
Equations of motion $\text{Div}\,\mathbf{F}\cdot\mathbf{S} + \rho_o(\mathbf{P} - \boldsymbol{a})$	3
Evolution laws $\mathbf{B}_{\dot{m}} = \frac{\partial H}{\partial \dot{\mathbf{m}}}$	n

7.6 Plastic Porosity

Introduce the following notation

$$\boldsymbol{\varepsilon}^p = \varepsilon^p_{ii} = \text{tr}\,\boldsymbol{\varepsilon}^p \tag{7.31}$$

and consider ϕ^e as the porosity relative to the state after a complete elastic unloading process from the present state.

Define the plastic porosity

$$\phi^p = \phi^e\left(1+\varepsilon^p_{ii}\right) - \phi_o \tag{7.32}$$

In infinitesimal transformations $\boldsymbol{\varepsilon}^p$ represent the irreversible volume dilatation occurring after complete unloading. By the volume transport formula (1.8) and the expression (1.20),

$$dV^e = \left(1+\varepsilon^p_{ii}\right)dV_r \tag{7.33}$$

where dV^e is the volume of the elementary system after a complete elastic unloading process from the present state. Eqs. (7.32) and (7.33) give

$$\phi^p dV_r = \phi^o dV^e - \phi_o dV_r \tag{7.34}$$

Porosity ϕ^e is not a convenient state variable since never refers to the same configuration. Note that the plastic porosity ϕ^p can be also defined directly from Eq. (7.34). The relation (7.3) can be written in the form

$$\left(1-\phi_o\right)\boldsymbol{\varepsilon}_m = \left(1-\phi\right)\boldsymbol{\varepsilon} - \left(\phi-\phi_o\right) \tag{7.35}$$

where $\boldsymbol{\varepsilon}_m$ is the average matrix material volume dilatation. By (7.3) and the above definitions we have

$$\left(1-\phi_o\right)\boldsymbol{\varepsilon}^p_m = \left(1-\phi^e\right)\varepsilon^p_{ii} - \left(\phi^e-\phi_o\right) \tag{7.36}$$

where $\boldsymbol{\varepsilon}^p_m$ is the permanent volume dilatation of the matrix material, after a complete elastic unloading process of the elementary system. The quantity $\boldsymbol{\varepsilon}^p_m$ is not measurable at the macroscopic level adopted here for the description of porous material. The expressions (7.32) and (7.36) give

$$\boldsymbol{\varepsilon}^p = \phi^p + \left(1-\phi_o\right)\boldsymbol{\varepsilon}^p_m \tag{7.37}$$

The relation (7.37) shows that the plastic volume dilatation $\boldsymbol{\varepsilon}^p$ consists of two parts. The first – the plastic porosity ϕ^p corresponds to the irreversible change in volume of the connected porous space. The second part shows the permanent volume dilatation $\boldsymbol{\varepsilon}^p_m$ of the matrix material and is quantified through the term $(1-\phi_o)\,\boldsymbol{\varepsilon}^p_m$ where the factor $(1-\phi_o)$ takes into account the volume part occupied by the matrix. The relation (7.37) is of interest only for some results derived about the macroscopic behaviour of the porous material from the matrix material behaviour. The expression (7.37) for an incompressible matrix material ($\boldsymbol{\varepsilon}^p_m = 0$) shows, that the plastic volume dilatation ε^p_{ii} (7.31) is equal to the plastic porosity ϕ^p, because of irreversible variation of the porous space.

The experimental determinations of the plastic variables $\boldsymbol{\varepsilon}^p$ and ϕ^p show that these are internal state variables, because these variables are not accessible to direct observation. In experimental investigations we can measure only $d\varepsilon$. The increments $d\boldsymbol{\varepsilon}^p$ and $d\phi^p$ of the internal variables are obtained by subtracting values of external variable increments that correspond to opposite variations of the loading. The external variable $\boldsymbol{\varepsilon}$ is directly measurable and the internal plastic variables $\boldsymbol{\varepsilon}^p$ and ϕ^p are measured only as integrals of increments.

8 State Equations

The thermodynamic states of porous material are characterized by the external variables T, $\boldsymbol{\varepsilon}$, and the internal variables $\boldsymbol{\varepsilon}^p$, ϕ^p and m_I (I = 1, ..., N).

$$\Psi = \Psi\,(T, \boldsymbol{\varepsilon}, \boldsymbol{\varepsilon}^p, \phi^p, m_I) \tag{8.1}$$

where Ψ is the free energy. The variables m_I are internal variables characterizing the hardening state. As stated in section 6.3, the state equations in infinitesimal transformation are

$$S = -\frac{\partial \Psi}{\partial T} \qquad \boldsymbol{\sigma} = \frac{\partial \Psi}{\partial \boldsymbol{\varepsilon}} \tag{8.2}$$

The expression for free energy Ψ limited to a second-order expansion with respect to variables T, $\boldsymbol{\varepsilon}$, $\boldsymbol{\varepsilon}^p$, and ϕ^p is

$$\begin{aligned}\Psi = {} & \boldsymbol{\sigma}^o\,(\boldsymbol{\varepsilon} - \boldsymbol{\varepsilon}^p) - S_o\theta + \frac{1}{2}\,(\boldsymbol{\varepsilon} - \boldsymbol{\varepsilon}^p)\,\mathbf{C}\,(\boldsymbol{\varepsilon} - \boldsymbol{\varepsilon}^p) \\ & - \theta\mathbf{A}\,(\boldsymbol{\varepsilon} - \boldsymbol{\varepsilon}^p) - \frac{1}{2}\frac{C_\varepsilon}{T_o}\theta^2 + U(m)\end{aligned} \tag{8.3}$$

with $\theta = T - T_o$. The state equations obtained from Eqs. (8.2) and (8.3) are

$$\boldsymbol{\sigma} = \boldsymbol{\sigma}^o + \mathbf{C}\,(\boldsymbol{\varepsilon} - \boldsymbol{\varepsilon}^p) - \mathbf{A}\theta \tag{8.4}$$

$$S = S_o + \mathbf{A}\,(\boldsymbol{\varepsilon} - \boldsymbol{\varepsilon}^p) + C_\varepsilon\theta/T_o \tag{8.5}$$

$$\phi^p = \varepsilon^p_{ii} \tag{8.6}$$

The expressions (8.4) – (8.5) are identical to Eqs. (6.11) – (6.12). The third above equation i.e. Eq. (8.6) represents the changes of plastic porosity in the case of incompressible matrix material.

Since ϕ^p can be obtained from plastic strain $\boldsymbol{\varepsilon}^p$ all expressions from section 6 can be applied.

8.1 The Poroplastic Flow Rule

The poroplastic flow rule describes how the plastic phenomena in porous material occur. The poroplastic flow rule does not differ from that described in Section 6. The only difference is the additional term $d\phi^p = d\varepsilon^p_{ii}$ added to the flow rule for plastic material.

Consider the loading point ($\boldsymbol{\sigma}$) on the boundary of the domain of elasticity $\mathcal{E}_D$, ($F = dF = 0$). The infinitesimal loading increment $d\boldsymbol{\sigma}$ corresponding to ($\boldsymbol{\sigma}$) is said to be a neutral loading increment. The increment $d\boldsymbol{\varepsilon}$ may not be elastic. This is expressed by

$$d\boldsymbol{\varepsilon}^p = d\phi^p = 0 \quad \text{if} \quad F(\boldsymbol{\sigma}) < 0 \quad \text{or if} \quad F(\boldsymbol{\sigma}) = 0 \quad \text{and} \quad dF = \frac{\partial F}{\partial \boldsymbol{\sigma}} d\boldsymbol{\sigma} < 0$$
$$d\boldsymbol{\varepsilon} = d\boldsymbol{\varepsilon}^e + d\boldsymbol{\varepsilon}^p \quad \text{if} \quad F = dF = 0 \tag{8.7}$$

The notation dF assumes that the function F is differentiable with respect to its arguments, at the considered loading point ($\boldsymbol{\sigma}$) on the boundary of the domain $\mathcal{E}_D$.

The subdifferential of F with respect to ($\boldsymbol{\sigma}$) at the present loading point ($\boldsymbol{\sigma}$) is denoted by ∂F. The subdifferential ∂F at any regular point on the boundary of domain $\mathcal{E}_D$ defined by $F(\boldsymbol{\sigma}) = 0$ reduces to the gradient of F with respect to ($\boldsymbol{\sigma}$) and thus corresponds geometrically to the outward normal $\partial F/\partial \boldsymbol{\sigma}$ to the boundary at this point. In the case of a singular point, the boundary of domain $\mathcal{E}_D$ being a convex loading function shows that $\alpha \partial F$ with $\alpha \geq 0$ is the set of outward normals to the boundary of $\mathcal{E}_D$.

An evolution of the plastic strain is indicated by Eq. (8.7) occurs. The values of $d\boldsymbol{\varepsilon}^p$ are not all admissible, because they must ensure the non-negativeness of the intrinsic dissipation Π_1. The intrinsic dissipation reduces to the plastic work rate for ideal plastic materials, so the non-negativeness of intrinsic dissipation, requires the relation on the plastic increments $d\boldsymbol{\varepsilon}^p$

$$\boldsymbol{\sigma}\, d\boldsymbol{\varepsilon}^p \geq 0 \tag{8.8}$$

Let

$$d\boldsymbol{\varepsilon}^p \in d\lambda\, \mathcal{G}(\boldsymbol{\sigma}) \quad d\phi^p = d\varepsilon^p_{ii} \qquad d\lambda \geq 0 \tag{8.9}$$

where $d\lambda$ is a non-negative scalar factor called the plastic multiplier. The plastic multiplier $d\lambda$ is non-negative. The set $\mathcal{G}(\boldsymbol{\sigma})$ of directions in the loading point space $\{\boldsymbol{\sigma}\}$ represents the set of thermodynamically admissible directions for the vector ($d\boldsymbol{\varepsilon}^p$), which ensure the non-negativeness of its scalar product $\boldsymbol{\sigma}\, d\boldsymbol{\varepsilon}^p$. The admissible directions for vector ($d\boldsymbol{\varepsilon}^p$) are independent of the plastic criterion.

The plastic flow rule derived from the non-negativeness of intrinsic dissipation for an ideal plastic material can be summarized by

$$d\boldsymbol{\varepsilon}^p \in d\lambda\ \mathcal{G}(\boldsymbol{\sigma}),\ d\phi^p = d\varepsilon^p_{ii}\quad d\lambda \geq 0 \ \text{ where } \begin{cases} d\lambda \geq 0 \text{ if } F = 0 \text{ and } dF = 0 \\ d\lambda = 0 \text{ if } F < 0 \text{ or } dF < 0 \end{cases} \tag{8.10}$$

If the plastic increments $d\boldsymbol{\varepsilon}^p$ is non-zero, the loading is said to be plastic.

The evolution to hardening materials is elastic with no change of the hardening state, if in the space $\{\boldsymbol{\sigma} \times \boldsymbol{\eta}\}$ the present extended loading point $(\boldsymbol{\sigma}, \boldsymbol{\eta})$ lies inside the fix domain $\mathbb{E}$ $(F < 0)$ or on its boundary $F = 0$ and leaves it $(dF < 0)$. The above is expressed by

$$(d\boldsymbol{\varepsilon}^p, d\mathbf{m}) \in d\lambda\ \mathcal{G}(\boldsymbol{\sigma}, \boldsymbol{\eta}),\qquad d\phi^p = d\varepsilon^p_{ii}\qquad d\lambda \geq 0$$

$$\text{where } \begin{cases} d\lambda \geq 0 \text{ if } F = 0 \text{ and } dF = 0 \\ d\lambda = 0 \text{ if } F = 0 \text{ or } dF = 0 \end{cases} \tag{8.11}$$

We say that the loading is plastic if the increments $(d\boldsymbol{\varepsilon}^p, d\mathbf{m})$ are non-zero. The scalar $d\lambda$ appearing in Eq. (8.11) is the plastic multiplier, and $\mathcal{G}(\boldsymbol{\sigma}, \boldsymbol{\eta})$ is the set of thermodynamically admissible directions for the increments $(d\boldsymbol{\varepsilon}^p, d\mathbf{m})$ ensuring the non-negativeness of the intrinsic dissipation

$$\boldsymbol{\sigma}\, d\boldsymbol{\varepsilon}^p + \boldsymbol{\eta} \cdot d\mathbf{m} \geq 0 \tag{8.12}$$

The increment $d\mathbf{m}$ of the hardening variable is written in the form

$$d\mathbf{m} = d\lambda\ \mathcal{G}_{\mathbf{m}}(\boldsymbol{\sigma}, \boldsymbol{\eta}) \tag{8.13}$$

where $\mathcal{G}_{\mathbf{m}}(\boldsymbol{\sigma}, \boldsymbol{\eta})$ represents the directions actually realized among the permissible directions for $d\mathbf{m}$. The loading function in hardening plasticity depends on the hardening force $\boldsymbol{\eta}$. Then the expression on dF is

$$dF = d_{\boldsymbol{\eta}}F + d_{\boldsymbol{\sigma}}F \tag{8.14}$$

with

$$d_{\boldsymbol{\eta}}F = \frac{\partial F}{\partial \boldsymbol{\sigma}}\, d\boldsymbol{\sigma} \tag{8.15}$$

$$d_{\boldsymbol{\sigma}}F = \frac{\partial F}{\partial \boldsymbol{\eta}}\, d\boldsymbol{\eta} \tag{8.16}$$

where $d_{\boldsymbol{\eta}}F$ is the differential of function F for the assumed constant hardening force $\boldsymbol{\eta}$. The expression (8.16) can be written as

$$d_{\boldsymbol{\sigma}}F = -H\, d\lambda$$

where H is the hardening modulus defined by

$$H = -\frac{\partial F}{\partial \boldsymbol{\eta}} \cdot \frac{\partial \boldsymbol{\eta}}{\partial \mathbf{m}} \cdot \frac{\partial \mathbf{m}}{\partial \lambda} = \frac{\partial F}{\partial \boldsymbol{\eta}} \cdot \frac{\partial^2 U}{\partial \mathbf{m}^2} \cdot \mathcal{G}_{\mathbf{m}}(\boldsymbol{\sigma}, \boldsymbol{\eta}) \tag{8.17}$$

The definition (8.16) of $d_{\boldsymbol{\sigma}}F$ and the relation (8.17) show that H has a stress unit. The plastic flow rule is specified, to examine the specific case of hardening (H > 0) and the case of softening (H < 0). In the case of hardening

$$\begin{aligned} &\left(d\boldsymbol{\varepsilon}^{p}, d\mathbf{m}\right) \in \frac{\partial_{\boldsymbol{\eta}} F}{H} \mathcal{G}(\boldsymbol{\sigma}, \boldsymbol{\eta}) \qquad d\phi^{p} = d\varepsilon^{p}_{ii} \qquad \text{if } F = 0 \text{ and } d_{\boldsymbol{\eta}}F > 0 \\ &d\boldsymbol{\varepsilon}^{p} = d\phi^{p} = d\mathbf{m} = 0 \qquad \text{if } F < 0 \text{ or if } F = 0 \text{ and } d_{\boldsymbol{\eta}}F < 0 \end{aligned} \tag{8.18}$$

In the case of softening

$$\begin{aligned} &\left(d\boldsymbol{\varepsilon}^{p}, d\mathbf{m}\right) \in \frac{\partial_{\boldsymbol{\eta}} F}{H} \mathcal{G}(\boldsymbol{\sigma}, \boldsymbol{\eta}) \qquad d\phi^{p} = d\varepsilon^{p}_{ii} \qquad \text{or } d\boldsymbol{\varepsilon}^{p} = d\mathbf{m} = 0 \\ &\qquad \text{if } F = 0 \text{ and } d_{\boldsymbol{\eta}}F < 0 \\ &d\boldsymbol{\varepsilon}^{p} = d\phi^{p} = d\mathbf{m} = 0 \text{ if } F < 0 \quad \text{or} \quad \text{if } F = d_{\boldsymbol{\eta}}F = 0 \end{aligned} \tag{8.19}$$

The expressions (8.18) and (8.19) for hardening or softening materials indicate that for the loading point ($\boldsymbol{\sigma}$) being remaining on the boundary of elasticity domain E_D, the evolution is purely elastic without a change in the hardening state (i.e. $d\boldsymbol{\varepsilon}^p = d\phi^p = d\mathbf{m} = d\boldsymbol{\eta} = 0$). The loading increment $d\boldsymbol{\sigma}$ is said then to be neutral.

8.2 The Associated Flow Rule

If material satisfies the maximal plastic work hypothesis then we say that the material is standard. The plastic criterion is then convex and the flow rule is normal, and we say that the flow rule is associated with the criterion of plasticity.

Consider an ideal plastic standard material. Eqs. (8.10) and (8.19) lead to write the flow rule as

$$d\boldsymbol{\varepsilon}^{p} \in d\lambda\, \partial F(\boldsymbol{\sigma}) \qquad d\phi^{p} = d\varepsilon^{p}_{ii}\; d\lambda \geq 0$$

$$\text{where} \quad \begin{aligned} &d\lambda \geq 0 \quad \text{if } F = 0 \quad \text{and } dF = 0 \\ &d\lambda = 0 \quad \text{if } F < 0 \quad \text{and } dF < 0 \end{aligned} \tag{8.20}$$

and for any regular point on the boundary of the elasticity domain, in the case of plastic loading

$$d\boldsymbol{\varepsilon}^{p} = d\lambda \frac{\partial F}{\partial \sigma} \qquad d\phi^{p} = d\varepsilon^{p}_{ii} \qquad d\lambda \geq 0 \qquad \text{if } F = dF = 0 \tag{8.21}$$

Consider a hardening standard material. Eqs. (8.18), (8.19) and (8.21) give for hardening material (H > 0)

$$\begin{aligned} &(d\boldsymbol{\varepsilon}^{p}, d\mathbf{m}) \in \frac{d_{\boldsymbol{\eta}} F}{H} \partial\eta F(\boldsymbol{\sigma}, \boldsymbol{\eta}) \qquad \phi^{p} = d\varepsilon^{p}_{ii} \qquad \text{if } F = 0 \text{ and } d_{\boldsymbol{\eta}}F > 0 \\ &d\boldsymbol{\varepsilon}^{p} = d\phi^{p} = d\mathbf{m} = 0 \qquad \text{if } F < 0 \quad \text{or if } F = 0 \text{ and } d_{\boldsymbol{\eta}}F \leq 0 \end{aligned} \tag{8.22}$$

and for softening (H < 0)

$$(d\boldsymbol{\varepsilon}^p, d\mathbf{m}) \in \frac{d_\eta F(\boldsymbol{\sigma},\boldsymbol{\eta})}{H} \partial_\eta F(\boldsymbol{\sigma},\boldsymbol{\eta}) \qquad d\phi^p = d\varepsilon_{ii}^p \qquad \text{if } F = 0 \text{ and } d_\eta F < 0$$

$$\text{or } d\boldsymbol{\varepsilon}^p = d\phi^p = d\mathbf{m} = 0 \qquad \text{if } F < 0 \quad \text{or if } F = 0 \text{ and } d_\eta F = 0 \tag{8.23}$$

Consider a regular point on the boundary of the elasticity domain and non-zero plastic increments $d\boldsymbol{\varepsilon}^p$. Then we have

$$d\boldsymbol{\varepsilon}^p = \frac{d_\eta F}{H}\left(\frac{\partial F}{\partial \boldsymbol{\sigma}}\right) \tag{8.24}$$

8.3 The Non-associated Flow Rule

For many materials the hypothesis of normality of dissipative mechanisms is not associated with the flow rule. The flow rule of these materials is presented then by the non-associated potentials, which are different from the loading function F. Then we say that plastic materials are non-standard.

In a general case the loading function F and the non-associated potentials are convex. Introduce a function G different from the loading function F, but with the same arguments. The function G for the ideal plastic material is a non-associated potential, if

$$d\boldsymbol{\varepsilon}^P \in d\lambda\, \partial G(\boldsymbol{\sigma}) \qquad d\phi^p = d\varepsilon_{ii}^p \qquad d\lambda \geq 0$$

$$\text{where} \quad \begin{cases} d\lambda \geq 0 & \text{if} \quad F = 0 \quad \text{and} \quad dF = 0 \\ d\lambda = 0 & \text{if} \quad F < 0 \quad \text{or} \quad dF < 0 \end{cases} \tag{8.25}$$

In Eq. (8.25) $\partial G(\boldsymbol{\sigma})$ is the subdifferential of non-associated potential $G(\boldsymbol{\sigma})$ with respect to $\boldsymbol{\sigma}$. The flow rule in the case of plastic loading if G is without singular points is

$$d\boldsymbol{\varepsilon}^p = d\lambda \frac{\partial G}{\partial \boldsymbol{\sigma}} \qquad d\phi^p = d\varepsilon_{ii}^p \qquad d\lambda \geq 0 \qquad \text{if } F = dF = 0 \tag{8.26}$$

The flow rule for a hardening material (H > 0) is

$$d\boldsymbol{\varepsilon}^p \in \frac{d_\eta F}{H} \partial_\eta\, G(\boldsymbol{\sigma},\boldsymbol{\eta}) \qquad d\phi^p = d\varepsilon_{ii}^p \qquad \text{if } F = 0 \quad \text{and} \quad d_\eta F > 0$$

$$d\boldsymbol{\varepsilon}^p = d\phi^p = d\mathbf{m} = 0 \qquad \text{if } F < 0 \quad \text{or if } F = 0 \quad \text{and} \quad d_\eta F < 0$$

and for softening (H < 0)

$$d\boldsymbol{\varepsilon}^p \in \frac{d_\eta F}{H} \partial_\eta\, G(\boldsymbol{\sigma},\boldsymbol{\eta}), d\phi^p = d\varepsilon_{ii}^p \qquad \text{if } F = 0 \quad \text{and} \quad d_\eta F < 0$$

$$d\boldsymbol{\varepsilon}^p = d\phi^p = d\mathbf{m} = 0 \qquad \text{if } F < 0 \quad \text{or if } F = d_\eta F = 0 \tag{8.27}$$

In the above equations $\partial_\eta G$ is the subdifferential of a non-associated potential $G(\boldsymbol{\sigma}, \boldsymbol{\eta})$ with respect to $\boldsymbol{\sigma}$ at a constant hardening force $\boldsymbol{\eta}$. If G is without points, in the case of plastic loading the flow rule is

$$d\boldsymbol{\varepsilon}^p = \frac{d_\eta F}{H} \frac{\partial G}{\partial \boldsymbol{\sigma}} \tag{8.28}$$

The non-associated potentials can be extended to hardening variables. Consider the subdifferential $\partial G(\boldsymbol{\sigma}, \boldsymbol{\eta})$ of the non-associated potential $G(\boldsymbol{\sigma}, \boldsymbol{\eta})$ with respect to the extended forces $(\boldsymbol{\sigma}, \boldsymbol{\eta})$. If G is without singular points, in the case of plastic loading the extended flow rule is

$$\left(d\boldsymbol{\varepsilon}^p, d\mathbf{m}\right) = \frac{d_\eta F}{H}\left(\frac{\partial G}{\partial \boldsymbol{\sigma}}, \frac{\partial G}{\partial \boldsymbol{\eta}}\right) \qquad d\phi^p = d\varepsilon^p_{ii} \tag{8.29}$$

The potential G expressing the evolution for hardening variables may differ from this used in the flow rule. Denote it by G*

$$\left(d\boldsymbol{\varepsilon}^p, d\mathbf{m}\right) = \frac{d_\eta F}{H}\left(\frac{\partial G^*}{\partial \boldsymbol{\sigma}}, \frac{\partial G}{\partial \boldsymbol{\eta}}\right) \qquad d\phi^p = d\varepsilon^p_{ii} \tag{8.30}$$

For standard material G* = F and by Eq. (8.30)

$$\left(d\boldsymbol{\varepsilon}^p, d\mathbf{m}\right) = \frac{d_\eta F}{H}\left(\frac{\partial F}{\partial \boldsymbol{\sigma}}, \frac{\partial G}{\partial \boldsymbol{\eta}}\right) \qquad d\phi^p = d\varepsilon^p_{ii} \tag{8.31}$$

The above developed hardening laws indicate that the direction $\mathcal{G}_\mathbf{m}$ is by identified with $\partial G/\partial \boldsymbol{\eta}$.

The hardening modulus H is expressed in the form

$$H = \frac{\partial F}{\partial \boldsymbol{\eta}} \cdot \frac{\partial^2 U}{\partial \mathbf{m}^2} \cdot \frac{\partial G}{\partial \boldsymbol{\eta}} = \frac{\partial F}{\partial \boldsymbol{\eta}} \cdot \mathrm{H} \cdot \frac{\partial G}{\partial \boldsymbol{\eta}} \tag{8.32}$$

Consider a potential G without singular points for an ideal plastic material. According to the flow rule (8.25) the non-negativeness (6.24) of the intrinsic dissipation Π_1 leads to the following restriction upon the potential

$$\boldsymbol{\sigma} \frac{\partial G}{\partial \boldsymbol{\sigma}} \geq 0 \qquad \text{if } F = 0 \tag{8.33}$$

For a potential G without singular points and a hardening material following the flow rule (8.27) at a given hardening state, due to relation (8.13) the condition on the potential is

$$\boldsymbol{\sigma} \frac{\partial G}{\partial \boldsymbol{\sigma}} + \boldsymbol{\eta} \cdot \mathcal{G}_\mathrm{m} \geq 0 \qquad \text{if } F = 0 \tag{8.34}$$

Consider a potential G without singular points and a hardening material with the flow rule (8.29) and therefore $\mathcal{G}_m = \partial G / \partial \eta$. The condition upon G is

$$\sigma \frac{\partial G}{\partial \sigma} + \eta \cdot \frac{\partial G}{\partial \eta} \geq 0 \quad \text{if } F = 0 \tag{8.35}$$

Consider the non-associated potential G used to express the hardening law which is different from the non-associated potential G* used to express the flow rule. The condition that concern G and G* is

$$\sigma \frac{\partial G^*}{\partial \sigma} + \eta \cdot \frac{\partial G}{\partial \eta} \geq 0 \quad \text{if } F = 0 \tag{8.36}$$

and if F = G i.e. for standard material

$$\sigma \frac{\partial F}{\partial \sigma} + \eta \cdot \frac{\partial G}{\partial \eta} \geq 0 \quad \text{if } F = 0 \tag{8.37}$$

For the non-standard materials with the flow rule expressed by non-associated potentials inherent modeling problems is involved. The hypothesis of maximal plastic work is not followed and such materials do not satisfy the sufficient conditions ensuring thermodynamic stability, even if they exhibit hardening. The loading increment $d\boldsymbol{\sigma}$ and the response increment $d\boldsymbol{\varepsilon}$ are linked by the relation, which is not necessarily biunivocal.

8.4 Plasticity Models

For isotropic plastic material the loading function can be expressed as (Fig. 8.1)

$$F = h^2 + \psi_1 (3p)^2 - \psi_3 q^2 \tag{8.38}$$

where $\psi_1 = B/A$, $\psi_3 = 4\mu/A$. The parameters A, B and μ are described by the following authors:

$$\text{Gurson} \quad \begin{cases} A = 3 \\ B = (1-R)^2/8 \\ \mu = R^2 - R + 1 \end{cases} \tag{8.39}$$

$$\text{Green} \quad \begin{cases} A = 3 \\ B = 1/4\left[\ln(1-R)\right]^2 \\ \mu = 3\left[1-(1-R)^{\frac{2}{3}}\right]^2 \Big/ 3 - 2(1-R)^{\frac{1}{3}} \end{cases} \tag{8.40}$$

$$\text{Kuhn}\quad \begin{cases} A = 2 + R^2 \\ B = (1 - R^2)/3 \\ \mu = 1 \end{cases} \tag{8.41}$$

$$\text{Oyane}\quad \begin{cases} A = 3/R^2 \\ B = 1/\left[1 + \sqrt{R}/(1 - R)\right]^2 R^2 \\ \mu = R^2 \end{cases} \tag{8.42}$$

$$\text{Shima}\quad \begin{cases} A = 3 \\ B = 2.49\,(1 - R)^{0.514}/9R^5 \\ \mu = R \end{cases} \tag{8.43}$$

In the above expressions the variable R is the relative density and $R = 1 - \phi$.
Consider the form (6.149) of loading function with $n = 2$

$$F = \frac{3h^2}{2m^2} + \frac{1}{2}\left(p + \frac{1}{2}\eta\right)^2 - \frac{1}{8}\eta^2 \tag{8.44}$$

where η is a hardening force.

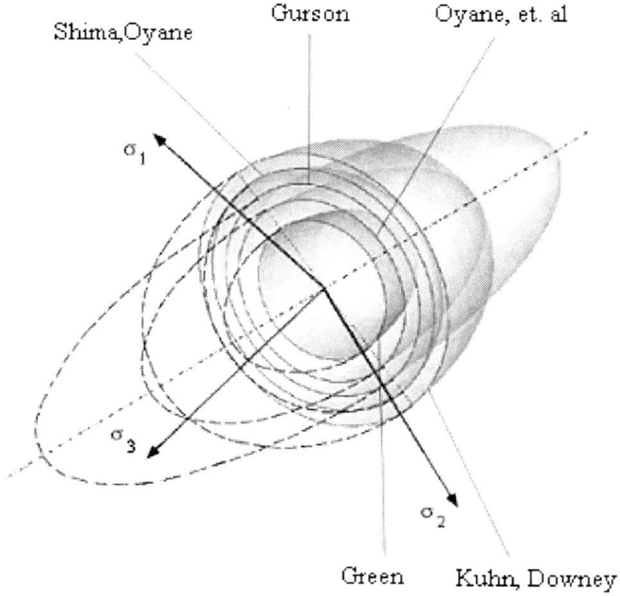

Fig. 8.1. Geometrical representations of yield locus for porous metals

The loading function (8.44) is convex. A material corresponding to this function exhibits isotropic hardening. The yield function defined by (8.44) has no singular points. In the principal stress space $\{\sigma_i\}$, the elasticity domain corresponds to an ellipsoid of revolution about the trisector. The length of the ellipsoid axis along the trisector is $\sqrt{3}\eta$. Any deviatoric plane, normal to the trisector, intersects the yield surface along a circle (Fig. 8.2).

Consider the associated potential. Then the material is standard. If the plastic loading occurs, by (6.159) and (6.160) with G = F and by (8.44) we get the flow rule

$$d\boldsymbol{\varepsilon}^p = d\lambda\left[\frac{3}{2m^2}\mathbf{s}+\frac{1}{3}(p+\eta)\,\mathbf{1}\right] \tag{8.45}$$

$$d\lambda = \frac{d_\eta F}{H} \geq 0 \qquad d_\eta F = (p+\eta)\,dp + \frac{3}{2m^2}\mathbf{s}\,d\mathbf{s} \tag{8.46}$$

In the principal stress space $\{\sigma_i\}$, let (pl) be the plane defined by $p + \eta = 0$. This plane passes through the ellipsoid center and is orthogonal to the trisector (see Fig. 8.2). It is a symmetry plane that divides the ellipsoid into two equal halves. For the points on a circle (cr), which is the intersection of the ellipsoid by the plane (pl), the plastic volume dilatation increment $d\varepsilon^p_{ii}$ and the plastic porosity increment are zero. Plastic evolution of $d\phi^p$ occurs without any change in volume and plastic porosity.

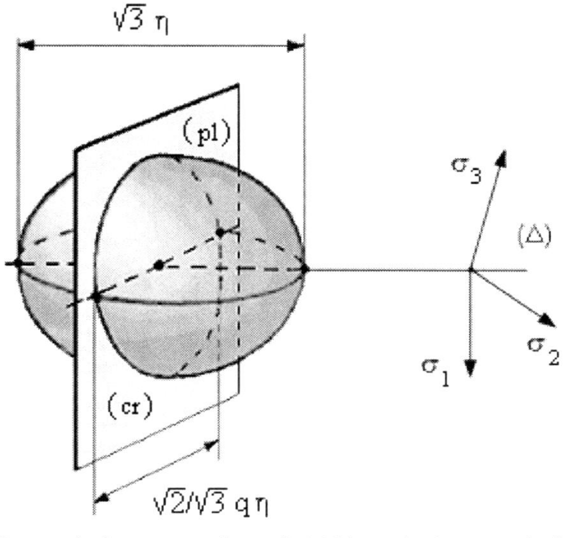

Fig. 8.2. Geometrical representations of yield locus in the space $\{\sigma_i\}$

Part V

Viscoplasticity

9 Viscoplastic Behaviour

9.1 Introduction

The models of plasticity analyzed in the preceding sections do not take into account viscous effects. The behaviour of material is then independent of loading rates and the evolution of the elementary system is considered as a sequence of equilibrium states. Considering viscous phenomena the response to an infinitesimal loading variation is partially delayed. The viscoplastic models developed in this section take into account both viscosity and plasticity effects. For simplicity only isothermal evolutions are discussed.

9.2 Dissipation Potential

Consider a non-negative function H $(\dot{m}_1, \dot{m}_2, ..., \dot{m}_N)$ of arguments $\dot{m}_I$. The evolution relations are based on the hypothesis of normality of a dissipative mechanism consisting of establishing the existence of a set of internal variables m_I and the function H of their rates

$$\mathbf{B}_{\dot{m}} = \frac{\partial H}{\partial \dot{\mathbf{m}}} \tag{9.1}$$

where $\mathbf{B}_{\dot{m}}$ are the thermodynamical forces associated with internal variables $\dot{m}_I$. The function H is said to be the dissipation potential. H is a convex function with respect to its arguments. The function H by its convexity satisfies the relation

$$\left(\dot{\mathbf{m}} - \dot{\mathbf{m}}\right) \cdot \frac{\partial H(\dot{m})}{\partial \dot{\mathbf{m}}} \geq H(\dot{\mathbf{m}}) - H(\dot{\mathbf{m}}^*) \tag{9.2}$$

for any couple of the set of internal variable rates $\left\{\dot{\mathbf{m}} = (\dot{m}_1, \dot{m}_2, ..., \dot{m}_N), (\dot{m}_1^*, \dot{m}_2^*, ..., \dot{m}_N^*)\right\}$.
By (9.2) with $\dot{\mathbf{m}}^* = 0$

$$\dot{\mathbf{m}} \cdot \frac{\partial H(\dot{\mathbf{m}})}{\partial \dot{\mathbf{m}}} \geq H(\dot{\mathbf{m}}) - H(0) \tag{9.3}$$

The instrinsic dissipation Π_1 defined by (4.52) can be expressed by using Eq. (9.2) as a function of only the arguments of H. If the dissipation potential H is a quadratic form of its arguments $\dot{m}_I$ then $\Pi_1 = 2H$.

The Legendre - Fenchel transform H* of the dissipation potential H with respect to all its arguments $\dot{m}_I$ is

$$H^*(\mathbf{B}_{\dot{m}}) = \mathbf{B}_{\dot{m}}\dot{\mathbf{m}} - H(\dot{\mathbf{m}}) \tag{9.4}$$

By (9.1) and (9.4) the following relation is obtained

$$\dot{\mathbf{m}} = \frac{\partial H^*}{\partial \mathbf{B}_{\dot{m}}} \tag{9.5}$$

The hypothesis of normality of the dissipative mechanism may concern only a limited number of internal variables. The above relations apply then only to these variables. The hypothesis of normality of the dissipative mechanism ensures only the non-negativeness of the corresponding part of the intrinsic dissipation.

9.3 State Equations

A thermodynamic state in a viscoplastic model is characterized by the external variable $\boldsymbol{\varepsilon}$, the internal variable $\boldsymbol{\varepsilon}^v$ and the hardening variables $\mathbf{m}$. Then the free energy ψ is expressed as

$$\Psi = \Psi(\boldsymbol{\varepsilon}, \boldsymbol{\varepsilon}^v, \mathbf{m}) \tag{9.6}$$

The isothermal state equation is

$$\boldsymbol{\sigma} = \frac{\partial \Psi}{\partial \boldsymbol{\varepsilon}} \tag{9.7}$$

The above equation is based on the normality of the external variable $\boldsymbol{\varepsilon}$ with regard to the whole set of state variables. The free energy Ψ limited to a second order expansion has the form

$$\Psi = \boldsymbol{\sigma}^o(\boldsymbol{\varepsilon} - \boldsymbol{\varepsilon}^v) + W(\boldsymbol{\varepsilon} - \boldsymbol{\varepsilon}^v) + U(\mathbf{m}) \tag{9.8}$$

The expression (9.8) is obstructed under the hypothesis of separativity of energies and irrespective of any additive constant. The energy frozen U ($\mathbf{m}$) is a function of hardening state variables $\mathbf{m}$. The expression for the reduced potential W is

$$W(\boldsymbol{\varepsilon} - \boldsymbol{\varepsilon}^v) = \frac{1}{2}(\boldsymbol{\varepsilon} - \boldsymbol{\varepsilon}^v)\,\mathbf{C}\,(\boldsymbol{\varepsilon} - \boldsymbol{\varepsilon}^v) \tag{9.9}$$

By (9.7), (9.8) and (9.9) the state equations are

$$\boldsymbol{\sigma} = \boldsymbol{\sigma}^o + \mathbf{C}(\boldsymbol{\varepsilon} - \boldsymbol{\varepsilon}^{vp}) \tag{9.10}$$

The state equation (9.10) is the same as the plastic state equations provided that the plastic strain is replaced by viscoplastic ones. Eq. (9.10) can be written in the inverted form

$$\boldsymbol{\varepsilon} - \boldsymbol{\varepsilon}^{vp} = \mathbf{C}^{-1}\,(\boldsymbol{\sigma} - \boldsymbol{\sigma}^{o}) \tag{9.11}$$

Introduce W interpreted as the reduced potential

$$\boldsymbol{\sigma} - \boldsymbol{\sigma}^{o} = \frac{\partial W}{\partial\left(\boldsymbol{\varepsilon} - \boldsymbol{\varepsilon}^{vp}\right)} \tag{9.12}$$

$$\boldsymbol{\varepsilon} - \boldsymbol{\varepsilon}^{vp} = \frac{\partial W^{*}}{\partial\left(\boldsymbol{\sigma} - \boldsymbol{\sigma}^{o}\right)} \tag{9.13}$$

with $W^{*}\,(\boldsymbol{\sigma} - \boldsymbol{\sigma}^{o})$ as the Legendre-Fenchel transform of a reduced potential W with respect to $\boldsymbol{\varepsilon} - \boldsymbol{\varepsilon}^{vp}$

$$W^{*}\,(\boldsymbol{\sigma} - \boldsymbol{\sigma}^{o}) = \frac{1}{2}\,(\boldsymbol{\sigma} - \boldsymbol{\sigma}^{o})\,\mathbf{C}^{-1}\,(\boldsymbol{\sigma} - \boldsymbol{\sigma}^{o}) \tag{9.14}$$

9.4 Evolution Laws

By Eqs. (9.8) and (9.10), the intrinsic dissipation Π_1 (6.24) is expressed as

$$\Pi_1 = \Pi^{vp} - \frac{\partial U}{\partial \mathbf{m}} \cdot \dot{\mathbf{m}} \geq 0 \tag{9.15}$$

$$\Pi^{vp} = \boldsymbol{\sigma}\,\dot{\boldsymbol{\varepsilon}}^{vp} \tag{9.16}$$

The following formulae hold

$$\boldsymbol{\eta} = -\frac{\partial U}{\partial \mathbf{m}} \qquad \mathbf{m} = -\frac{\partial U^{*}}{\partial \boldsymbol{\eta}} \tag{9.17}$$

where U^{*} is the Legendre-Fenchel transform of frozen energy U defined by (6.27). Then the intrinsic dissipation (9.15) reads

$$\Pi_1 = \boldsymbol{\sigma}\,\dot{\boldsymbol{\varepsilon}}^{vp} + \boldsymbol{\eta} \cdot \dot{\mathbf{m}} \geq 0 \tag{9.18}$$

The rates of irreversible strain in the viscoplastic case are non-zero only if the loading point ($\boldsymbol{\sigma}$) is exterior to the current elasticity domain. The interior of this elastic domain is defined by $F\,(\boldsymbol{\sigma}, \boldsymbol{\eta}) < 0$, F being the viscoplastic loading function. We have

$$\dot{\boldsymbol{\varepsilon}}^{vp} = 0 \quad \text{if} \quad F\,(\boldsymbol{\sigma}, \boldsymbol{\eta}) \leq 0 \qquad \dot{\boldsymbol{\varepsilon}}^{vp} \neq 0 \quad \text{if} \quad F\,(\boldsymbol{\sigma}, \boldsymbol{\eta}) > 0 \tag{9.19}$$

In the absence of hardening variables **m** i.e. if the elasticity domain is invariant, we say that material is ideal viscoplastic. The expression of viscoplastic rates is specified in the case of viscoplastic loading. The flow rule describes it with respect to the non-negativeness of intrinsic dissipation.

9.5 Viscoplastic Material

Consider standard viscoplastic material. The rates of viscoplastic strain for standard ideal plastic material is derived from a dissipation potential H* ($\boldsymbol{\sigma}$)

$$\dot{\boldsymbol{\varepsilon}}^{vp} = \frac{\partial H^*(\boldsymbol{\sigma})}{\partial \boldsymbol{\sigma}} \tag{9.20}$$

In Eq. (9.20) H* ($\boldsymbol{\sigma}$) is a continuous non-negative function of its arguments

$$H^*(\boldsymbol{\sigma}) = 0 \quad \text{if} \quad F(\boldsymbol{\sigma}) \leq 0 \tag{9.21}$$

and

$$H^*(\boldsymbol{\sigma}) > 0 \text{ and convex with respect to its arguments if } F(\boldsymbol{\sigma}) > 0$$

The rates of viscoplastic strain for standard hardening viscoplastic material are derived from a dissipation potential H* ($\boldsymbol{\sigma}, \boldsymbol{\eta}$). The flow rule is

$$\dot{\boldsymbol{\varepsilon}}^{vp} = \frac{\partial H^*(\boldsymbol{\sigma}, \boldsymbol{\eta})}{\partial \boldsymbol{\sigma}} \tag{9.22}$$

In Eq. (9.22) H* ($\boldsymbol{\sigma}, \boldsymbol{\eta}$) is a continuous non-negative function with respect to $\boldsymbol{\sigma}$ such that

$$\begin{aligned} &H^*(\boldsymbol{\sigma}, \boldsymbol{\eta}) = 0 \quad \text{if} \quad F(\boldsymbol{\sigma}, \boldsymbol{\eta}) \leq 0 \\ &H^*(\boldsymbol{\sigma}, \boldsymbol{\eta}) > 0 \text{ and convex with respect to } \boldsymbol{\sigma} \text{ if } F(\boldsymbol{\sigma}, \boldsymbol{\eta}) > 0 \end{aligned} \tag{9.23}$$

The rates of hardening state variables **m**, are derived from the dissipation potential H* ($\boldsymbol{\sigma}, \boldsymbol{\eta}$). The flow rule is

$$\dot{\boldsymbol{\varepsilon}}^{vp} = \frac{\partial H^*(\boldsymbol{\sigma}, \boldsymbol{\eta})}{\partial \boldsymbol{\sigma}} \qquad \dot{\mathbf{m}} = \frac{\partial H^*(\boldsymbol{\sigma}, \boldsymbol{\eta})}{\partial \boldsymbol{\eta}} \tag{9.24}$$

In Eq. (9.24) H* ($\boldsymbol{\sigma}, \boldsymbol{\eta}$) is a continuous non-negative function of its arguments, satisfying

$$\begin{aligned} &H^*(\boldsymbol{\sigma}, \boldsymbol{\eta}) = 0 \quad \text{if} \quad F(\boldsymbol{\sigma}, \boldsymbol{\eta}) \leq 0 \\ &H^*(\boldsymbol{\sigma}, \boldsymbol{\eta}) > 0 \text{ and convex if } F(\boldsymbol{\sigma}, \boldsymbol{\eta}) > 0 \end{aligned} \tag{9.25}$$

9.6 Stability in Viscoplasticity

Consider the stability with respect to stress ($\boldsymbol{\sigma}^o$) of a relaxed equilibrium state to which the elasticity domain E_D^o corresponds. For standard material the stability is ensured by viscoplastic behaviour. Consider the expression (6.82) used for the stability analysis

$$\int_V \left[\left(\boldsymbol{\sigma} - \boldsymbol{\sigma}^o\right)\dot{\boldsymbol{\varepsilon}}\right] dV \leq 0 \tag{9.26}$$

In the case of viscoplasic material

$$\int_V \left(\boldsymbol{\sigma} - \boldsymbol{\sigma}^o\right)\left(\dot{\boldsymbol{\varepsilon}} - \dot{\boldsymbol{\varepsilon}}^{vp}\right) dV \leq 0 \tag{9.27}$$

By (9.12) – (9.14) we get

$$\frac{d}{dt}\int_V W^* \, dV \leq -\int_V \left(\boldsymbol{\sigma} - \boldsymbol{\sigma}^o\right)\dot{\varepsilon}^{vp} dV \tag{9.28}$$

By (9.20), (9.21), (9.28) and relations (9.24), (9.25) we get

$$\frac{d}{dt}\int_V W^* \, dV \leq 0 \tag{9.29}$$

The existence of a Lyapunov functional with regard to stress is ensured by the strict convexity of W*. The state equations (9.12), (9.13), together with relations (9.20), (9.21) in the ideal case and the relation (9.22), (9.23) in the hardening case, ensure material stability with regard to $\boldsymbol{\varepsilon}$.

9.7 Viscoplastic Models

In order to describe a viscoplastic process the loading functions F ($\boldsymbol{\sigma}$) and F ($\boldsymbol{\sigma}$, $\boldsymbol{\eta}$) should be defined. These functions are the same as for plastic material. In the case of standard material the flow rule is defined by the viscoplastic dissipation potential H*. Consider the following expression for potential H*

$$H^* = \frac{1}{2\mu}|F|^2 \tag{9.30}$$

where F ($\boldsymbol{\sigma}$) and F ($\boldsymbol{\sigma}$, $\boldsymbol{\eta}$) are convex functions with regard to $\boldsymbol{\sigma}$ and $|F|$ is the positive part of F. By Eqs. (9.24) and (9.30) the flow rule is

$$\dot{\boldsymbol{\varepsilon}}^{vp} = \frac{1}{\mu}|F|\frac{\partial F}{\partial \boldsymbol{\sigma}} \tag{9.31}$$

In the case of non-associated viscoplastic flow the used function G differs from the loading function F and

$$\dot{\boldsymbol{\varepsilon}}^{vp} = \frac{1}{\mu}|F|\frac{\partial G}{\partial \boldsymbol{\sigma}} \tag{9.32}$$

The exponent 2 in expression (9.30) can be replaced by an exponent n. This exponent is not necessarily an integer. It must be superior to unity for the potential H* to remain convex. In the case of hardening materials in addition to the flow rule the hardening law is specified

$$\dot{\mathbf{m}} = \frac{1}{\mu}|F|\frac{\partial F}{\partial \boldsymbol{\eta}} \tag{9.33}$$

The non-associated hardening law can also be used

$$\dot{\mathbf{m}} = \frac{1}{\mu}|F|\frac{\partial G}{\partial \boldsymbol{\eta}} \tag{9.34}$$

The function G in the above expression may be different from the function G of the non-associated flow rule (9.32). Consider moreover the following associated and non-associated hardening laws

$$\dot{\mathbf{m}} = \frac{1}{\mu}|F|\frac{\partial F}{\partial \boldsymbol{\eta}} - \mathbf{N}\cdot\mathbf{m} \tag{9.35}$$

$$\dot{\mathbf{m}} = \frac{1}{\mu}|F|\frac{\partial G}{\partial \boldsymbol{\eta}} - \mathbf{N}\cdot\mathbf{m} \tag{9.36}$$

where ${}^{T}\mathbf{N}$ and $\mathbf{m}\cdot\mathbf{N}\cdot\mathbf{m} \geq 0$. In the case of isotropic hardening with the loading function given by (6.149) when one hardening variable **m** is involved, the matrix **N** reduces to a positive scalar N. Assuming the different forms of dissipation potential H* and the loading function F or G we get the flow rule associated with these notions. Using the Huber-Mises loading function F = h – q given by Eq. (6.170) where $\alpha = 0$ and the dissipation potential of the form $H^* = \frac{1}{2\mu}F$ we get the following flow rule

$$\dot{\boldsymbol{\varepsilon}}^{vp} = \frac{1}{2\mu}\mathbf{s} \tag{9.37}$$

Part VI

Discontinuous Fields

10 Surfaces of Discontinuity

10.1 The Jump Operator

The surfaces of discontinuity exist in metals. The formulation of discontinuity surfaces in metal forming plasticity and the formation of discontinuity surfaces of strain rates fields in metal forming have to be discussed. In this section the general relations in metal forming imposed by the physical laws of conservation on the discontinuities are derived.

Denote by **X** the coordinate of the material particle in the reference configuration, and **x** (**X**, t) its current coordinates. Let $\mathcal{L}$ be a surface that separates the material in its current state in region 1 and region 2. If the coordinates x_i, as its derivatives up to the order n-1 with respect to the initial coordinates X_α are continuous across surface $\mathcal{L}$,while certain derivatives of order n are discontinuous across $\mathcal{L}$, then $\mathcal{L}$ is said to be a surface of discontinuity of order n. The definition can be extended to any field $\mathcal{H}$ (**X**, t), other than **x** (**X**, t), then the surface is a surface of discontinuity of order n with respect to field $\mathcal{H}$. Denoting by the index i = 1 or 2 the region to which the quantity is attached the jump undergone by the quantity $\mathcal{H}$, when passing through $\mathcal{L}$ from region 2 to region 1, is denoted by

$$[\![\mathcal{H}]\!] = \mathcal{H}_1 - \mathcal{H}_2 \tag{10.1}$$

where the operator $[\![\ ]\!]$ is called the jump operator.

10.2 Discontinuity of Stress

In infinitesimal transformation consider the material body B (Fig. 10.1.) with local coordinate system **X**. Let in the body B be surface of stress discontinuity $\mathcal{L}_\sigma$, which divides this body into two regions denoted by 1 and 2.

The surface of discontinuity $\mathcal{L}_\sigma$ divides the surroundings ΔB of the point **1** laying on this surface into two parts ΔB^1 and ΔB^2. The states of stresses in both

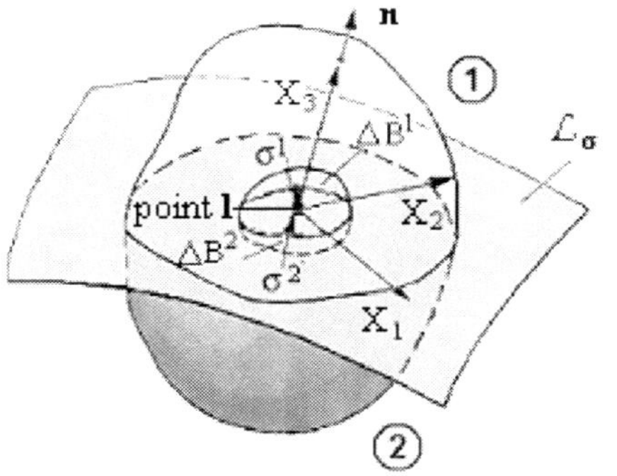

Fig. 10.1. The body with surface of stress discontinuity $\mathcal{L}_\sigma$

parts of these surroundings are denoted by $\boldsymbol{\sigma}^1$ and $\boldsymbol{\sigma}^2$ (Fig. 10.1). For the normal **n** to the surface of discontinuity $\mathcal{L}_\sigma$ at point **1** laying on this surface we assign two vectors (Fig. 10.2); from the site of region 1 the vector

$$\boldsymbol{t}^1 = \boldsymbol{\sigma}^1 \cdot \mathbf{n} \tag{10.2}$$

and from the site of region 2 the vector

$$\boldsymbol{t}^2 = \boldsymbol{\sigma}^2 \cdot \mathbf{n} \tag{10.3}$$

For the surroundings ΔB of the point **1** on the surface of discontinuity $\Delta\mathcal{L}$, the equilibrium condition of reactions for both regions is

$$\int_{\Delta\mathcal{L}} \boldsymbol{t}^1 \mathrm{d}\mathcal{L} + \int_{\Delta\mathcal{L}} \boldsymbol{t}^2 \mathrm{d}\mathcal{L} = 0 \tag{10.4}$$

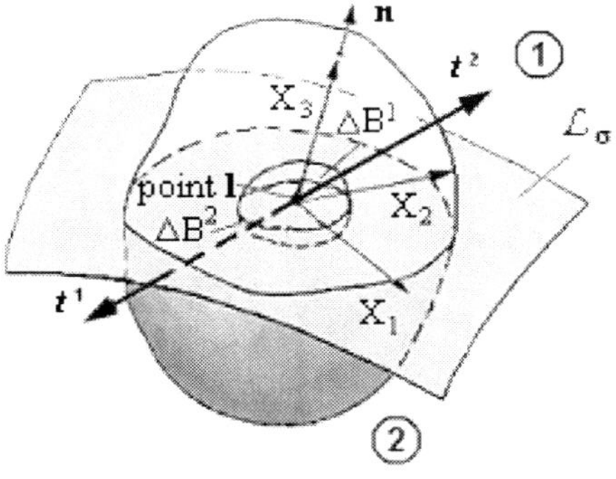

Fig. 10.2. The surroundings of the point **1** divided by surface of stress discontinuity $\mathcal{L}_\sigma$

and hence

$$\boldsymbol{t}^1 + \boldsymbol{t}^2 = 0 \tag{10.5}$$

If we go to the local coordinates **X** (Fig. 10.2) where the direction of axis X_3 is identical with the direction of normal **n**, then for this system **n** = [0, 0, 1] and the relations (10.2) and (10.3) are reduced to

$$t_j^1 = -\sigma_{3j}^1 \tag{10.6}$$

$$t_j^2 = \sigma_{3j}^2 \tag{10.7}$$

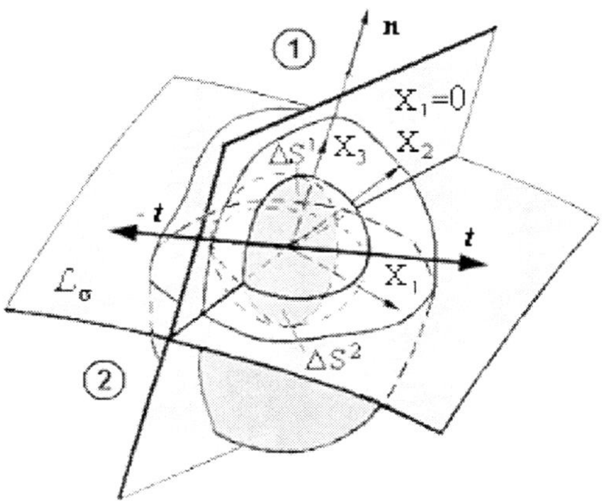

Fig. 10.3. The stress vector $\boldsymbol{t}$ in the cross-section by surface $X_1 = 0$

Substituting Eqs. (10.6) and (10.7) into Eq. (10.5) we get the desired condition for stress on the surface of discontinuity $\mathcal{L}_\sigma$ as

$$\sigma_{3j}^1 = \sigma_{3j}^2 \qquad j = 1, 2, 3 \tag{10.8}$$

The state of stress on the surface of discontinuity $\mathcal{L}_\sigma$ is described as

$$\left[\sigma_{ij}^1\right] = \begin{bmatrix} \sigma_{11}^1 & \sigma_{12}^1 & \sigma_{13} \\ \sigma_{21}^1 & \sigma_{22}^1 & \sigma_{23} \\ \sigma_{31} & \sigma_{32} & \sigma_{33} \end{bmatrix} \tag{10.9}$$

$$\left[\sigma_{ij}^2\right] = \begin{bmatrix} \sigma_{11}^2 & \sigma_{12}^2 & \sigma_{13} \\ \sigma_{21}^2 & \sigma_{22}^2 & \sigma_{23} \\ \sigma_{31} & \sigma_{32} & \sigma_{33} \end{bmatrix} \tag{10.10}$$

For cross – sections by a surface other than surface of discontinuity, the equilibrium condition (10.5) is satisfied independently for each of the regions ΔB^1, ΔB^2. This situation is shown in Fig. 10.3 where the cross-section is carried out by the plane $X_1 = 0$.

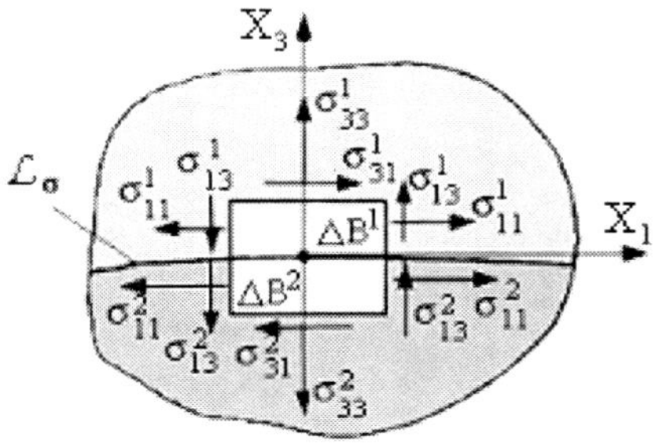

Fig. 10.4. Cross-section by the surface of discontinuity for plane state of stress; σ_2 is tangent to the surface of discontinuity $\mathcal{L}_\sigma$

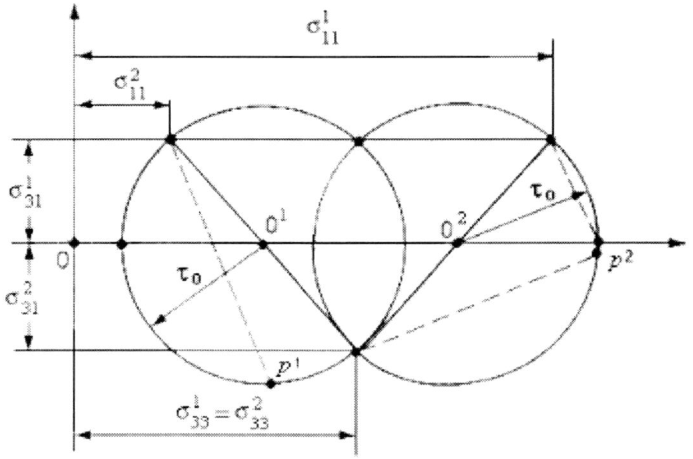

Fig. 10.5. Interpretation of Fig. 10.4 by Mohr's circles

In the particular case when X_2 is the principal direction, then $\sigma_{21} = 0$ and only the normal components of stress σ_{11} and σ_{22} undergo the jump on the surface $\mathcal{L}_\sigma$ (Fig. 10.4). In the case, when in both sides of the surface $\mathcal{L}_\sigma$ the material is in plastic range, the state of stress in the space $\{\sigma, \tau\}$ is represented by the Mohr's circles in Fig. 10.5. From this figure follows the simple interpretation of the jump of stress components.

$$[\![\sigma_{12}]\!] = \sigma_{12}^1 - \sigma_{12}^2$$

$$[\![\sigma_{11}]\!] = \sigma_{11}^1 - \sigma_{11}^2 \tag{10.11}$$

$$[\![\sigma_{22}]\!] = \sigma_{22}^1 - \sigma_{22}^2$$

The poles of Mohr's circles p^1 and p^2 allow us to determine the remaining principal directions in both regions ΔB^1 and ΔB^2.

10.3 Discontinuity of Velocity

Consider a body B (Fig. 10.6.) with surface of velocity discontinuity $\mathcal{L}_V$. This surface divides the body into two regions denoted by 1 and 2. Let **X** be the local coordinate system with an origin at point **1** on the surface of discontinuity $\mathcal{L}_V$ with axis X_3 normal to this surface (Fig. 10.6). The surface of discontinuity $\mathcal{L}_V$ divides the surroundings of the point **1** into two regions ΔB^1 and ΔB^2, which move with the velocities $\mathbf{V}^1$ and $\mathbf{V}^2$ in such a way that the compactness of the medium is not violated. The compactness condition requires that the velocities normal to the surface of discontinuity $\mathcal{L}_V$ are equal

$$\mathbf{V}^1 \cdot \mathbf{e}_3 = \mathbf{V}^2 \cdot \mathbf{e}_3 \tag{10.12}$$

and hence in our case

$$V_3^1 = V_3^2 \tag{10.13}$$

where $\mathbf{e}_3$ is the base vector of axis X_3.

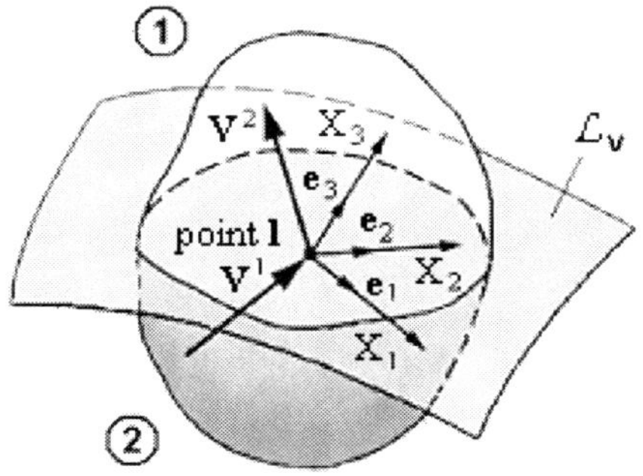

Fig. 10.6. The body B with surface of velocity discontinuity $\mathcal{L}_V$

From the above it follows that the differences in velocity components are permissible only in the directions tangent to the surface of discontinuity. The surface of velocity discontinuity is physically the layer in which the change of tangent velocity to this surface occurs. If the thickness of this layer tends to zero, then the components of tangent velocity to the surface $\mathcal{L}_V$ change with a jump. This jump change of velocity is called the velocity discontinuity. This discontinuity is expressed as

$$[\![\mathbf{v}]\!] = \mathbf{v}^1 - \mathbf{v}^2 \tag{10.14}$$

10.4 The Virtual Work Theorem for Discontinuous Material

By recalling the theorem of virtual work rate for quasistatic small perturbations and not considering body forces, the virtual strain work rate is equal to the virtual work rate of surface forces **t** exerted on the boundary A = ∂V of the domain V

$$\int_V \sigma\,\dot{\varepsilon}^*\,dV = \int_{\partial V} \mathbf{t}\cdot\mathbf{v}^*\,d(\partial V) \tag{10.15}$$

Let $\mathcal{L}$ be the surface of stress discontinuity of the body B of the domain V with the boundary surface ∂V consisting of a surface with the given velocity ∂V_v and with the given surface tractions ∂V_t (Fig. 10.7). Consider both regions of the body B, i.e. B^1 and B^2 separately where the superscripts 1 and 2 are related to regions 1 or 2, respectively. The regions B^1 and B^2 have a common surface. Let it be a surface of stress discontinuity $\mathcal{L}_\sigma$ (Fig. 10.8.).

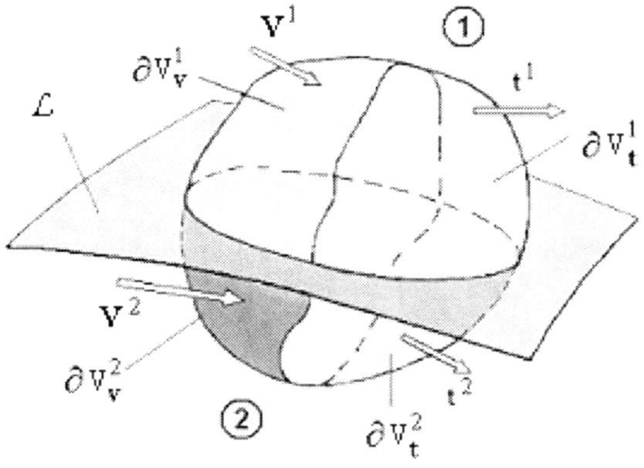

Fig. 10.7. The body B and surfaces ∂V_V and ∂V_t divided by the surface of discontinuity $\mathcal{L}$

The state of stress within the body fulfills the equilibrium equations. In regions B^1 and B^2 surface forces $\boldsymbol{\sigma}^1 \cdot \mathbf{n}^1$ and $\boldsymbol{\sigma}^2 \cdot \mathbf{n}^2$ are continuous. The normal outward vectors $\mathbf{n}^1$ and $\mathbf{n}^2$ fulfill the expression

$$\mathbf{n}^1 = -\mathbf{n}^2 \tag{10.16}$$

so the following relation holds

$$\boldsymbol{\sigma}^1 \cdot \mathbf{n}^1 = -\boldsymbol{\sigma}^2 \cdot \mathbf{n}^2 \tag{10.17}$$

The virtual work theorem for each region of the body can be written as

$$\int_{V^1} \sigma\, \dot{\varepsilon}\, dV = \int_{\partial V_t^1} \mathbf{t}^* \mathbf{V}\, d(\partial V) + \int_{\partial V_v^1} \mathbf{t}\, \mathbf{V}^*\, d(\partial V) + \int_{\mathcal{L}} \mathbf{t}^1\, \mathbf{V}\, d\mathcal{L} \tag{10.18}$$

$$\int_{V^2} \sigma\, \dot{\varepsilon}\, dV = \int_{\partial V_t^2} \mathbf{t}^* \mathbf{V}\, d(\partial V) + \int_{\partial V_v^2} \mathbf{t}\, \mathbf{V}^*\, d(\partial V) + \int_{\mathcal{L}} \mathbf{t}^2\, \mathbf{V}\, d\mathcal{L} \tag{10.19}$$

By adding the above expression we get the virtual work theorem for the whole body

$$\int_{V^1+V^2} \sigma\, \dot{\varepsilon}\, dV = \int_{\partial V_t^1 + \partial V_t^2} \mathbf{t}^* \mathbf{V}\, d(\partial V) + \int_{\partial V_v^1 + \partial V_v^2} \mathbf{t}\, \mathbf{V}^*\, d(\partial V) \tag{10.20}$$

which is the virtual work theorem for the whole body B without the discontinuity of stress. In Eq. (10.20) $\mathbf{t}^* = \mathbf{t}^*(\mathbf{X})$ are the surface tractions given on ∂V_t and $\mathbf{V}^* = \mathbf{V}^*(\mathbf{X})$ is the velocity given on ∂V_V. The theorem of virtual work has been analyzed here neglecting the body forces.

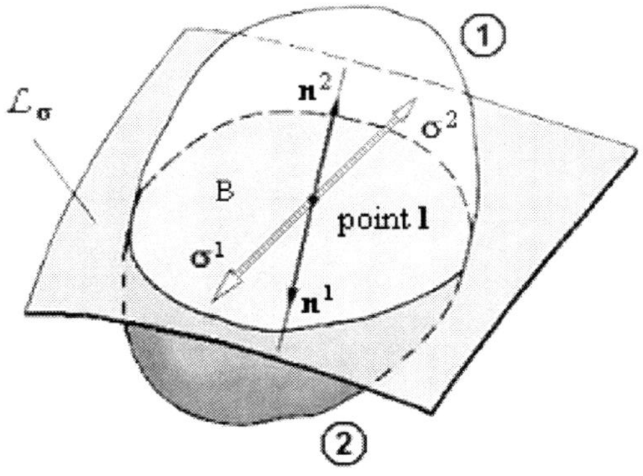

Fig. 10.8. The body B and the surface of discontinuity $\mathcal{L}_\sigma$

Consider the surface of velocity discontinuity $\mathcal{L}_v$ under the assumption that the jump of velocity is at the plane tangent to the surface of discontinuity (Fig. 10.9).

The work dissipated in the surface of discontinuity $\mathcal{L}_v$ is given by the scalar product of **t** and the jump of displacement velocity $[\![\mathbf{V}]\!]$.

$$\int_{\mathcal{L}_V} \mathbf{t}\,[\![\mathbf{V}]\!]\,d\mathcal{L} \tag{10.21}$$

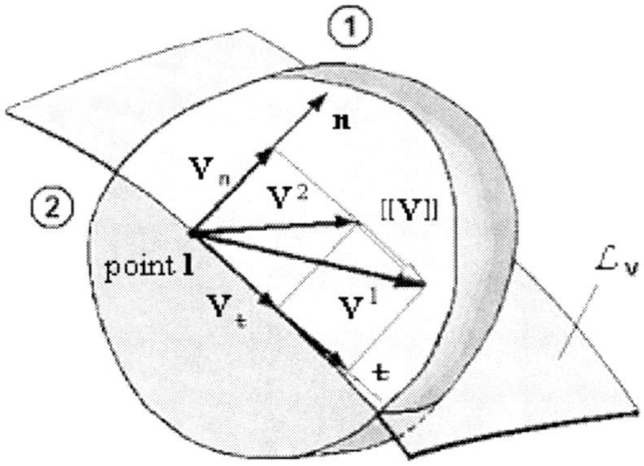

Fig. 10.9. The body B and the surface of discontinuity $\mathcal{L}_v$

It is the work dissipated by external forces, so the expression (10.21) is added to the left hand side of Eq. (10.20). We get

$$\int_V \sigma\,\dot{\varepsilon}\,dV + \int_{\mathcal{L}_V} \mathbf{t}\,[\![\mathbf{V}]\!]\,d\mathcal{L} = \int_{\partial V_t} \mathbf{t}^*\,\mathbf{V}\,d(\partial V) + \int_{\partial V_V} \mathbf{t}\;\;\mathbf{V}^*\,d(\partial V) \tag{10.22}$$

10.5 Strain in the Discontinuity Field

The strain rate tensor can be written as

$$\dot{\boldsymbol{\varepsilon}} = \frac{1}{2}\left([\![\mathbf{V}]\!]\otimes\mathbf{n} + \mathbf{n}\otimes[\![\mathbf{V}]\!]\right)\delta_{\mathcal{L}} \tag{10.23}$$

where $\delta_{\mathcal{L}}$ is the Dirac δ-function relative to the surface $\mathcal{L}$ and **n** is the unit normal vector to surface $\mathcal{L}$ oriented in the direction of region 2 (Fig. 10.10). A flow rule defined through a non-associated potential G is

$$\dot{\boldsymbol{\varepsilon}}^p = \dot{\lambda}\frac{\partial G}{\partial \sigma} \qquad \dot{\lambda} \geq 0 \tag{10.24}$$

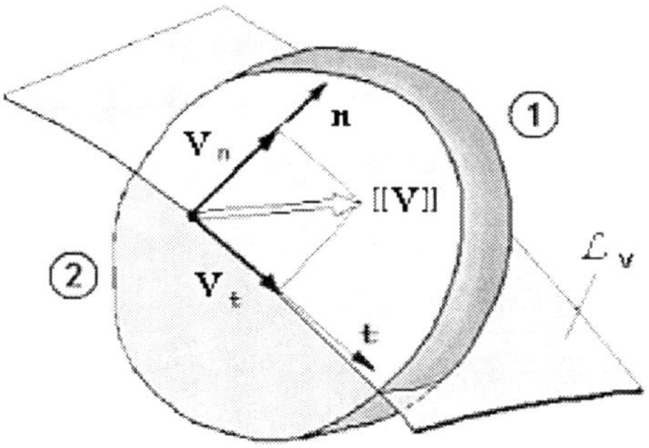

Fig. 10.10. The decomposition of the discontinuity of the material velocity

By (10.23) and (10.24) we get the relation

$$\dot{\varepsilon}^{p} = \frac{1}{2}\left([\![\mathbf{V}]\!] \otimes \mathbf{n} + \mathbf{n} \otimes [\![\mathbf{V}]\!]\right)\delta_{\mathcal{L}} \qquad \dot{\lambda} = \dot{\Lambda}\delta_{\mathcal{L}} \qquad \dot{\Lambda} \geq 0 \tag{10.25}$$

The flow rule will involve the discontinuity $[\![\mathbf{V}]\!]$. Introduce the vector **t** defined as the vector tangent to the discontinuity surface at the considered point, and lying in the plane contacting both the normal **n** and the discontinuity of velocity $[\![\mathbf{V}]\!]$ (Fig. 10.10).

The singular part of the plastic strain rate tensor is

$$\dot{\varepsilon}^{p} = \left(\mathbf{V}_{n}\mathbf{n} \otimes \mathbf{n} + \frac{1}{2}\mathbf{V}_{t}\left(\mathbf{t} \otimes \mathbf{n} + \mathbf{n} \otimes \mathbf{t}\right)\right)\delta_{\mathcal{L}} \tag{10.26}$$

where $\mathbf{V_n}$ and $\mathbf{V_t}$ are the normal and the tangent components of the velocity discontinuity

$$[\![\mathbf{V}]\!] = \mathbf{V}_{n}\mathbf{n} + \mathbf{V}_{t}\,\mathbf{t} \tag{10.27}$$

The equivalent of plastic distortion $\gamma^{p}_{e\sigma}$ defined by (6.164) and corresponding to the singular part of the plastic strain rate (10.26) is

$$\dot{\gamma}^{p}_{eq} = \sqrt{\frac{4}{3}\mathbf{V}_{n}^{2} + \mathbf{V}_{t}^{2}}\;\delta_{\mathcal{L}} \tag{10.28}$$

The tensor, which is multiplied by surface δ-function $\delta_{\mathcal{L}}$ in Eq. (10.26) corresponds to a tensor of plane plastic rate. The axis of this plane strain, orthogonal to **n** and **t**, is located in the plane tangent to the discontinuity surface. The intrinsic dissipated work rate Π_1, due to plastic effects associated with the singular parts of the plastic rates, is

$$\Pi_{1} = \Pi^{p} = [\![\mathbf{V}]\!] \cdot \mathbf{t} \tag{10.29}$$

per unit of discontinuity surface $d\mathcal{L}$. According to the second law it has to be non-negative.

10.6 Models of Plasticity with Discontinuities

In ideal plasticity, the flow rule (10.24) for non-singular points of potential G is given as

$$\dot{\varepsilon}^{p} = \dot{\lambda}\left(\frac{\mathbf{s}}{2h} + \frac{\delta}{3}\mathbf{1}\right) \tag{10.30}$$

which corresponds to Eq. (6.198). From Eq. (10.30) it follows that the principal axes of stress tensor and that of the plastic strain rate tensor coincide. Let $\mathbf{t}$ be the unit vector such that the set of orthogonal vectors **n**, **t** and forms a direct trihedron. Since the plastic strain rate (10.25) corresponds to a plane strain rate with an axis oriented along the vector $\mathbf{t}$, the flow rule (10.30) entails that one of the principal stress axes is oriented among $\mathbf{t}$. Let σ_2 be the principal stress corresponding to this direction. The two other principal directions, with principal stresses σ_1 and σ_3, therefore lie in plane (**n**, **t**). Expressing that the plastic strain along the vector $\mathbf{t}$ is zero, and using the relation (6.149), which combines h and the principal stresses, it follows that

$$\sigma_2 = \frac{1}{2}(\sigma_1 + \sigma_3) - h\delta \qquad h = \frac{\sigma_1 - \sigma_3}{2\sqrt{1 - \delta^{2/3}}} \tag{10.31}$$

where $\sigma_1 \geq \sigma_3$. By (6.220) the dilatancy factor δ satisfies inequality $0 \leq \delta \leq \sqrt{3}/2$. From Eq. (10.31) it follows that

$$\sigma_1 \geq \sigma_2 \geq \sigma_3 \tag{10.32}$$

The principal stress σ_3 along the vector $\mathbf{t}$ is the intermediary principal stress. By Eqs. (10.25), (10.26) and (10.30) we get

$$\mathbf{V_n} = \mathrm{tg}\vartheta \left| \mathbf{V_t} \right| \tag{10.33}$$

where

$$\delta = \frac{3\sin\vartheta}{\sqrt{3(3 + \sin^2\vartheta)}} \qquad \mathrm{tg}\vartheta = \frac{\delta}{\sqrt{1 - \frac{4}{3}\delta^2}}, \qquad 0 \leq \vartheta \leq \frac{\Pi}{2} \tag{10.34}$$

which is consistent with the inequality $0 \leq \delta \leq \sqrt{3}/2$. The absolute value of the normal discontinuity is therefore equal to that of the tangent discontinuity multiplied by the factor $\mathrm{tg}\vartheta$. The latter depends only on the plastic dilatancy factor δ, which aggresses with the purely plastic origin of the discontinuity. In the plane (**n**, **t**) normal to the direction of the intermediary principal stress σ_2 the discontinuity of the material velocity $[\![\mathbf{V}]\!]$ makes with the vector **n** an angle equal to $\Pi/2-\vartheta$. There are two possible directions for $[\![\mathbf{V}]\!]$, symmetrical with respect to the direction oriented along **n**. The associated motion corresponds to the separation in two parts of the body along the discontinuity surface, since the discontinuity velocity $[\![\mathbf{V}]\!]$ is oriented in the same direction as the vector **n**. The principal values $\dot{\varepsilon}_1^p$ and $\dot{\varepsilon}_2^p$ of the plastic strain rate tensor defined in the space (**n**, **t**) by Eq. (10.26) can be derived from Eq. (10.33) to give

$$\dot{\varepsilon}_1^p=\frac{1}{2}\left|\mathbf{V}_\mathbf{t}\right|\operatorname{tg}\left(\frac{\Pi}{4}+\frac{\vartheta}{2}\right)\delta_{\mathcal{L}}\quad \dot{\varepsilon}_3^p=-\frac{1}{2}\left|\mathbf{V}_\mathbf{t}\right|\operatorname{tg}\left(\frac{\Pi}{4}-\frac{\vartheta}{2}\right)\delta_{\mathcal{L}} \tag{10.35}$$

The factor term of the δ-function $\delta_{\mathcal{L}}$ is positive for one principal value, and negative for the other. The frame formed of the principal strain rate directions $\mathbf{e}_1$, $\mathbf{e}_2$, and $\mathbf{e}_3$, coinciding with that of the principal stress directions (isotropic material), is obtained through a relation of the frame **n**, **t** and **ŧ** for an angle $\Pi/4-\vartheta/2$ around vector **t**, i.e. the direction of the intermediary principal stress σ_2. Owing to the non-negativity of the undetermined factor involved in Eq. (10.25), the principal strain rates $\dot{\varepsilon}_1^p$ and $\dot{\varepsilon}_3^p$ are associated with the stresses σ_1 and σ_2. By Eqs. (10.25), (10.30), (10.31), (10.35) and (10.36) these principal strain rates can be written as

$$\dot{\varepsilon}_1^p=\dot{\Lambda}\,\delta\,\frac{1+\sin\vartheta}{2\sin\theta}\,\delta_{\mathcal{L}}\quad \dot{\varepsilon}_3^p=-\dot{\Lambda}\,\delta\,\frac{1-\sin\vartheta}{2\sin\theta}\,\delta_{\mathcal{L}}\qquad \dot{\Lambda}\geq 0 \tag{10.36}$$

The main results obtained above are illustrated in Fig. 10.11.

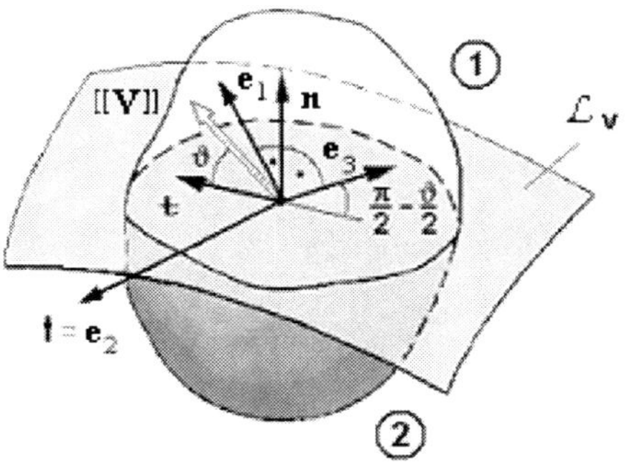

Fig. 10.11. Velocity discontinuity for standard ideal plastic material

10.7 Hardening Materials

In the case of hardening materials the discontinuity of velocity cannot appear. It is because no scalar remains undetermined in the formulation of their constitutive equations. The localization of a deformation in hardening materials in quasi-static experiments is modeled by considering discontinuity surfaces of a higher order. Introduce the polarization vector p such that for vector field $\mathbf{g}$

$$[\![\mathrm{div}\,\mathbf{g}]\!] = p \cdot \mathbf{n} \tag{10.37}$$

$$[\![\mathrm{grad}\,\mathbf{g}]\!] = p \otimes \mathbf{n} \tag{10.38}$$

From the above we get

$$[\![\mathrm{grad}\,\mathbf{V}]\!] = p \otimes \mathbf{n} \tag{10.39}$$

By (1.17) we have

$$[\![\dot{\varepsilon}]\!] = \frac{1}{2}(p \otimes \mathbf{n} + \mathbf{n} \otimes p) \tag{10.40}$$

Introduce the vector $\mathbf{t}$, defined as the tangent vector to the discontinuity surface at the considered point, located in the plane that contains the normal $\mathbf{n}$ and the polarization vector p. The decomposition of the jump of the strain rate tensor is

$$p = p_{\mathrm{n}}\,\mathbf{n} + p_{\mathrm{t}}\mathbf{t} \tag{10.41}$$

$$[\![\dot{\varepsilon}]\!] = \left\{ p_{\mathrm{n}}\,\mathbf{n} \otimes \mathbf{n} + \frac{1}{2} p_{\mathrm{t}} (\mathbf{t} \otimes \mathbf{n} + \mathbf{n} \otimes \mathbf{t}) \right\} \tag{10.42}$$

In quasi-static state the momentum balance equation is

$$\mathrm{div}\,(\boldsymbol{\sigma} - \boldsymbol{\sigma}^{\mathrm{o}}) = 0 \tag{10.43}$$

By (10.43) it follows that

$$[\![\dot{\boldsymbol{\sigma}} \cdot \mathbf{n}]\!] = 0 \tag{10.44}$$

For plastic loading the incremental equation of plasticity is

$$[\![\dot{\boldsymbol{\sigma}}]\!] = \mathbf{C}_{\mathrm{o}}^{\mathrm{T}} \cdot [\![\dot{\varepsilon}]\!] \tag{10.45}$$

where the tangent module $\mathbf{C}_{\mathrm{o}}^{\mathrm{T}}$ is

$$\mathbf{C}_{\mathrm{o}}^{\mathrm{T}} = \mathbf{C}_{\mathrm{o}} - \mathbf{K}_{\mathrm{o}} \mathbf{H}_{\mathrm{o}} \otimes \mathbf{F}_{\mathrm{o}} \tag{10.46}$$

and

$$\mathbf{K}_{o}^{-1} = \mathbf{H} + \frac{\partial F}{\partial \boldsymbol{\sigma}} \mathbf{C}_{o} \frac{\partial G}{\partial \boldsymbol{\sigma}} \tag{10.47}$$

$$\mathbf{H}_{o} = \mathbf{C}_{o} \frac{\partial G}{\partial \boldsymbol{\sigma}} \tag{10.48}$$

$$\mathbf{F}_{o} = \frac{\partial F}{\partial \boldsymbol{\sigma}} \mathbf{C}_{o} \tag{10.49}$$

The expressions (10.40), (10.44) and (10.45) require that the polarization vector p satisfies

$$\left(\mathbf{n} \cdot \mathbf{C}_{o}^{T} \cdot \mathbf{n}\right) \cdot p = 0 \tag{10.50}$$

Localization of deformation becomes possible if the above equation admits a non-zero solution for the polarization vector p

$$\det\left(\mathbf{n} \cdot \mathbf{C}_{o}^{T} \cdot \mathbf{n}\right) = 0 \tag{10.51}$$

Part VII

Numerical Solution Methods

11 Numerical Solutions

11.1 Introduction

Most of the problems considered in the book can only be solved by using numerical procedures. In this section the general data concerning the numerical solving methods are given. The non-linearity of the constitutive equations and the time-depending process require the solution technique by means of a step-by-step method.

11.2 Uniqueness of Solution

The state equation of elasticity for a small isothermal perturbation is expressed as

$$\boldsymbol{\sigma} = \boldsymbol{\sigma}_o + \mathbf{C} \cdot \boldsymbol{\varepsilon} \tag{11.1}$$

or in inverted form

$$\boldsymbol{\varepsilon} = \mathbf{C}^{-1} (\boldsymbol{\sigma} - \boldsymbol{\sigma}^o) \tag{11.2}$$

The functional norm $\|\cdot\|$ of any stress field $\boldsymbol{\sigma}$ is defined by

$$\|\boldsymbol{\sigma}\| = \left(\int_V \frac{1}{2} \boldsymbol{\sigma} \cdot \mathbf{C}^{-1} \cdot \boldsymbol{\sigma} \, dV \right)^{\frac{1}{2}} \tag{11.3}$$

The functional norm (11.3) is the integral over a domain V of the isothermal reduced potential W*($\boldsymbol{\sigma}$) defined by (6.86) in the case of infinitesimal transformations. The potential W* can be used to define a norm, provided that it is a quadratic form of its arguments. This is the case if the material satisfies the sufficient condition of stability (6.89).

In an elastic case, the relations (11.1) and (11.2) must hold for two possible solutions ($\boldsymbol{\sigma}$, $\boldsymbol{\varepsilon}$) and ($\boldsymbol{\sigma}^*$, $\boldsymbol{\varepsilon}^*$).

By the virtual work theorem (6.81) it follows that

$$\int_V (\boldsymbol{\sigma} - \boldsymbol{\sigma}*) \cdot (\dot{\boldsymbol{\varepsilon}} - \dot{\boldsymbol{\varepsilon}}*) dV \leq 0 \tag{11.4}$$

From the above we have

$$\frac{d}{dt} \|\boldsymbol{\sigma} - \boldsymbol{\sigma}*\|^2 \leq 0 \tag{11.5}$$

By (11.5), the distance between two solutions $\boldsymbol{\sigma}(t)$ and $\boldsymbol{\sigma}^*(t)$ cannot increase. This distance is non-negative and equal to zero at $t = 0$, $\boldsymbol{\sigma}(0) = \boldsymbol{\sigma}^*(0) = \boldsymbol{\sigma}_o$ and remains zero as time goes by i.e. $\boldsymbol{\sigma}(t) = \boldsymbol{\sigma}^*(t)$ which proves the uniqueness of the solution. By inequality

$$\frac{1}{2} \int_V \boldsymbol{\sigma} \cdot \mathbf{C}^{-1} \cdot \boldsymbol{\sigma} \, dV > \lambda_{inf} \|\boldsymbol{\sigma}\|^2 \tag{11.6}$$

the uniqueness in the sense of distance associated with the scalar product

$$\|\boldsymbol{\sigma}\| = \left(\frac{1}{2} \int_V \boldsymbol{\sigma} \cdot \boldsymbol{\sigma} dV \right)^{\frac{1}{2}} \tag{11.7}$$

is also ensured. In Eq. (11.6) λ_{inf} is the smallest of the eigenvalues associated with the positively defined symmetric form $\mathbf{C}_o$.

By (11.2) the solution concerning the strain field $\boldsymbol{\varepsilon}$ is also unique. By the relation that links displacement $\mathbf{u}$ and strain $\boldsymbol{\varepsilon}$, the uniqueness of the solution with respect to $\mathbf{u}$ is proved. In the case of an ideal plastic material the state equations are written in the form

$$\boldsymbol{\sigma} = \boldsymbol{\sigma}_o + \mathbf{C} (\boldsymbol{\varepsilon} - \boldsymbol{\varepsilon}^p) \tag{11.8}$$

and its inversion is

$$\boldsymbol{\varepsilon} - \boldsymbol{\varepsilon}^p = \mathbf{C}^{-1} (\boldsymbol{\sigma} - \boldsymbol{\sigma}_o) \tag{11.9}$$

Let $\boldsymbol{\sigma}(t)$ and $\boldsymbol{\sigma}^*(t)$ be two possible solutions in stress of the evolution problem. Consider the two possible solutions $(\boldsymbol{\sigma}, \boldsymbol{\varepsilon}, \boldsymbol{\varepsilon}^p)$ and $(\boldsymbol{\sigma}^*, \boldsymbol{\varepsilon}^*, \boldsymbol{\varepsilon}^{p*})$ of state equations (11.8) and (11.9). We have

$$\frac{d}{dt} \|\boldsymbol{\sigma} - \boldsymbol{\sigma}*\|^2 \leq -\int_V (\boldsymbol{\sigma} - \boldsymbol{\sigma}*) \cdot (\dot{\boldsymbol{\varepsilon}}^p - \dot{\boldsymbol{\varepsilon}}^p *) dV \tag{11.10}$$

If material is standard, the hypothesis of maximal plastic work (6.54) is satisfied. Then for two solutions $(\boldsymbol{\sigma}, \boldsymbol{\varepsilon}, \boldsymbol{\varepsilon}^p)$ and $(\boldsymbol{\sigma}^*, \boldsymbol{\varepsilon}^*, \boldsymbol{\varepsilon}^{p*})$

$$(\boldsymbol{\sigma} - \boldsymbol{\sigma}^*) \cdot \dot{\boldsymbol{\varepsilon}}^{p} \geq 0 \qquad (\boldsymbol{\sigma}^* - \boldsymbol{\sigma}) \cdot \dot{\boldsymbol{\varepsilon}}^{p*} \geq 0 \tag{11.11}$$

By (11.10) and (11.11) it follows that (11.4) holds for an ideal plastic standard material. Hence, for such materials, the solution in stress is unique. This does not ensure the uniqueness of the solution in displacement. For an ideal plastic standard material in the case of plastic loading, the flow rule reads

$$\dot{\boldsymbol{\varepsilon}}^{p} = \lambda \frac{\partial F}{\partial \sigma} \qquad \dot{\lambda} \geq 0 \tag{11.12}$$

due to the undetermined character of plastic multiplier $\dot{\lambda}$ involved in Eq. (11.12), the uniqueness of the solution in strain $\boldsymbol{\varepsilon}$ can not be derived from the uniqueness of the solution in stress through the constitutive equations of plasticity.

Since the solution in stress $\boldsymbol{\sigma}$ is unique for an ideal plastic standard material, the plastic multiplier $\lambda^{o} = d\lambda/dt$ considered as a function of time is unique. Hence by (11.9) and (11.12) the solution in the strain rate is unique. By relation

$$\dot{\boldsymbol{\varepsilon}} = \frac{1}{2}\left(\text{grad}\,\mathbf{V} + {}^{T}\text{grad}\,\mathbf{V}\right) \tag{11.13}$$

which combines the material velocity and the strain rate tensor, the uniqueness of the material velocity $\mathbf{V}$ is ensured. This field depends in a continuous way on the time function λ^{o} ($d\lambda/dt$), which is unique. The preceding developments result in the uniqueness of the solution in the material displacement $\mathbf{u}$.

11.3 Time Discretization

Assuming the incremental nature of the equations of elasto-plasticity, the time is first discretized by

$$t_n = n \cdot \Delta t \tag{11.14}$$

where n is the time step.

$$S_{n-1} = (\boldsymbol{\sigma}_{n-1}, \mathbf{u}_{n-1}) \tag{11.15}$$

The solution S_{n-1} at time t_{n-1} is assumed as known. The solution S_n at time t_n is obtained as

$$S_n = S_{n-1} + \Delta_n S \quad \text{where } \Delta_n S = (\Delta_n \boldsymbol{\sigma}, \Delta_n \mathbf{u}). \tag{11.16}$$

The problem discretized with respect to time satisfies the discretized momentum equation

$$\text{div}\, \Delta_n \boldsymbol{\sigma} = 0 \tag{11.17}$$

and the compatibility equation

$$2\,\Delta_n\,\boldsymbol{\varepsilon} = \mathrm{grad}\,\Delta_n\,\mathbf{u} + {}^{T}\mathrm{grad}\,\Delta_n\,\mathbf{u} \tag{11.18}$$

and the discretized boundary conditions

$$\begin{aligned} \Delta_n\,\boldsymbol{\sigma}\,\mathbf{n} &= \Delta_n\,\mathbf{t}^* \qquad \text{on } \partial V_t \\ \Delta_n\,\mathbf{u} &= \Delta_n\,\mathbf{u}^* \qquad \text{on } \partial V_u \end{aligned} \tag{11.19}$$

11.4 The Theorem of Virtual Work

A theorem of virtual work in finite stress increments $\Delta_n\boldsymbol{\sigma}$ and in virtual finite displacement increments $\Delta_n\mathbf{u}^*$ can be derived from the time-discretized equations in a similar way to the derivation of the theorem of virtual work (6.81) from the non-discretized equations of the problem

$$\int_V \Delta_n\boldsymbol{\sigma}\,\Delta_n\boldsymbol{\varepsilon}^*\,dV - \int_{\partial V} \Delta_n\,\mathbf{t}\,\Delta_n\mathbf{u}^*\,d(\partial V) = 0 \tag{11.20}$$

In (11.20) $\Delta_n\mathbf{u}$ has the dimension of a displacement, and the integrals in Eq. (11.20) represent work quantities. The theorem of virtual work for finite increments is the time-discretized theorem as of the virtual work rate. Consider two fields of kinematically admissible finite displacement increments $\Delta_n\mathbf{u}$ and $\Delta_n\mathbf{u}^o$. The difference $\Delta_n\mathbf{u} - \Delta_n\mathbf{u}^o$ is kinematically admissible with zero finite displacement increments imposed on the boundary ∂V_u. By (11.20) applied to these fields we get

$$\int_V \Delta_n\boldsymbol{\sigma}\left(\Delta_n\boldsymbol{\varepsilon} - \Delta_n\boldsymbol{\varepsilon}^o\right)dV - \int_{\partial V_t} \Delta_n\,\mathbf{t}\left(\Delta_n\mathbf{u} - \Delta_n\mathbf{u}^o\right)d(\partial V_t) = 0 \tag{11.21}$$

for every statically admissible $\Delta_n\,\boldsymbol{\sigma}$ and $\Delta_n\mathbf{u}$ and kinematically admissible $\Delta_n\mathbf{u}^o$. Consider the statically admissible finite increment fields of stress $\Delta_n\boldsymbol{\sigma}$ and $\Delta_n\boldsymbol{\sigma}^o$. Their differences $\Delta_n\boldsymbol{\sigma}$ and $\Delta_n\boldsymbol{\sigma}^o$ are also statically admissible with zero data imposed on ∂V_t. By (11.20) applied to fields of finite increments

$$\Delta_n\,\mathbf{u}^* = \Delta_n\,\mathbf{u} - \Delta_n\,\mathbf{u}^o \tag{11.22}$$

and to fields of auto-equilibrated stress $(\Delta_n\,\boldsymbol{\sigma} - \Delta_n\,\boldsymbol{\sigma}^o)$

$$\int_V \left(\Delta_n\boldsymbol{\sigma} - \Delta_n\boldsymbol{\sigma}^o\right)\left(\Delta_n\boldsymbol{\varepsilon} - \Delta_n\boldsymbol{\varepsilon}^o\right)dV = 0 \tag{11.23}$$

for every statically admissible $\Delta_n\,\boldsymbol{\sigma}$, $\Delta_n\,\boldsymbol{\sigma}^o$ and kinematically admissible $\Delta_n\mathbf{u}$, $\Delta_n\mathbf{u}^o$. By expression (11.23) the sum of the virtual works of auto-equilibrated finite increments of all the forces, which are developed in fields of virtual displacements kinematically admissible with zero data, is zero.

11.5 Variational Formulation

In elastic state the discretized constitutive equations are

$$\Delta_n \boldsymbol{\sigma} = \mathbf{C}_o \, \Delta_n \boldsymbol{\varepsilon} \tag{11.24}$$

The expression (11.24) can be written

$$\Delta_n \boldsymbol{\sigma} = \frac{\partial W_n^*}{\partial(\Delta_n \boldsymbol{\varepsilon})} \tag{11.25}$$

where $W_n^*(\Delta_n \boldsymbol{\varepsilon})$ is the potential of finite increments

$$W_n^*(\Delta_n \boldsymbol{\varepsilon}) = \frac{1}{2} \Delta_n \boldsymbol{\varepsilon} \, \mathbf{C}_o \, \Delta_n \boldsymbol{\varepsilon} \tag{11.26}$$

Consider the finite increment solution for the displacement field $\Delta_n \mathbf{u}$ and another finite increment displacement field $\Delta_n \mathbf{u}^o$ that is different from $\Delta_n \mathbf{u}$, but kinematically admissible. The field $\Delta_n \mathbf{u}$ is associated by (11.18) with finite increment of strain field $\Delta_n \boldsymbol{\varepsilon}$. Denote by $\Delta_n \boldsymbol{\varepsilon}^o$ the associated finite strain increment with $\Delta_n \mathbf{u}^o$.

By (11.21) and (11.25) we get

$$\int_V \frac{\partial W_n^*}{\partial(\Delta_n \boldsymbol{\varepsilon})} \left(\Delta_n \boldsymbol{\varepsilon}^o - \Delta_n \boldsymbol{\varepsilon}\right) dV - \int_{\partial V} \Delta_n \mathbf{t} \left(\Delta_n \mathbf{u}^o - \Delta_n \mathbf{u}\right) d(\partial V) = 0 \tag{11.27}$$

Since the elastic moduli $\mathbf{C}_o$ corresponds to a symmetric positive quadratic form, the potential $W_n^*(\Delta_n \boldsymbol{\varepsilon})$ is strictly convex with respect to $\Delta_n \boldsymbol{\varepsilon}$. By the property of strictly convex functions (6.72) we have

$$W_n^*\left(\Delta_n \boldsymbol{\varepsilon}^o\right) - W_n^*(\Delta_n \boldsymbol{\varepsilon}) > \frac{\partial W_n^*}{\partial(\Delta_n \boldsymbol{\varepsilon})} \left(\Delta_n \boldsymbol{\varepsilon}^o - \Delta_n \boldsymbol{\varepsilon}\right) \tag{11.28}$$

and by (11.27) it implies

$$\begin{aligned} &\int_V W_n^*\left(\Delta_n \boldsymbol{\varepsilon}^o\right) dV - \int_{\partial V} \Delta_n \mathbf{t} \left(\Delta_n \mathbf{u}^o\right) d(\partial V) > \\ &> \int_V W_n^*(\Delta_n \boldsymbol{\varepsilon}) \, dV - \int_{\partial V} \Delta_n \mathbf{t} \left(\Delta_n \mathbf{u}\right) d(\partial V) \end{aligned} \tag{11.29}$$

Define the functional $\mathcal{H}_n(\Delta_n \mathbf{u})$

$$\mathcal{H}_n(\Delta_n \mathbf{u}) = \int_V W_n^*(\Delta_n \boldsymbol{\varepsilon})\, dV - \int_{\partial V} \Delta_n \mathbf{t} \Delta_n \mathbf{u}\, d(\partial V) \tag{11.30}$$

By (11.29) the solution $(\Delta_n \mathbf{u})$ satisfies

$$\mathcal{H}_n(\Delta_n \mathbf{u}) < \mathcal{H}_n(\Delta_n \mathbf{u}^o) \tag{11.31}$$

for kinematically admissible fields $\Delta_n \mathbf{u}^o$.

The solution $\Delta_n\mathbf{u}$ minimizes with respect to $\Delta_n\mathbf{u}$. The finite increment solution $\Delta_n \mathbf{u}$ of the discretized problem from step n – 1 to step n, is unique because the functional $\mathcal{H}_n(\Delta_n\mathbf{u})$ is strictly convex with respect to $\Delta_n\mathbf{u}$. Let $\delta\mathbf{u} = \Delta_n\mathbf{u} - \Delta_n\mathbf{u}^o$, $2\delta\boldsymbol{\varepsilon} = \text{grad}\ \delta\mathbf{u} + {}^t\text{grad}\ \delta\mathbf{u}$. By (11.24), (11.25) and (11.27) the solution $(\Delta_n\mathbf{u})$ satisfies

$$\int_V \delta\boldsymbol{\varepsilon}\,(\mathbf{C}_o\ \Delta_n \boldsymbol{\varepsilon})\, dV - \int_{\partial V_t} \Delta_n\ \mathbf{t}\,\delta\,\mathbf{u}\, d(\partial V) = 0 \tag{11.32}$$

for all kinematically admissible variations $\delta\boldsymbol{\varepsilon}$ with zero boundary conditions imposed on ∂V_u. Eq. (11.32) is the explicit formulation of the extremum condition

$$\delta\, \mathcal{H}_n\ (\Delta_n\ \mathbf{u}) = 0 \tag{11.33}$$

The variational theorem in discretized elasticity is formulated as follows. In a discretized problem the solution in finite increments $\Delta_n\mathbf{u}$ minimizes with respect to $\Delta_n\mathbf{u}$ the functional $\mathcal{H}_n\ (\Delta_n\mathbf{u})$ in a unique way.

The variational result concerning the finite increment solutions in stress $\Delta_n\boldsymbol{\sigma}$ can also be derived.

Introduce the functional $\mathcal{T}_n\ (\Delta_n\ \boldsymbol{\sigma})$

$$\mathcal{T}_n(\Delta_n \boldsymbol{\sigma}) = \int_V W_n^*(\Delta_n \boldsymbol{\sigma}) - \int_{\partial V_u} \Delta_n\ \mathbf{u}^* \Delta_n \boldsymbol{\sigma} \cdot \mathbf{n}\, d(\partial V) \tag{11.34}$$

with

$$W_n^*(\Delta_n \boldsymbol{\sigma}) = \frac{1}{2} \Delta_n \boldsymbol{\sigma}\ \mathbf{C}_o^{-1}\ \Delta_n \boldsymbol{\sigma} \tag{11.35}$$

The finite increment solution in stress $\Delta_n\boldsymbol{\sigma}$ minimizes in a unique way the functional $\mathcal{T}_n\ (\Delta_n\ \boldsymbol{\sigma})$. In plasticity the discretized state equation is

$$\Delta_n \boldsymbol{\sigma} = \mathbf{C}_o\ (\Delta_n \boldsymbol{\varepsilon} - \Delta_n \boldsymbol{\varepsilon}^p) \tag{11.36}$$

For a hardening material, the state equation between hardening variable $\mathbf{m}$ and hardening force $\boldsymbol{\eta}$ is added to the above equations

$$\Delta_n \mathbf{m} = - \mathrm{H} \cdot \Delta_n \boldsymbol{\eta} \tag{11.37}$$

The flow rule (11.12) for ideal plastic material is discretized in an implicit way

$$\Delta_n \boldsymbol{\varepsilon}^p = \Delta\lambda \frac{\partial F}{\partial \boldsymbol{\sigma}} \left(\boldsymbol{\sigma}_{n-1} + \Delta_n \boldsymbol{\sigma} \right) \quad \Delta\lambda \geq 0 \quad \text{if } F\,(\boldsymbol{\sigma}_{n-1} + \Delta_n \boldsymbol{\sigma}) = 0 \tag{11.38}$$

In plasticity an iterative scheme needs to be applied because the problem of determining the finite increment solution Δ_n S = ($\Delta_n \boldsymbol{\sigma}$, $\Delta_n \mathbf{u}$) is non-linear. In time-discretized problems the solution Δ_n S is obtained by successive iterations within the same time step. Δ_n S is determined as the limit of a series of function $\Delta_{n,k}$ S as the number k of iterations tends to infinity. The iterative field equations are

$$\operatorname{div} \Delta_{n,k} \boldsymbol{\sigma} = 0 \tag{11.39}$$

$$2\, \Delta_{n,k} \boldsymbol{\varepsilon} = \operatorname{grad} \Delta_{n,k} \boldsymbol{\varepsilon} = \operatorname{grad} \Delta_{n,k} \mathbf{u} + {}^{T}\operatorname{grad} \Delta_{n,k} \mathbf{u} \tag{11.40}$$

The constitutive equation is

$$\Delta_{n,k} \boldsymbol{\sigma} = \mathbf{C}_o \left(\Delta_{n,k} \boldsymbol{\varepsilon} - \Delta_{n,k-1} \boldsymbol{\varepsilon}^p \right) \tag{11.41}$$

with initial conditions

$$\Delta_{n,k-1} \boldsymbol{\varepsilon}^p \quad \text{with} \quad \Delta_{n,0} \boldsymbol{\varepsilon}^p = 0 \tag{11.42}$$

The finite stress $\Delta_{n,k} \boldsymbol{\sigma}$ and the plastic finite increments $\Delta_{n,k} \boldsymbol{\varepsilon}^p$ are determined and are combined through the discretized constitutive equation (11.41). The flow rule for ideal standard plastic material is

$$\Delta_{n,k} \boldsymbol{\varepsilon}^p = \Delta\lambda \frac{\partial F}{\partial \boldsymbol{\sigma}} \left(\boldsymbol{\sigma}_{n-1} + \Delta_{n,k} \boldsymbol{\sigma} \right) \Delta\lambda \geq 0 \quad \text{if } F\,(\boldsymbol{\sigma}_{n-1} + \Delta_{n,k} \boldsymbol{\sigma}) = 0 \tag{11.43}$$

A variational formulation of the linear problem is needed to apply the iterative method. Rewrite the discretized equation (11.41) in the form

$$\Delta_{n,k} \boldsymbol{\sigma} = \frac{\partial W^*_{n,k}}{\partial \left(\Delta_{n,k} \boldsymbol{\varepsilon} \right)} \tag{11.44}$$

$W^*_{n,k} \left(\Delta_{n,k} \boldsymbol{\varepsilon} \right)$ is the finite increment potential defined by

$$W^*_{n,k} \left(\Delta_{n,k} \boldsymbol{\varepsilon} \right) = \frac{1}{2} \left(\Delta_{n,k} \boldsymbol{\varepsilon} - \Delta_{n,k-1} \boldsymbol{\varepsilon}^p \right) \mathbf{C}_o \left(\Delta_{n,k} \boldsymbol{\varepsilon} - \Delta_{n,k-1} \boldsymbol{\varepsilon}^p \right) \tag{11.45}$$

The potential $W^*_{n,k} \left(\Delta_{n,k} \boldsymbol{\varepsilon} \right)$ is strictly convex with respect to $\Delta_{n,k} \boldsymbol{\varepsilon}$ analogously as the potential defined by (11.26). The variational approach can be applied to the problem. The solution $\Delta_{n,k} \mathbf{u}$ is the unique one which, among all solutions kinematically and statically admissible, minimizes with respect to $\Delta_{n,k} \mathbf{u}$ the functional $\mathcal{H}_{n,k}$ ($\Delta_{n,k} \mathbf{u}$) defined by

$$\mathcal{H}_{n,k}\left(\Delta_{n,k}\mathbf{u}\right)=\int_V W^*_{n,k}\left(\Delta_{n,k}\boldsymbol{\varepsilon}\right)dV-\int_{\partial V_t}\Delta_n\,\mathbf{t}\,\Delta_{n,k}\mathbf{u}\,d(\partial V) \tag{11.46}$$

The solution ($\Delta_{n,k}\,\mathbf{u}$) satisfies

$$\int_V \delta\boldsymbol{\varepsilon}\left[\mathbf{C}_o\left(\Delta_{n,k}\boldsymbol{\varepsilon}-\Delta_{n,k-1}\boldsymbol{\varepsilon}^p\right)\right]dV-\int_{\partial V_t}\Delta_n\,\mathbf{t}\,\delta\mathbf{u}\,d(\partial V)=0 \tag{11.47}$$

for any variation $\delta\boldsymbol{\varepsilon}$ kinematically admissible with zero boundary conditions imposed on ∂V_n. Equation (11.48) expresses the extremum condition $\delta\mathcal{H}_{n,\,k}=0$. By (11.23) in letting $\Delta_n(\cdot)=\Delta_{n,\,k}(\cdot)$ and $\Delta_n(\cdot)^o=\Delta_{n,\,1}(\cdot)$ we get

$$\int_V\left(\Delta_{n,k}\boldsymbol{\sigma}-\Delta_{n,l}\boldsymbol{\sigma}\right)\left(\Delta_{n,k}\boldsymbol{\varepsilon}-\Delta_{n,l}\boldsymbol{\varepsilon}\right)dV=0 \tag{11.48}$$

11.6 Geometrical Interpretation

Rewrite the discretized state equations (11.16) and (11.41) as the forms

$$\boldsymbol{\sigma}_{n,\,k}=\boldsymbol{\sigma}_{n-1}+\Delta_{n,\,k}\,\boldsymbol{\sigma} \tag{11.49}$$

$$\Delta_{n,\,k}\,\boldsymbol{\sigma}\;=\mathbf{C}\,(\Delta_{n,k}\,\boldsymbol{\varepsilon}-\Delta_{n,k-1}\,\boldsymbol{\varepsilon}^p) \tag{11.50}$$

The inversion of the relation (11.50) is

$$\Delta_{n,k}\,\boldsymbol{\varepsilon}=\mathbf{C}^{-1}\cdot\Delta_{n,k}\,\boldsymbol{\sigma}+\Delta_{n,k-1}\,\boldsymbol{\varepsilon}^p \tag{11.51}$$

Introduce the scalar product $<\boldsymbol{\sigma},\boldsymbol{\sigma}*>$ in the space of the vectorial field $\boldsymbol{\sigma}$

$$<\boldsymbol{\sigma},\boldsymbol{\sigma}*>=\frac{1}{2}\int_V\boldsymbol{\sigma}\cdot\mathbf{C}^{-1}\cdot\boldsymbol{\sigma}*dV \tag{11.52}$$

By Eq. (11.48) the orthogonality relation holds

$$<\Delta_{n,\,k}\,\boldsymbol{\sigma}-\Delta_{n,\,1}\,\boldsymbol{\sigma},\;\mathbf{C}\cdot(\Delta_{n,k}\,\boldsymbol{\varepsilon}-\Delta_{n,1}\,\boldsymbol{\varepsilon})>=0 \tag{11.53}$$

At the first iteration (i.e. for k = 1) $\Delta_{n,k-1\,=\,0}\,\boldsymbol{\varepsilon}^p$ is equal to zero, and the field $\boldsymbol{\sigma}_{n,1}$ is equal to $\boldsymbol{\sigma}_{n-1}+\mathbf{C}\cdot\Delta_{n,1}\,\boldsymbol{\varepsilon}$. The field $\boldsymbol{\sigma}_{n,1}$ belongs to the vectorial subspace of fields $\boldsymbol{\sigma}=\boldsymbol{\sigma}_{n-1}+\mathbf{C}\cdot\Delta_{n,k}\,\boldsymbol{\varepsilon}$ with the kinematically admissible field $\Delta_{n,k}\,\boldsymbol{\varepsilon}$, and to the vectorial subspace of fields $\boldsymbol{\sigma}=\boldsymbol{\sigma}_{n-1}+\Delta_{n,\,k}\,\boldsymbol{\sigma}$ with the statically admissible field $\Delta_{n,\,k}\,\boldsymbol{\sigma}$. In the case of ideal plastic material the loading function F = F ($\boldsymbol{\sigma}$) is convex with respect to its argument $\boldsymbol{\sigma}$. The domain defined in the space of vector fields $\boldsymbol{\sigma}(\mathbf{x})$ by $F(\boldsymbol{\sigma})\le 0$ is convex.

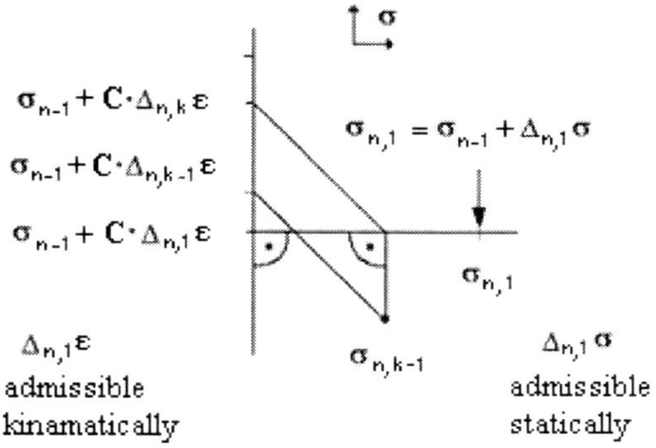

Fig. 11.1. Relations of orthogonality in the space $\{\boldsymbol{\sigma}\}$

By Eq. (11.44) we have

$$\Delta_{n,k}\boldsymbol{\varepsilon}^{p} = \Delta\lambda \frac{\partial F(\boldsymbol{\sigma}_{n,k})}{\partial \boldsymbol{\sigma}_{n,k}} \quad F(\boldsymbol{\sigma}_{n,k}) \leq 0 \quad \Delta\lambda \geq 0\Delta\lambda \;\; F(\boldsymbol{\sigma}_{n,k}) = 0 \tag{11.54}$$

The discretized state equations (11.50) and (11.54) give

$$\boldsymbol{\sigma}_{n,k} - \boldsymbol{\sigma}_{n,k-1} = \mathbf{C}\cdot\Delta_{n,k}\,\boldsymbol{\varepsilon} - \mathbf{C}\cdot\Delta_{n,k-1}\,\boldsymbol{\varepsilon} \tag{11.55}$$

The property of orthogonality (11.53) and the property described by (11.55) are illustrated in Fig. 11.1.

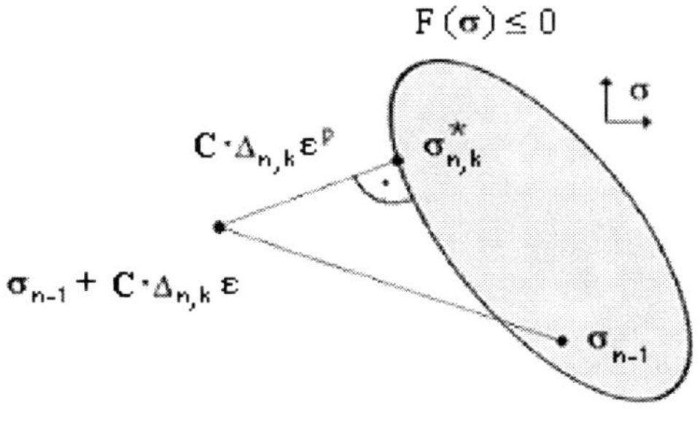

Fig. 11.2. Interpretation of the solution $\boldsymbol{\sigma}_{n,k}$ for standard material

If $\boldsymbol{\sigma}_{n-1} + \mathbf{C}\cdot\Delta_{n,k}\,\boldsymbol{\varepsilon}$ is outside the convex domain defined by $F(\boldsymbol{\sigma}) \leq 0$ i.e. if solution $\Delta_{n,k}\,\boldsymbol{\varepsilon}^{p}$ is a non-zero, Eqs. (11.49) and (11.55) indicate that the solution $\boldsymbol{\sigma}_{n,k}$ of the problem is in the sense of scalar product (11.52) the orthogonal projection of the field $\boldsymbol{\sigma}_{n-1} + \mathbf{C}\cdot\Delta_{n,k}\,\boldsymbol{\varepsilon}$ onto the convex surface of the region defined by $F(\boldsymbol{\sigma}) \leq 0$. This is shown in Fig. 11.2. The uniqueness of this construction is due to the convexity of F and the condition $\Delta\lambda \geq 0$ in Eq. (11.54). The geometrical interpretation of the iterative solving method for an ideal standard plastic material is shown in Fig. 11.3.

The orthogonality relations (11.53) and the relation (11.55) are satisfied in the case of non-standard material, because they are independent of the flow rule. The geometrical interpretation illustrated in Fig. 11.2 does not hold, because it is based on the normality of the flow rule. Consider the flow rule expressed by a non-associated potential G. Then

$$\Delta_{n,k}\boldsymbol{\varepsilon}^{p} = \Delta\lambda \frac{\partial G(\boldsymbol{\sigma}_{n,k})}{\partial \boldsymbol{\sigma}_{n,k}} \quad F(\boldsymbol{\sigma}_{n,k}) \le 0 \Delta\lambda \ge 0 \quad \Delta\lambda\, F(\boldsymbol{\sigma}_{n,k}) = 0 \tag{11.56}$$

Define the unit vector **n** by

$$\mathbf{n} = \frac{\mathbf{C}\cdot\dfrac{\partial G}{\partial \boldsymbol{\sigma}}}{\sqrt{\left\langle \dfrac{\partial G}{\partial \boldsymbol{\sigma}}, \dfrac{\partial G}{\partial \boldsymbol{\sigma}} \right\rangle}} \tag{11.57}$$

where the norm defined by scalar product (11.52) is introduced. The solution $\mathbf{C} \cdot \Delta_{n,k}\, \boldsymbol{\varepsilon}^{p}$ of the problem (11.49) – (11.50) and (11.56) is pointed in the direction of vector $\mathbf{n}_{n,\,k}$ defined by (11.57) with $\boldsymbol{\sigma} = \boldsymbol{\sigma}_{n,\,k}$ (Fig. 11.4).

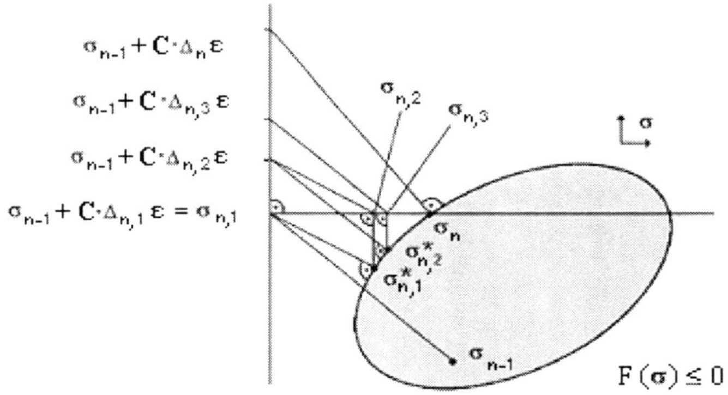

Fig. 11.3. Iterative solution method for standard material

Consider a non-standard material and a flow rule expressed by a non-associated potential G. The geometrical construction of the solution given in Fig. 11.3 has to be modified according to Fig. 11.4. If positive hardening takes place, most of the usual criteria with non-associated potentials allow the construction of a unique solution according to Fig. 11.4. The analysis should be specific to any particular case considered. If negative hardening takes place, elastic unloading cannot be distinguished from plastic loading. This implies that the solution is non-unique, intrinsic to the material behaviour.

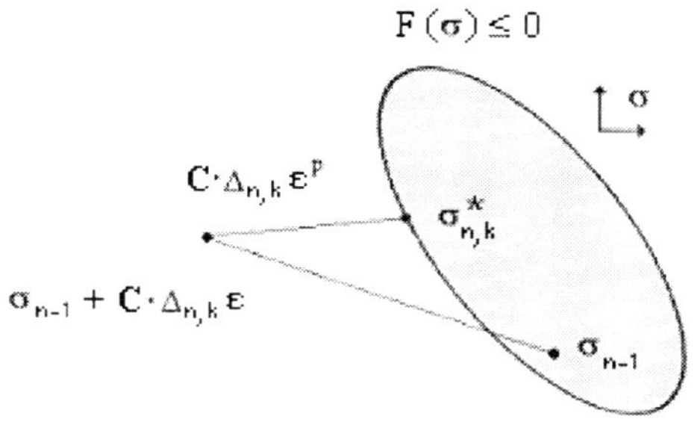

Fig. 11.4. Solution $\boldsymbol{\sigma}_{n,k}$ for non- standard material

11.7 Convergence

Assume the existence of the increment solution $d\boldsymbol{\sigma}$ of the non-discretized problem. In a general case the convergence of the iteration scheme is then satisfied. A convergence criterion required for an iterative scheme can be provided by the plastic criterion. Define the positive scalar ξ_k in the case of an ideal plastic material by

$$\xi_k = \max_{x \in V}\left\{\frac{1}{C}\left\langle\left\langle F\left(\boldsymbol{\sigma}_{n-1} + \Delta_{n,k}\boldsymbol{\sigma}\right)\right\rangle\right\rangle\right\} \tag{11.58}$$

In the case of the hardening

$$\xi_k = \max_{x \in V}\left\{\frac{1}{C}\left\langle\left\langle F\left(\boldsymbol{\sigma}_{n-1} + \Delta_{n,k}\boldsymbol{\sigma},\ \boldsymbol{\eta}_{n-1} + \Delta_{n,k}\boldsymbol{\eta}\right)\right\rangle\right\rangle\right\} \tag{11.59}$$

where $\langle\langle \cdot \rangle\rangle$ is for the positive parts of the term between brackets, and C represents a positive normalization constant. It has the same unit as the loading function. It is assumed to be any critical parameter, for instance, the limit in pure shear stress. The parameter ξ_k represents the distance between the loading point and the current elasticity domain at iteration k. The convergence criterion is

$$\xi_k \leq \xi \tag{11.60}$$

where ξ is a small positive real. The value of ξ can vary depending on the problem considered. In practice ξ is assumed to be 10^{-2}.

12 Numerical Models of Plasticity

12.1 The Thermo-Elasto-Plastic Finite Element Model

Consider the virtual work equation for a finite element assemblage for a thermo-elastic-plastic material model at time t + Δt (step n+1)

$$\int_V \mathbf{B}_L^T \boldsymbol{\sigma}_{n+1} dV = \mathbf{R}_{n+1} \tag{12.1}$$

$$\boldsymbol{\sigma}_{n+1} = \mathbf{C}_{n+1}\left(\boldsymbol{\varepsilon}_{n+1} - \boldsymbol{\varepsilon}_{n+1}^{p} - \boldsymbol{\varepsilon}_{n+1}^{\theta}\right) \tag{12.2}$$

$$\boldsymbol{\varepsilon}_n^p = \Lambda_n \mathbf{D}\boldsymbol{\sigma}_n \tag{12.3}$$

$$\boldsymbol{\varepsilon}_{n+1}^{\theta} = \alpha\left(\theta_{n+1} - \theta_R\right)\boldsymbol{\delta} \tag{12.4}$$

where

$$\boldsymbol{\varepsilon}_{n+1} = \mathbf{B}\mathbf{U}_{n+1} \tag{12.5}$$

and $\mathbf{B}$ is the total strain-displacement transformation matrix, $\mathbf{U}_{n+1}$ is the nodal point displacement vector, $\mathbf{R}_{n+1}$ is nodal point external load vector, $\mathbf{D}$ is the deviatoric stress operator matrix and $\boldsymbol{\delta}^T$ is [1,1,1,0,0,0]. Substituting Eqs. (12.2) and (12.5) into Eq. (12.1) we get

$$\mathbf{K}\mathbf{U}_{n+1} = \mathbf{R}_{n+1} + \int_V \mathbf{B}^T \mathbf{C}_{n+1}\left(\boldsymbol{\varepsilon}^p + \boldsymbol{\varepsilon}^{\theta}\right) dV \tag{12.6}$$

where

$$\mathbf{K} = \int_V \mathbf{B}^T \mathbf{C}_{n+1} \mathbf{B}\, dV \tag{12.7}$$

is the elastic stiffness matrix.

12.2 The Rigid-Viscoplastic Finite Element Model

In a rigid-viscoplastic model the flow rule with the omitted superscript vp is given in the form (9.37)

$$\dot{\boldsymbol{\varepsilon}} = \frac{1}{2\mu}\mathbf{s} \tag{12.8}$$

The theorem of virtual work (10.15) is of the form

$$\int_V \boldsymbol{\sigma}\,\dot{\boldsymbol{\varepsilon}}\,dV = \int_{\partial V} \mathbf{t}\cdot\mathbf{V}\,d(\partial V) \tag{12.9}$$

introducing the deviatoric stress operator **D**

$$\boldsymbol{\sigma} = \mathbf{D}^{-1}\cdot\mathbf{s} \tag{12.10}$$

we get from (12.8)

$$\dot{\boldsymbol{\varepsilon}} = \frac{1}{2\mu}\mathbf{D}\cdot\boldsymbol{\sigma} \tag{12.11}$$

or in inverted form

$$\boldsymbol{\sigma} = \mathbf{P}\cdot\dot{\boldsymbol{\varepsilon}} \tag{12.12}$$

where $\mathbf{P} = 2\mu\mathbf{D}^{-1}$. Substituting (12.11) into Eq. (12.9) we obtain

$$\mathbf{K}_v\cdot\mathbf{V} - \mathbf{Q} = 0 \tag{12.13}$$

where

$$\mathbf{K}_v = \int_V {}^{T}\mathbf{B}\,\mathbf{D}^{-1}\,2\mu\,\mathbf{B}\,dV \tag{12.14}$$

is the stiffness matrix
and

$$\mathbf{Q} = \int_{\partial V} \mathbf{N}\,\mathbf{t}\,d(\partial V) \tag{12.15}$$

is the mechanical load vector.

12.3 The Rigid-Poroplastic Finite Element Model

The loading function for porous material is assumed in the form (8.45)

$$F = A\,h^2 + \frac{(3-A)(3p)^2}{3} - q^2 \tag{12.16}$$

For standard material Eq. (8.46) gives the flow rule

$$\dot{\boldsymbol{\varepsilon}} = \dot{\lambda}\left[\mathrm{A}\,\mathbf{s} + 3(3-\mathrm{A})\mathrm{p}\,\mathbf{1}\right] \tag{12.17}$$

with

$$\dot{\lambda} = \frac{\mathrm{d_q F}}{\mathrm{H}} \qquad \mathrm{d_q F} = 9\,(3-\mathrm{A})\,\mathrm{p\,dp} + \mathrm{A}\,\mathbf{s}\,\mathrm{d}\mathbf{s} \tag{12.18}$$

By (12.14), (12.15) and the virtual work theorem (12.9) we finally get

$$\mathbf{K}_\phi \cdot \mathbf{V} = \mathbf{Q} \tag{12.19}$$

where

$$\mathbf{Q} = \int_V \mathbf{N}\,\mathbf{t}\,\mathrm{d}(\partial \mathrm{V}) \tag{12.20}$$

$$\mathbf{K}_\phi \int_V \dot{\lambda}^{\mathrm{T}}\mathbf{B}\,\mathbf{D}\mathbf{B}\,\mathrm{dV} \tag{12.21}$$

and

$$\mathbf{D} = \begin{bmatrix} \frac{4-\mathrm{A}}{\mathrm{A}(3-\mathrm{A})} & \frac{\mathrm{A}-2}{\mathrm{A}(3-\mathrm{A})} & \frac{\mathrm{A}-2}{\mathrm{A}(3-\mathrm{A})} & 0 \\ \frac{\mathrm{A}-2}{\mathrm{A}(3-\mathrm{A})} & \frac{4-\mathrm{A}}{\mathrm{A}(3-\mathrm{A})} & \frac{\mathrm{A}-2}{\mathrm{A}(3-\mathrm{A})} & 0 \\ \frac{\mathrm{A}-2}{\mathrm{A}(3-\mathrm{A})} & \frac{\mathrm{A}-2}{\mathrm{A}(3-\mathrm{A})} & \frac{4-\mathrm{A}}{\mathrm{A}(3-\mathrm{A})} & 0 \\ 0 & 0 & 0 & \frac{1}{\mathrm{A}} \end{bmatrix} \tag{12.22}$$

12.4 The Finite Differences Model

The matrix forms of equations for finite differences approximations are the same as for finite element models. The only difference is in the forms of matrices and vectors components. The application of finite differences to a solution of metal forming plasticity problems is the rarity, so a detailed description of this solution has been omitted.

Part VIII

Sensitivity in Metal Forming Plasticity

13 Sensitivity

13.1 Introduction

The theory of sensitivity originates form purely mathematical studies of the influence of coefficient variations on differential equations. It was much later that the theory of sensitivity became the subject of studies in the field of metal forming.

In metal forming a set of equations that defines the relationship between external loads, prescribed displacements, stresses, etc. in material body are considered. A sensitivity of the metal forming process to variations of its parameters is one of the most important aspects necessary for a proper understanding of the process.

The properties of the metal forming process are characterized by some parameters **h** called the design parameters. Typical parameters are dimensions of metallic body, material constants, geometric parameters defining the over all shape of the material. The change of these parameters influences on the final metal forming process. Sensitivity of metal forming is a measure of the change in the metal forming response under the change in the design parameters. By the metal forming response we mean any quantity that may be used to characterize process behaviour. In this section the sensitivity in metal forming is limited to rigid-viscoplastic and elasto-plastic material models.

The two basic methodologies for solving sensitivity problems in metal forming called the direct differentiation method and the adjoint system method are discussed.

The design sensitivity analysis consists of non-shape and sensitivity techniques. The shape design sensitivity analysis is based on the material derivative approach and control volume approach.

13.2 Notations and Terminology

Consider the functional

$$\mathcal{G}(\mathbf{h}) = G[\mathbf{V}(\mathbf{h}), \mathbf{h}] \tag{13.1}$$

where **V** is the velocity vector treated as dependent on the metal forming parameters **h**. The dependence **V** = **V** (**h**) on **h** is implicit since the solution **V** for a pre-set value of **h** is given.

The total derivative of the functional $\mathcal{G}$ with respect to **h** gives

$$\frac{\mathrm{d}\mathcal{G}}{\mathrm{d}\mathbf{h}} = \frac{\mathrm{d}G}{\mathrm{d}\mathbf{h}} = \frac{\partial G}{\partial \mathbf{V}}\frac{\mathrm{d}\mathbf{V}}{\mathrm{d}\mathbf{h}} + \frac{\partial G}{\partial \mathbf{h}} \tag{13.2}$$

The term $\frac{\partial G}{\partial \mathbf{V}}$ is the partial derivative of G with respect to **V**, and $\frac{\mathrm{d}G}{\mathrm{d}\mathbf{h}}$ is the absolute derivative of G with respect to **h**.

The derivatives $\frac{\partial G}{\partial \mathbf{V}}$ and $\frac{\partial G}{\partial \mathbf{h}}$ in Eq. (13.2) are explicit, while the derivatives $\frac{\mathrm{d}\mathbf{V}}{\mathrm{d}\mathbf{h}}$ are implicit.

Consider a function $\mathcal{G}$ of the form

$$\mathcal{G}(\mathbf{h}) = G[\Delta\mathbf{V}(\mathbf{h}), \boldsymbol{\sigma}(\mathbf{h}), \mathbf{h}] \tag{13.3}$$

The total derivative of $\mathcal{G}$ given by (13.3) is

$$\frac{\mathrm{d}\mathcal{G}}{\mathrm{d}\mathbf{h}} = \frac{\partial G}{\partial \Delta\mathbf{V}}\frac{\partial \Delta\mathbf{V}}{\partial \mathbf{h}} + \frac{\partial G}{\partial \boldsymbol{\sigma}}\frac{\mathrm{d}\boldsymbol{\sigma}}{\mathrm{d}\mathbf{h}} + \frac{\partial G}{\partial \mathbf{h}} \tag{13.4}$$

The derivatives $\frac{\partial G}{\partial \Delta\mathbf{V}}, \frac{\partial G}{\partial \boldsymbol{\sigma}}$ and $\frac{\partial G}{\partial \mathbf{h}}$ in Eq. (13.4) are explicit, while $\frac{\partial \Delta\mathbf{V}}{\partial \mathbf{h}}$ is implicit. By assumption a collection of known variables $\frac{\mathrm{d}\boldsymbol{\sigma}}{\mathrm{d}\mathbf{h}}$ is not treated as the design derivative.

The different variations of $\mathcal{G}$ can be considered. The symbol $\delta\mathcal{G}$ stands for its arbitrary variation as known in variational calculus. A special case of the variation $\delta\mathcal{G}$ denoted by $\bar{\delta}\mathcal{G}$ is the variation of $\mathcal{G}$ with respect to changes in **h**. The total design variation of $\mathcal{G}$ is

$$\begin{aligned} \bar{\delta}\mathcal{G} &= \frac{\mathrm{d}\mathcal{G}}{\mathrm{d}\mathbf{h}}\delta\mathbf{h} = \frac{\partial G}{\partial \mathbf{h}}\delta\mathbf{h} + \frac{\partial G}{\partial \mathbf{V}}\bar{\delta}\mathbf{V} \\ &= \frac{\partial G}{\partial \mathbf{h}}\delta\mathbf{h} + \frac{\partial G}{\partial \mathbf{V}}\frac{\mathrm{d}\mathbf{V}}{\partial \mathbf{h}}\delta\mathbf{h} = \left(\frac{\partial G}{\partial \mathbf{h}} + \frac{\partial G}{\partial \mathbf{V}}\frac{\mathrm{d}\mathbf{V}}{\partial \mathbf{h}}\right)\delta\mathbf{h} \end{aligned} \tag{13.5}$$

The notation $\bar{\delta}\mathbf{V}$ used in (13.5) indicates on variation of the argument **V** as its total design variation, $\bar{\bar{\delta}}\mathcal{G} = \frac{\partial G}{\partial \mathbf{h}}\delta\mathbf{h}$ is the variation of $\mathcal{G}$ related to the explicit dependence of $\mathcal{G}$ on **h** called the explicit design variation of $\mathcal{G}$ and

$\tilde{\delta} G = \frac{\partial G}{\partial \mathbf{V}} \frac{\mathrm{d}\mathbf{V}}{\mathrm{d}\mathbf{h}} \delta \mathbf{h}$ is the variation of G related to the implicit dependence of G on **h**, called the implicit design variation of G. By the above considerations the following expression holds

$$\bar{\delta} G = \bar{\bar{\delta}} G + \tilde{\delta} G \tag{13.6}$$

13.3 Discretized Equation of Rigid-Viscoplasticity

The discretization of rigid-viscoplasticity equations (12.13) gives

$$\mathbf{K}_{\mathrm{V}}(\mathbf{h})\, \mathbf{V}(\mathbf{h}) = \mathbf{Q}(\mathbf{h}) \tag{13.7}$$

where the forms of matrix $\mathbf{K}_{\mathrm{V}}$ and the vector $\mathbf{Q}$ are given by (12.14) – (12.15). By differentiating Eq. (13.7) with respect to **h**, after re-ordering the terms, gives

$$\mathbf{K}_{\mathrm{V}} \frac{\mathrm{d}\mathbf{V}}{\mathrm{d}\mathbf{h}} = \frac{\mathrm{d}\mathbf{Q}}{\mathrm{d}\mathbf{h}} - \frac{\mathrm{d}\mathbf{K}}{\mathrm{d}\mathbf{h}} \mathbf{V} \tag{13.8}$$

The expression (13.8) is used to calculate the derivatives $\frac{\mathrm{d}\mathbf{V}}{\mathrm{d}\mathbf{h}}$. If the solution vector **V** is known the right-hand side can be easily calculated. The method described above is known as the direct differentiation method. Another technique to the direct differentiation method is known as the adjoint system method. It is based on introducing an adjoint vector variable $\boldsymbol{\lambda} = \boldsymbol{\lambda}(\mathbf{h})$ defined as the solution of the set of linear equations

$$\mathbf{K}_{\mathrm{V}}(\mathbf{h})\boldsymbol{\lambda}(\mathbf{h}) = \left(\frac{\partial G}{\partial \mathbf{V}}\right)^{\mathrm{T}} \tag{13.9}$$

i.e. as

$$\boldsymbol{\lambda}(\mathbf{h}) = \mathbf{K}^{-1}(\mathbf{h}) \left(\frac{\partial G}{\partial \mathbf{V}}\right)^{\mathrm{T}} \tag{13.10}$$

where G is the function given by (13.1). By the symmetry of the stiffness matrix and substituting $\frac{\partial G}{\partial \mathbf{V}}$ from Eq. (13.9), Eq. (13.2) may be rewritten as

$$\frac{\mathrm{d}G}{\mathrm{d}\mathbf{h}} = \frac{\partial G}{\partial \mathbf{h}} + \boldsymbol{\lambda}^{\mathrm{T}} \mathbf{K}_{\mathrm{V}} \frac{\mathrm{d}\mathbf{V}}{\partial \mathbf{h}} \tag{13.11}$$

or, by (13.8)

$$\frac{\mathrm{d}G}{\mathrm{d}\mathbf{h}}=\frac{\partial G}{\partial \mathbf{h}}+\boldsymbol{\lambda}^{\mathrm{T}}\left(\frac{\mathrm{d}\mathbf{Q}}{\partial \mathbf{h}}-\frac{\mathrm{d}\mathbf{K}_{\mathrm{V}}}{\mathrm{d}\mathbf{h}}\mathbf{V}\right) \tag{13.12}$$

The vector of Lagrange multiplier $\boldsymbol{\lambda}$ by considering an extended constraint function is introduced as

$$G^{*}\,[\boldsymbol{\lambda}\,(\mathbf{h}),\,\mathbf{V}\,(\mathbf{h}),\,\mathbf{h}] = G\,[\mathbf{V}\,(\mathbf{h}),\,\mathbf{h}] + \boldsymbol{\lambda}^{\mathrm{T}}\,(\mathbf{h})\,\mathbf{R}\,[\mathbf{V}\,(\mathbf{h}),\,\mathbf{h}] \tag{13.13}$$

where

$$\mathbf{R}\,[\mathbf{V}\,(\mathbf{h}),\,\mathbf{h}] = \mathbf{Q}\,(\mathbf{h}) - \mathbf{K}_{\mathrm{V}}\,(\mathbf{h})\,\mathbf{V}\,(\mathbf{h}) = 0 \tag{13.14}$$

is the residual of the matrix equation. By differentiating Eq. (13.13) with respect to $\mathbf{h}$ we get

$$\frac{\mathrm{d}G^{*}}{\mathrm{d}\mathbf{h}}=\frac{\partial G}{\partial \mathbf{h}}+\frac{\partial G}{\partial \mathbf{V}}\frac{\mathrm{d}\mathbf{V}}{\partial \mathbf{h}}+\frac{\mathrm{d}\boldsymbol{\lambda}^{\mathrm{T}}}{\mathrm{d}\mathbf{h}}\left(\mathbf{Q}-\mathbf{K}_{\mathrm{V}}\,\mathbf{V}\right)+\boldsymbol{\lambda}^{\mathrm{T}}\left(\frac{\mathrm{d}\mathbf{Q}}{\mathrm{d}\mathbf{h}}-\frac{\mathrm{d}\mathbf{K}_{\mathrm{V}}}{\mathrm{d}\mathbf{h}}\mathbf{V}-\mathbf{K}_{\mathrm{V}}\frac{\mathrm{d}\mathbf{V}}{\partial \mathbf{h}}\right) \tag{13.15}$$

The state equation holds both at current $\mathbf{h}$ and perturbed design $\mathbf{h} + \delta\mathbf{h}$

$$\mathbf{R}\,[\mathbf{V}\,(\mathbf{h}+\delta\mathbf{h}),\,\mathbf{h}+\delta\mathbf{h}] = \mathbf{Q}\,(\mathbf{h}+\delta\mathbf{h}) - \mathbf{K}_{\mathrm{V}}\,(\mathbf{h}+\delta\mathbf{h})\,\mathbf{V}\,(\mathbf{h}+\delta\mathbf{h}) = 0 \tag{13.16}$$

then also

$$\frac{\mathrm{d}\mathbf{R}}{\mathrm{d}\mathbf{h}}=\frac{\mathrm{d}\mathbf{Q}}{\mathrm{d}\mathbf{h}}-\frac{\mathrm{d}\mathbf{K}_{\mathrm{V}}}{\mathrm{d}\mathbf{h}}\mathbf{V}-\mathbf{K}_{\mathrm{V}}\frac{\mathrm{d}\mathbf{V}}{\mathrm{dh}}=0 \tag{13.17}$$

By (13.15) we have

$$\frac{\mathrm{d}G^{*}}{\mathrm{d}\mathbf{h}}=\frac{\mathrm{d}G}{\mathrm{d}\mathbf{h}} \tag{13.18}$$

The expression (13.15) can be rewritten as

$$\frac{\mathrm{d}G^{*}}{\mathrm{d}\mathbf{h}}=\frac{\partial G}{\partial \mathbf{h}}+\boldsymbol{\lambda}^{\mathrm{T}}\left(\frac{\mathrm{d}\mathbf{Q}}{\mathrm{d}\mathbf{h}}-\frac{\mathrm{d}\mathbf{K}_{\mathrm{V}}}{\mathrm{d}\mathbf{h}}\mathbf{V}\right)+\left(\frac{\partial G}{\partial \mathbf{V}}-\boldsymbol{\lambda}^{\mathrm{T}}\,\mathbf{K}_{\mathrm{V}}\right)\frac{\mathrm{d}\mathbf{V}}{\mathrm{d}\mathbf{h}} \tag{13.19}$$

The Lagrange multiplies can be assumed to ensure the functional G^{*} to be stationary with respect to the variable $\mathbf{V}$. Then

$$\delta G^{*}=\frac{\partial G^{*}}{\partial \mathbf{V}}\delta\mathbf{V}=\left(\frac{\partial G}{\partial \mathbf{V}}-\boldsymbol{\lambda}^{\mathrm{T}}\,\mathbf{K}_{\mathrm{V}}\right)\delta\mathbf{V}=0 \tag{13.20}$$

or

$$\frac{\partial G^{*}}{\partial \mathbf{V}}\delta\mathbf{V}=\frac{\partial G}{\partial \mathbf{V}}-\boldsymbol{\lambda}^{\mathrm{T}}\,\mathbf{K}_{\mathrm{V}}=0 \tag{13.21}$$

Selecting $\boldsymbol{\lambda}$ to eliminate from Eq. (13.19) the coefficient at $\frac{\mathrm{d}\mathbf{V}}{\mathrm{d}\mathbf{h}}$ and by using Eqs. (13.18), (13.20) the relationship (13.19) becomes

$$\frac{\mathrm{d}G^*}{\mathrm{d}\mathbf{h}} = \frac{\partial G}{\partial \mathbf{h}} + \boldsymbol{\lambda}^{\mathrm{T}}\left(\frac{\mathrm{d}\mathbf{Q}}{\mathrm{d}\mathbf{h}} - \frac{\mathrm{d}\mathbf{K}_{\mathrm{V}}}{\mathrm{d}\mathbf{h}}\mathbf{V}\right) \tag{13.22}$$

which is identical to Eq. (13.12).

Consider the state equation given in the form

$$\mathbf{R}\,(\mathbf{V}\,(\mathbf{h}),\,\mathbf{h}) = 0 \tag{13.23}$$

The extended functional (13.13) differentiated with respect to $\mathbf{h}$ is

$$\frac{\mathrm{d}G^*}{\mathrm{d}\mathbf{h}} = \frac{\partial G}{\partial \mathbf{h}} + \frac{\partial G}{\partial \mathbf{V}}\frac{\mathrm{d}\mathbf{V}}{\mathrm{d}\mathbf{h}} + \frac{\mathrm{d}\boldsymbol{\lambda}^{\mathrm{T}}}{\mathrm{d}\mathbf{h}}\mathbf{R} + \boldsymbol{\lambda}^{\mathrm{T}}\left(\frac{\partial \mathbf{R}}{\partial \mathbf{h}} + \frac{\partial \mathbf{R}}{\partial \mathbf{V}}\right)\frac{\mathrm{d}\mathbf{V}}{\mathrm{d}\mathbf{h}} \tag{13.24}$$

The third term on the right-hand side can be dropped on account of the condition $\mathbf{R} = 0$. From the above we have

$$\frac{\mathrm{d}G^*}{\mathrm{d}\mathbf{h}} = \frac{\partial G}{\partial \mathbf{h}} + \boldsymbol{\lambda}^{\mathrm{T}}\frac{\partial \mathbf{R}}{\partial \mathbf{h}} + \left(\frac{\partial G}{\partial \mathbf{V}} + \boldsymbol{\lambda}^{\mathrm{T}}\frac{\partial \mathbf{R}}{\partial \mathbf{V}}\right)\frac{\mathrm{d}\mathbf{V}}{\mathrm{d}\mathbf{h}} \tag{13.25}$$

By selecting $\boldsymbol{\lambda}$ in such a way as to make G^* stationary with respect to the variable $\mathbf{V}$,

$$\frac{\partial G^*}{\partial \mathbf{V}} = \frac{\partial G}{\partial \mathbf{V}} + \boldsymbol{\lambda}^{\mathrm{T}}\frac{\partial \mathbf{R}}{\partial \mathbf{V}} = 0 \tag{13.26}$$

we can eliminate from the right-hand side of Eq. (13.26) the last term so that

$$\frac{\mathrm{d}G^*}{\mathrm{d}\mathbf{h}} = \frac{\partial G}{\partial \mathbf{h}} + \boldsymbol{\lambda}^{\mathrm{T}}\frac{\partial \mathbf{R}}{\partial \mathbf{h}} \tag{13.27}$$

We may write

$$\frac{\partial G^*}{\partial \mathbf{h}} = \frac{\partial G}{\partial \mathbf{h}} + \boldsymbol{\lambda}^{\mathrm{T}}\frac{\partial \mathbf{R}}{\partial \mathbf{h}} \tag{13.28}$$

since the right-hand side of Eq. (13.27) contains the terms independent of $\frac{\mathrm{d}\mathbf{V}}{\mathrm{d}\mathbf{h}}$ or $\frac{\mathrm{d}\boldsymbol{\lambda}}{\mathrm{d}\mathbf{h}}$.

By Eq. (13.24) the following relation holds

$$\frac{\mathrm{d}G^*}{\mathrm{d}\mathbf{h}} = \frac{\mathrm{d}G}{\mathrm{d}\mathbf{h}} \tag{13.29}$$

which is due to the fact that the state equation holds true at both the current and perturbed designs

$$\frac{d\mathbf{R}}{d\mathbf{h}} = \frac{\partial \mathbf{R}}{\partial \mathbf{h}} + \frac{\partial \mathbf{R}}{\partial \mathbf{V}}\frac{d\mathbf{V}}{d\mathbf{h}} = 0 \tag{13.30}$$

By Eqs. (13.27) and (13.29) the fundamental adjoint system method is obtained

$$\frac{dG}{d\mathbf{h}} = \frac{\partial G}{\partial \mathbf{h}} + \boldsymbol{\lambda}^{T}\frac{\partial \mathbf{R}}{\partial \mathbf{h}} \tag{13.31}$$

which gives

$$\frac{dG}{d\mathbf{h}} = \frac{\partial G^*}{\partial \mathbf{h}} \tag{13.32}$$

From the abcve for the given constraint functional $G[\mathbf{V}(\mathbf{h}), \mathbf{h}]$ and the function $\mathbf{R}[\mathbf{V}(\mathbf{h}), \mathbf{h}] = 0$ the sensitivity gradient may be obtained as

$$\frac{dG}{d\mathbf{h}} = \frac{\partial G^*}{\partial \mathbf{h}}$$

in which

$$G^* = G + \boldsymbol{\lambda}^{T}\mathbf{R} \tag{13.33}$$

and $\boldsymbol{\lambda}$ satisfies the equation

$$\frac{\partial G}{\partial \mathbf{V}} + \boldsymbol{\lambda}^{T}\frac{\partial \mathbf{R}}{\partial \mathbf{V}} = 0 \tag{13.34}$$

The main point in the above theorem is that the total design derivative $\frac{dG}{d\mathbf{h}}$ of the original functional G is replaced by the explicit design derivative $\frac{dG^*}{d\mathbf{h}}$ of the extended functional G^* at the cost of having to compute the Lagrange multiplier vector $\boldsymbol{\lambda}$.

13.4 Continuous Formulation for Rigid-Viscoplasticity

Consider the theorem of virtual work (10.15) for the rigid-viscoplastic material model written in the form

$$\begin{aligned} &\int_{\partial V} \mathbf{t}\,\delta\mathbf{V}\,d(\partial V) - \int_{V} \mathbf{P}\mathbf{A}^{m}\mathbf{V}\mathbf{A}^{n}\,\delta\mathbf{V}\,dV = 0 \\ &\int_{\partial V} t_i\delta V_i d(\partial V) - \int_{V} P_{ijkl}A_{ijm}A_{kln}V_m\delta V_n dV = 0 \end{aligned} \tag{13.35}$$

where matrix operator $\mathbf{A}$ is introduced by the expression

$$\dot{\boldsymbol{\varepsilon}} = \mathbf{A}^{k}\mathbf{V} \qquad \dot{\varepsilon}_{ij} = A_{ijk} V_k \tag{13.36}$$

The stress-strain rate relation is assumed in the form (12.12) as

$$\boldsymbol{\sigma} = \mathbf{P}\dot{\boldsymbol{\varepsilon}} \qquad \sigma_{ij} = P_{ijkl}\dot{\varepsilon}_{kl} \tag{13.37}$$

Consider the response functional

$$\mathcal{G}(\mathbf{h}) = G(\boldsymbol{\sigma}, \mathbf{V}, \mathbf{h}) = \int_V g(\boldsymbol{\sigma}, \mathbf{V}, \mathbf{h})\, dV + \int_{\partial V} g^{\sigma}(\mathbf{V}, \mathbf{h})\, d(\partial V) \tag{13.38}$$

in which dependence of all the functions on their arguments is explicitly given. The total design variation applied to (13.38) is

$$\overline{\delta}\mathcal{G} = \int_V \overline{\overline{\delta}} g\, dV + \int_{\partial V} \overline{\overline{\delta}} g^{\sigma} d(\partial V) + \int_V \widetilde{\delta} g\, dV + \int_{\partial V} \widetilde{\delta} g^{\sigma} d(\partial V) \tag{13.39}$$

where

$$\overline{\overline{\delta}} g = \frac{\partial g}{\partial h}\delta h \tag{13.40}$$

$$\overline{\overline{\delta}} g^{(\sigma)} = \frac{\partial g^{(\sigma)}}{\partial \mathbf{h}}\delta \mathbf{h} \tag{13.41}$$

$$\widetilde{\delta} g = \frac{\partial g}{\partial \boldsymbol{\sigma}}\overline{\delta}\boldsymbol{\sigma} + \frac{\partial g}{\partial \mathbf{V}}\overline{\delta}\mathbf{V} = g_{,\boldsymbol{\sigma}}\overline{\delta}\boldsymbol{\sigma} + g_{,\mathbf{V}}\overline{\delta}\mathbf{V} \tag{13.42}$$

$$\widetilde{\delta} g^{(\sigma)} = \frac{\partial g^{(\sigma)}}{\partial \mathbf{V}}\overline{\delta}\mathbf{V} = g^{(\sigma)}_{,\mathbf{V}}\overline{\delta}\mathbf{V} \tag{13.43}$$

The arguments $\boldsymbol{\sigma}$ and $\mathbf{V}$ of g are treated as independent so that no relation between $\overline{\delta}\boldsymbol{\sigma}$ and $\overline{\delta}\mathbf{V}$ is used. The total design variation of the Eq. (13.35) gives

$$\begin{gathered}\int_V \mathbf{P}\,\mathbf{A}^{m}\,\overline{\delta}\,\mathbf{V}\mathbf{A}^{n}\,\delta\,\mathbf{V}\,dV = \int_{\partial V} \overline{\overline{\delta}}\,\mathbf{t}\,\delta\,\mathbf{V}\,d(\partial V) - \int_V \overline{\overline{\delta}}\,\mathbf{P}\,\mathbf{A}^{m}\mathbf{V}\mathbf{A}^{n}\,\delta\,\mathbf{V}\,dV \\ \int_V P_{ijkl}A_{ijm}A_{kln}\overline{\delta}V_m\delta V_n\,dV = \int_{\partial V} \overline{\overline{\delta}}t_i\delta V_i\,d(\partial V) - \int_V \overline{\overline{\delta}}P_{ijkl}A_{ijm}A_{kln}V_m\delta V_n\,dV\end{gathered} \tag{13.44}$$

in which advantage was taken of the relations

$$\overline{\delta}\,\mathbf{P} = \overline{\overline{\delta}}\,\mathbf{P} \tag{13.45}$$

$$\overline{\delta}\,\mathbf{t} = \overline{\overline{\delta}}\,\mathbf{t} \tag{13.46}$$

$$\overline{\delta}\,\mathbf{A} = 0 \tag{13.47}$$

The expression (13.44) is solved using the same technique as for solving the equilibrium problem (13.35) with the different terms of the right-hand side. The design derivatives of stresses are

$$\begin{aligned}\bar{\delta}\boldsymbol{\sigma} &= \bar{\delta}(\mathbf{P}\dot{\boldsymbol{\varepsilon}}) = \delta(\mathbf{P}\mathbf{A}^{m}\mathbf{V}) = \underbrace{\bar{\bar{\delta}}\,\mathbf{P}\mathbf{A}^{m}\mathbf{V}}_{\bar{\bar{\delta}}\boldsymbol{\sigma}} + \underbrace{\mathbf{P}\mathbf{A}^{m}\,\bar{\delta}\,\mathbf{V}}_{\bar{\delta}\boldsymbol{\sigma}} \\ \bar{\delta}\sigma_{ij} &= \bar{\delta}(P_{ijkl}\dot{\varepsilon}_{kl}) = \bar{\delta}(P_{ijnl}A_{klm}V_{m}) = \underbrace{\bar{\bar{\delta}}P_{ijkl}A_{klm}V_{m}}_{\bar{\bar{\delta}}\sigma_{ij}} + \underbrace{P_{ijkl}A_{klm}\bar{\delta}V_{m}}_{\bar{\delta}\sigma_{ij}}\end{aligned} \tag{13.48}$$

13.5 Adjoint System Method for Rigid-Viscoplasticity

In the adjoint system method introduce the adjoint body with a displacement field $\mathbf{V}^{a}$ by the strain rate-velocity relation

$$\dot{\boldsymbol{\varepsilon}}^{(a)} = \mathbf{A}^{k}\mathbf{V}^{(a)} \qquad \dot{\boldsymbol{\varepsilon}}_{ij}^{(a)} = A_{ijk}V_{k}^{(a)} \tag{13.49}$$

the constitutive equation

$$\boldsymbol{\sigma}^{(a)} = \mathbf{P}\dot{\boldsymbol{\varepsilon}}^{(a)} \qquad \sigma_{ij}^{(a)} = P_{ijkl}\dot{\varepsilon}_{kl} \tag{13.50}$$

and the surface tractions

$$\mathbf{t} = \frac{\partial g^{\sigma}}{\partial \mathbf{V}} = g_{,V}^{\sigma} \qquad t_{i}^{(a)} = \frac{\partial g^{\sigma}}{\partial V_{i}} = g_{,V_{i}}^{\sigma} \tag{13.51}$$

and the zero boundary displacement on ∂V_{v}. The equilibrium for the adjoint system is expressed by the virtual work theorem as

$$\int_{V}\boldsymbol{\sigma}^{(a)}\delta\dot{\boldsymbol{\varepsilon}} = \int_{\partial V}\mathbf{t}^{(a)}\,\delta\,\mathbf{V}\,d(\partial V) \tag{13.52}$$

or

$$\int_{V}\boldsymbol{\sigma}^{(a)}\,\mathbf{A}^{k}\,\delta\,\mathbf{V} = \int_{V}g_{,V}\delta\,\mathbf{V}\,dV + \int_{\partial V}g_{,V}^{\sigma}\delta\,\mathbf{V}\,d(\partial V) \tag{13.53}$$

By using the standard derivations, the stress equilibrium equations for the adjoint system operated by the above variational statement can be shown

$$\operatorname{div}\boldsymbol{\sigma}^{(a)} + g_{,\mathbf{V}} = 0 \tag{13.54}$$

$$\boldsymbol{\sigma}^{(a)}\cdot\mathbf{n} = g_{,\mathbf{V}}^{\sigma} \tag{13.55}$$

In the case $\delta \mathbf{V} = \bar{\delta} \mathbf{V}$, Eq. (13.53) becomes

$$\int_V \boldsymbol{\sigma}^{(a)} \bar{\delta} \dot{\boldsymbol{\varepsilon}} dV = \int_V g_{,V} \bar{\delta} \mathbf{V} dV + \int_{\partial V} \bar{\delta} \mathbf{V} d(\partial V) \tag{13.56}$$

The integrand on the left-hand side of Eq. (13.56) can be transformed as follows

$$\begin{aligned} \boldsymbol{\sigma}^{(a)} \bar{\delta} \dot{\boldsymbol{\varepsilon}} &= \mathbf{P} \dot{\boldsymbol{\varepsilon}}^{(a)} \bar{\delta} \dot{\boldsymbol{\varepsilon}} = \left(\bar{\delta} \boldsymbol{\sigma} - \bar{\bar{\delta}} \mathbf{P} \dot{\boldsymbol{\varepsilon}} \right) \dot{\boldsymbol{\varepsilon}}^{(a)} \\ &= \bar{\delta} \boldsymbol{\sigma} \dot{\boldsymbol{\varepsilon}}^{(a)} - \bar{\delta} \boldsymbol{\sigma} g_{,\boldsymbol{\sigma}} - \bar{\bar{\delta}} \mathbf{P} \dot{\boldsymbol{\varepsilon}} \dot{\boldsymbol{\varepsilon}}^{(a)} \end{aligned} \tag{13.57}$$

which yields

$$\bar{\delta} \boldsymbol{\sigma} \dot{\boldsymbol{\varepsilon}}^{(a)} = \boldsymbol{\sigma}^{(a)} \bar{\delta} \dot{\boldsymbol{\varepsilon}} + \bar{\delta} \boldsymbol{\sigma} g_{,\boldsymbol{\sigma}} + \bar{\bar{\delta}} \mathbf{P} \dot{\boldsymbol{\varepsilon}} \dot{\boldsymbol{\varepsilon}}^{(a)} \tag{13.58}$$

The total variation of the virtual work equation is

$$\int_V \bar{\delta} \boldsymbol{\sigma} \delta \dot{\boldsymbol{\varepsilon}} = \int_V \bar{\bar{\delta}} \mathbf{t} \, \delta \mathbf{V} d(\partial V) \tag{13.59}$$

which for $\delta \mathbf{V} = \mathbf{V}^{(a)}$. Since $\mathbf{V}^{(a)}$ is kinematically admissible, Eq. (13.59) becomes

$$\int_V \bar{\delta} \boldsymbol{\sigma} \dot{\boldsymbol{\varepsilon}}^{(a)} dV = \int_V \bar{\bar{\delta}} \mathbf{t} \mathbf{V}^{(a)} d(\partial V) \tag{13.60}$$

By Eqs. (13.58) and (13.60) we obtain

$$\begin{aligned} &\int_V \left[\boldsymbol{\sigma}^{(a)} \bar{\delta} \dot{\boldsymbol{\varepsilon}} + g_{,\boldsymbol{\sigma}} \bar{\delta} \boldsymbol{\sigma} + \bar{\bar{\delta}} \mathbf{P} \dot{\boldsymbol{\varepsilon}} \left(\dot{\boldsymbol{\varepsilon}}^{(a)} - g_{,\boldsymbol{\sigma}} \right) \right] dV \\ &= \int_{\partial V} \bar{\bar{\delta}} \mathbf{t} \mathbf{V}^{(a)} d(\partial V) \end{aligned} \tag{13.61}$$

which by Eq. (13.53) is

$$\begin{aligned} &\int_V \left(g_{,\boldsymbol{\sigma}} - \dot{\boldsymbol{\varepsilon}}^{(a)} \right) \bar{\bar{\delta}} \mathbf{P} \dot{\boldsymbol{\varepsilon}} dV + \int_V \bar{\bar{\delta}} \mathbf{t} \mathbf{V}^{(a)} d(\partial V) \\ &= \int_V \left(g_{,\boldsymbol{\sigma}} \bar{\delta} \sigma + g_{,V} \bar{\delta} \mathbf{V} \right) dV + \int_{\partial V} g_{,V}^{\sigma} \bar{\delta} \mathbf{V} d(\partial V) \end{aligned} \tag{13.62}$$

The left-hand side of Eq. (13.62) can be computed directly provided the solution to both the equilibrium and adjoint problems have been found and the right-hand side matches exactly the value of $\tilde{\delta} \mathcal{G}$. The fundamental relationship of the adjoint system method is

$$\bar{\delta} \mathcal{G} = \bar{\bar{\delta}} \mathcal{G} + \int_V \left(g_{,\boldsymbol{\sigma}} - \dot{\boldsymbol{\varepsilon}}^{(a)} \right) \bar{\bar{\delta}} \mathbf{P} \dot{\boldsymbol{\varepsilon}} dV + \int_{\partial V} \bar{\bar{\delta}} \mathbf{t} \mathbf{V}^{(a)} d(\partial V) \tag{13.63}$$

By (13.63) the sensitivity function $\dfrac{d\mathcal{G}}{d\mathbf{h}}$ is computed by solving two boundary-value problems that are governed by the same differential operator but different right-hand side terms. By relations

$$\bar{\bar{\delta}} g = \frac{\partial g}{\partial \mathbf{h}} \delta \mathbf{h} + \frac{\partial g}{\partial \boldsymbol{\sigma}} \partial \boldsymbol{\sigma} \tag{13.64}$$

$$\tilde{\delta} g = \frac{\partial g}{\partial \boldsymbol{\sigma}} \tilde{\delta} \boldsymbol{\sigma} \tag{13.65}$$

The expression (13.62) can be employed in the form

$$\begin{aligned} &\int_V -\dot{\boldsymbol{\varepsilon}}^{(a)} \bar{\bar{\delta}} \mathbf{P} \dot{\boldsymbol{\varepsilon}} dV + \int_{\partial V} \bar{\bar{\delta}} \mathbf{t} \mathbf{V}^{(a)} d(\partial V) \\ &\quad = \int_V \left(g_{,\boldsymbol{\sigma}} \tilde{\delta} \boldsymbol{\sigma} + g_{,V} \bar{\delta} V \right) dV + \int_{\partial V} g_{,V}^{\sigma} \bar{\delta} V d(\partial V) \end{aligned} \tag{13.66}$$

which leads to the fundamental adjoint system method relation

$$\bar{\delta} \mathcal{G} = \partial \mathcal{G} - \int_V \dot{\boldsymbol{\varepsilon}}^{(a)} \bar{\bar{\delta}} \mathbf{P} \dot{\boldsymbol{\varepsilon}} dV + \int_V \bar{\bar{\delta}} \mathbf{t} \mathbf{V}^{(a)} d(\partial V) \tag{13.67}$$

The expression (13.67) can be written in the equivalent form as

$$\frac{d\mathcal{G}}{d\mathbf{h}} = \frac{\partial \mathcal{G}}{\partial \mathbf{h}} - \int_V \frac{\partial \mathbf{P}}{\partial \mathbf{h}} \mathbf{A}^m \mathbf{V}^{(a)} \mathbf{A}^n \mathbf{V} \, dV + \int_{\partial V} \frac{d\mathbf{t}}{d\mathbf{h}} \mathbf{V}^{(a)} d(\partial V) \tag{13.68}$$

13.6 Shape Sensitivity

The two basic methods of shape design sensitivity analysis are the material derivative approach and the control volume/reference domain approach. The discussion of the material derivative approach is omitted in this section. In this approach, the material derivative concept of continuum mechanics described in Section 1.9 is used to obtain variations of the field variables. Also, variations of the volume and surface integrals over a variable domain from the calculus of variations are used to obtain the design sensitivity expression for a response functional. In the control volume approach, all of the quantities and integrals are first transformed to a fixed reference domain. Then variations are taken to develop the design sensitivity expression. The material derivative approach is equivalent to the control volume approach.

13.6.1 The Control Volume Approach

In order to express any previous reference coordinates in terms of the design variables and the current reference coordinates $^r\mathbf{x}$ we define the mapping $\mathbf{x} = \mathbf{x}\,(^r\mathbf{x}\,,\mathbf{h})$. It enables us to calculate the Jacobian transformation matrix $F = F_{ij} = \partial x_i/\partial^r x_j$ and to relate the volume and surface differential elements between the actual and the reference configurations as

$$dV = dV^r J \tag{13.69}$$

$$d(\partial V) = d(\partial V^r)J_\partial \tag{13.70}$$

where $J = \|\mathbf{F}\|, J_\partial = \|\mathbf{F}^{-T}\,\mathbf{n}\|$ and $\mathbf{n}$ is the unit outward normal to the previous reference boundary. Substituting the transformations to the reference domain the virtual work theorem (12.9), the strain tensor (1.19) and its arbitrary variation become

$$\int_{\Omega^r}(\boldsymbol{\sigma}^T\delta\dot{\boldsymbol{\varepsilon}})J dV^r - \int_{\partial\Omega_t^r}\mathbf{t}^T\,\delta\,\mathbf{V} J_\partial d(\partial V^r) = 0 \tag{13.71}$$

where

$$\dot{\varepsilon}_{ij}(V_k) = \frac{1}{2}\left(\frac{\partial V_i}{\partial^r x_k}\frac{\partial^r x_k}{\partial x_j} + \frac{\partial V_j}{\partial^r x_k}\frac{\partial^r x_k}{\partial x_i}\right) \tag{13.72}$$

$$\delta(\dot{\varepsilon}_{ij}(V_k)) = \dot{\varepsilon}_{ij}(\delta V_k) = \frac{1}{2}\left(\frac{\partial(\delta V_i)}{\partial^r x_k}\frac{\partial^r x_k}{\partial x_j} + \frac{\partial(\delta V_j)}{\partial^r x_k}\frac{\partial^r x_k}{\partial x_i}\right) \tag{13.73}$$

Note that the strain rate tensor and its arbitrary variation (13.73) now depend explicitly on the design variables by the transformation matrix $\partial^r x_k/\partial x_i$. They also depend implicitly on the design variables because $V_{j,k}$ depends in that way.

13.6.2 Design Sensitivity Analysis

Consider a response functional subject to sensitivity analysis

$$\mathcal{G} = \int_V g(\boldsymbol{\sigma}, \dot{\boldsymbol{\varepsilon}}, \mathbf{V}, \mathbf{h})dV + \int_{\partial V_V} g^v(\hat{\mathbf{V}}, \mathbf{t}, \mathbf{h})d(\partial V_v) + \int_{\partial V_\sigma} g^{(\sigma)}(\mathbf{V}, \hat{\mathbf{t}}, \mathbf{h})d(\partial V_\sigma) \tag{13.74}$$

where $\hat{\mathbf{V}}$ and $\hat{\mathbf{t}}$ are the specified velocity and traction vectors imposed on the boundaries ∂V_v and ∂V_σ, respectively. The integrands in Eq. (13.74) are assumed to be known functions of their indicated arguments.

After substituting the design variation of the stress-strain rate and strain velocity relations to the design variation of the virtual work theorem, the design variation of the velocity field can be computed. The design variations of the stresses, strains, and the response functional $\mathcal{G}$ can be then computed using the velocity variations.

The notations $\bar{\delta}, \bar{\bar{\delta}}$ and $\tilde{\delta}$ are used to represent the total, explicit and implicit design variations, respectively. Thus $\bar{\delta}\mathcal{G}$ will represent

$$\bar{\delta}\mathcal{G} = \left(\frac{D\mathcal{G}}{D\mathbf{h}}\right)^T \delta\mathbf{h} \tag{13.75}$$

where $D\mathcal{G}/D\mathbf{h}$ is the desired design gradient. The total design variation of the response functional in Eq. (13.74) gives

$$\begin{aligned}\bar{\delta}\mathcal{G} = &\int_V \left(\bar{\delta}g\mathrm{J} + g\bar{\bar{\delta}}\mathrm{J}\right)dV^r + \int_{\partial V_v}\left(\bar{\delta}g^v \mathrm{J}_\partial + g^v\bar{\bar{\delta}}\mathrm{J}_\partial\right)d(\partial V_v)^r \\ &+ \int_{\partial V_\sigma}\left(\bar{\delta}g^{(\sigma)}\mathrm{J}_\partial + g^{(\sigma)}\bar{\bar{\delta}}\mathrm{J}_\partial\right)d(\partial V_\sigma)^r\end{aligned} \tag{13.76}$$

$$\bar{\delta}g = g_{,\boldsymbol{\sigma}}\bar{\delta}\boldsymbol{\sigma} + g_{,\dot{\boldsymbol{\varepsilon}}}\bar{\delta}\dot{\boldsymbol{\varepsilon}} + g_{,\mathbf{v}}\bar{\delta}\mathbf{V} + g_{,\mathbf{h}}\bar{\delta}\mathbf{h} \tag{13.77}$$

$$\bar{\delta}g^v = g^v_{,\hat{v}}\bar{\delta}\hat{\mathbf{V}} + g^v_{,t}\bar{\delta}\mathbf{t} + g^v_{,h}\delta\mathbf{h} \tag{13.78}$$

$$\bar{\delta}g^{(\sigma)} = g^\sigma_{,\hat{v}}\bar{\delta}\mathbf{V} + g^\sigma_{,t}\bar{\delta}\mathrm{t}_i + g^\sigma_{,h}\delta\mathbf{h} \tag{13.79}$$

where the foot that J and J_∂ depend only explicitly on design has been used. The total design variations of the stress and strain tensors are given as

$$\bar{\delta}\boldsymbol{\sigma} = \tilde{\delta}\boldsymbol{\sigma} + \bar{\bar{\delta}}\boldsymbol{\sigma} \tag{13.80}$$

$$\bar{\delta}\boldsymbol{\varepsilon} = \tilde{\delta}\boldsymbol{\varepsilon} + \bar{\bar{\delta}}\boldsymbol{\varepsilon} \tag{13.81}$$

13.6.3 Variation of Variables

Variation of variables can be expressed in terms of implicit and explicit forms as

$$\begin{gathered}\bar{\delta}\mathbf{x} = \bar{\bar{\delta}}\mathbf{x} \quad \bar{\delta}\mathbf{F} = \bar{\bar{\delta}}\mathbf{F} \quad \bar{\delta}\mathbf{V} = \tilde{\delta}\mathbf{V} \\ \bar{\delta}(\delta\mathbf{V}) = \bar{\bar{\delta}}(\delta\mathbf{V})\end{gathered} \tag{13.82}$$

By (13.72), (13.73) and (13.82) the implicit variation of the strain rate is

$$\begin{aligned}\bar{\delta}\dot{\varepsilon}_{ij} &= \frac{1}{2}\left[\tilde{\delta}\left(\frac{\partial V_i}{\partial^r x_k}\right)\frac{\partial^r x_k}{\partial x_j}+\tilde{\delta}\left(\frac{\partial V_j}{\partial^r x_k}\right)\frac{\partial^r x_k}{\partial x_i}\right]= \\ &= \frac{1}{2}\left[\frac{\partial(\tilde{\delta}V_i)}{\partial^r x_k}\frac{\partial^r x_k}{\partial x_j}+\frac{\partial(\tilde{\delta}V_j)}{\partial^r x_k}\frac{\partial^r x_k}{\partial x_i}\right]\end{aligned} \tag{13.83}$$

so

$$\tilde{\delta}\dot{\varepsilon}_{ij} = \dot{\varepsilon}_{ij}\left(\tilde{\delta}\,\mathbf{V}\right) \tag{13.84}$$

provided the variation is smooth enough to allow the permutation of the differentials. The explicit strain rate variation reads

$$\bar{\bar{\delta}}\dot{\varepsilon}_{ij} = \frac{1}{2}\left[\frac{\partial V_i}{\partial^r x_k}\bar{\bar{\delta}}\left(\frac{\partial^r x_k}{\partial x_j}\right)+\frac{\partial V_j}{\partial^r x_k}\bar{\bar{\delta}}\left(\frac{\partial^r x_k}{\partial x_i}\right)\right] \tag{13.85}$$

by Eq. (13.82) and $\bar{\delta}\left(\partial^r x_k / \partial x_j\right) = \bar{\bar{\delta}}\left(\partial^r x_k / \partial x_j\right)$.

From the formula for the derivative of a matrix in terms of the derivative of its inverse

$$\frac{\partial F_{ji}^{-1}}{\partial \mathbf{h}} = -F_{jk}^{-1}\frac{\partial F_{kl}}{\partial \mathbf{h}}F_{li}^{-1} \tag{13.86}$$

we get

$$\bar{\bar{\delta}}\left(\frac{\partial^r x_i}{\partial x_j}\right) = -\frac{\partial^r x_i}{\partial x_k}\frac{\partial^r x_l}{\partial x_j}\bar{\bar{\delta}}\left(\frac{\partial x_k}{\partial^r x_l}\right) \tag{13.87}$$

In the case of the stress tensor, by Eq. (12.12)

$$\tilde{\delta}\boldsymbol{\sigma} = \tilde{\delta}(\mathbf{P})\dot{\boldsymbol{\varepsilon}} + \mathbf{P}\,\tilde{\delta}\dot{\boldsymbol{\varepsilon}} \tag{13.88}$$

$$\bar{\bar{\delta}}\boldsymbol{\sigma} = \bar{\bar{\delta}}(\mathbf{P})\dot{\boldsymbol{\varepsilon}} + \mathbf{P}\,\bar{\bar{\delta}}\dot{\boldsymbol{\varepsilon}} \tag{13.89}$$

Similarly to (13.83) and (13.85) we can write

$$\tilde{\delta}(\delta\dot{\varepsilon}_{ij}) = \frac{1}{2}\left[\bar{\delta}\left(\frac{\partial(\delta V_i)}{\partial^r x_k}\right)\frac{\partial^r x_k}{\partial x_j}+\bar{\delta}\left(\frac{\partial(\delta V_j)}{\partial^r x_k}\right)\frac{\partial^r x_k}{\partial x_i}\right] \tag{13.90}$$

$$\bar{\bar{\delta}}(\delta\dot{\varepsilon}_{ij}) = \frac{1}{2}\left[\frac{\partial(\delta V_i)}{\partial^r x_k}\bar{\bar{\delta}}\left(\frac{\partial^r x_k}{\partial x_j}\right)+\frac{\partial(\delta V_j)}{\partial^r x_k}\bar{\bar{\delta}}\left(\frac{\partial^r x_k}{\partial x_i}\right)\right] \tag{13.91}$$

13.6.4 Sensitivities in the Virtual Work Theorem

The design variations of the virtual work theorem (10.15) is

$$\int_{V^r} \bar{\delta}\left(\boldsymbol{\sigma}^T \delta\dot{\boldsymbol{\varepsilon}} J\right) dV^r - \int_{\partial V_t^r} \bar{\delta}\left(\mathbf{t}^T \delta \mathbf{V} J_\partial\right) d\left(\partial V^r\right) = 0 \tag{13.92}$$

and by Eq. (13.82) we have

$$\begin{aligned} &\int_{V^r} \left[\bar{\delta}\boldsymbol{\sigma}^T \delta\dot{\boldsymbol{\varepsilon}} J + \boldsymbol{\sigma}^T \bar{\delta}(\delta\dot{\boldsymbol{\varepsilon}}) J + \boldsymbol{\sigma}^T \delta\dot{\boldsymbol{\varepsilon}} \bar{\delta} J\right] dV^r \\ &- \int_{\partial V_t^r} \left[\bar{\bar{\delta}}\left(\mathbf{t}^T J_\partial\right)\delta v + \mathbf{t}^t J_\partial \bar{\delta}(\delta \mathbf{V})\right] d\left(\partial V^r\right) = 0 \end{aligned} \tag{13.93}$$

and separating the implicit variation from the explicit one

$$\begin{aligned} &\left\{\int_{V^r} \boldsymbol{\sigma}^T \tilde{\delta}(\delta\dot{\boldsymbol{\varepsilon}}) J dV^r - \int_{\partial V^r} \mathbf{t}^T \bar{\delta}(\delta \mathbf{V}) J_\partial d\left(\partial V^r\right)\right\} \\ &+ \int_{V^r} \left[\bar{\delta}\boldsymbol{\sigma}^T \delta\dot{\boldsymbol{\varepsilon}} J + \boldsymbol{\sigma}^T \bar{\bar{\delta}}(\delta\dot{\boldsymbol{\varepsilon}}) J + \boldsymbol{\sigma}^T \delta\dot{\boldsymbol{\varepsilon}} \bar{\bar{\delta}} J\right] dV^r - \int_{\partial V^r} \bar{\bar{\delta}}\left(\mathbf{t}^T J_\partial\right)\delta \mathbf{V}\, d\left(\partial V^r\right) = 0 \end{aligned} \tag{13.94}$$

Assume the arbitrary variation $\bar{\delta}(\delta \mathbf{V})$ to be a kinematically admissible velocity field satisfying the smoothness requirements to ensure continuous derivatives. Since $\bar{\delta}(\delta\dot{\boldsymbol{\varepsilon}})$ is assumed to be compatible with $\bar{\delta}(\delta \mathbf{V})$ we get that $\tilde{\delta}(\delta\dot{\boldsymbol{\varepsilon}}) = \dot{\boldsymbol{\varepsilon}}\left(\bar{\delta}(\delta \mathbf{V})\right)$. Therefore $\tilde{\delta}(\delta\dot{\boldsymbol{\varepsilon}})$ is compatible with $\bar{\delta}(\delta \mathbf{V})$ and since $\bar{\delta}(\delta \mathbf{V})$ is arbitrary, the form in braces in Eq. (13.94) represents the virtual work theorem (10.15) and thus vanishes. By separating the stress variations we have

$$\begin{aligned} \int_{V^r} \tilde{\delta}\boldsymbol{\sigma}^T \delta\dot{\boldsymbol{\varepsilon}}\, J dV^r = &\int_{\partial V^r} \bar{\bar{\delta}}\, \mathbf{t}^T \left(J_\partial\right)\delta \mathbf{V}\, d\left(\partial V^r\right) \\ &- \int_{\partial V^r} \bar{\bar{\delta}}\left(\mathbf{t}^T J_\partial\right)\delta \mathbf{V}\, d\left(\partial V^r\right) \\ &- \int_{V^r} \left[\bar{\bar{\delta}}\boldsymbol{\sigma}^T \delta\dot{\boldsymbol{\varepsilon}} J + \boldsymbol{\sigma}^T \bar{\bar{\delta}}(\delta\dot{\boldsymbol{\varepsilon}}) J + \boldsymbol{\sigma}^T \delta\dot{\boldsymbol{\varepsilon}} \bar{\bar{\delta}} J\right] dV^r \end{aligned} \tag{13.95}$$

13.7 Sensitivity in Elasto-Plasticity

The virtual work theorem (13.71) in incremental form is

$$\int_{V^r} \boldsymbol{\sigma}_n \boldsymbol{\varepsilon}(\delta \mathbf{u}_n) J\, dV^r = \int_{V^r} \mathbf{t}_n\, \delta \mathbf{u}_n\, J_\partial d(\partial V^r) \tag{13.96}$$

where n denotes the step of the analysis. In the direct variational method, the total design variation of all the field variables is calculated by taking the design variation of the virtual work equations and the constitutive equations. The total design variations of the strain and its arbitrary variation are given as

$$\begin{aligned} \overline{\delta}\boldsymbol{\varepsilon}_n &= \overline{\overline{\delta}}\boldsymbol{\varepsilon}_n + \widetilde{\delta}\boldsymbol{\varepsilon}_n \\ \widetilde{\delta}\boldsymbol{\varepsilon}_n &= \boldsymbol{\varepsilon}(\widetilde{\delta}\, \mathbf{u}_n) \\ \overline{\delta}(\delta\boldsymbol{\varepsilon}_n) &= \varepsilon\left[\widetilde{\delta}(\delta\, \mathbf{u}_n)\right] + \overline{\overline{\delta}}\boldsymbol{\varepsilon}_n(\delta\, \mathbf{u}_n) \end{aligned} \tag{13.97}$$

The total design variations of the constitutive equation (6..21) in isothermal case are

$$\begin{aligned} \overline{\delta}\boldsymbol{\sigma}_n &= \overline{\overline{\delta}}\boldsymbol{\sigma}_n + \widetilde{\delta}\boldsymbol{\sigma}_n \\ \widetilde{\delta}\boldsymbol{\sigma}_n &= \mathbf{C}\left[\boldsymbol{\varepsilon}(\widetilde{\delta}\, \mathbf{u}_n) - \overline{\delta}\varepsilon_n^p\right] \\ \overline{\overline{\delta}}\boldsymbol{\sigma}_n &= \overline{\overline{\delta}}\, \mathbf{C}(\boldsymbol{\varepsilon}_n - \boldsymbol{\varepsilon}_n^p) + \mathbf{C}\, \overline{\overline{\delta}}\boldsymbol{\varepsilon}_n \end{aligned} \tag{13.98}$$

In order to calculate $\overline{\delta}\, \mathbf{u}_n$, the total design variation of the virtual equation (13.96) is computed as

$$\int_{V^r} \widetilde{\delta}\boldsymbol{\sigma}_n \boldsymbol{\varepsilon}(\delta\, \mathbf{u}_n) J dV^r = \int_{\partial V^r} \overline{\overline{\delta}}(\mathbf{t}_n\, J_\partial)\delta\, \mathbf{u}_n\, J_\partial d(\partial V^r) \tag{13.99}$$

The sensitivity equation (13.99) has the same operators as in the incremental equilibrium equation (13.96). The left – hand side of the sensitivity equation depends on the solution variable $\overline{\delta}\dot{\boldsymbol{\varepsilon}}^p$. This quantity must be calculated by the flow rule for the assumed material model.

Part IX

Stochastic Metal Forming Process

14 Stochastic Viscoplasticity

14.1 Introduction

Stochastic methods have recently become an area of research in metal forming. As the name suggests, these methods combine two crucial methodologies developed to deal with problems of metal forming: analytical or numerical analysis with the stochastic one.

The stochastic analysis in the broadest sense refers to the explicit treatment of uncertainty in any quantity entering the corresponding deterministic analysis. The exact values of these quantities are usually unknown because they cannot be precisely measured.

The stochastic approach to metal forming problems is important not only because of random material parameters, but particularly because of boundary problems appearing in these processes. Contact problems die-workpiece have exceptional random character and lead to determine the boundary forces in the contact die-workpiece considering the random character of friction between them.

Existing uncertain variations in parameters may have significant effects on such fundamental final characteristics, as strain and stress distributions, and they must affect the final design. Useful analytical tools for performing analysis of workpieces with uncertain properties are provided by the theory of random fields, which is an outgrowth of the probability theory.

This section is limited to equations typical for metal forming i.e. rigid-viscoplasticity. The necessity of performing numerical calculations while analysing the stochastic behaviour of manufacturing processes has now been widely recognized. Since uncertainties appear in the operators of governing equations, systems of equations are nonlinear as functions of random variables. The distinguishing feature of the stochastic methods, which is based on the perturbation approach, is treating probabilistic nonlinear problems with deterministic computational techniques that take full advantage of the mathematical properties of linear and nonlinear operators.

This section offers a specific treatment of metal forming problems for which we can use the term probabilistic numerical techniques.

The basic goal set in this section is getting the reader acquainted with the theoretical fundamentals of the method. The difficulty for users of these formulations is certainly experienced in the apparent complexity of programming and using the method.

14.2 The Stochastic Virtual Energy Principle

In this point a stochastic virtual energy principle for rigid - viscoplastic processes is discussed. In formulation of the stochastic virtual energy principle a combination of the theorem of virtual work and the second-order perturbation technique is applied.

Consider a workpiece under load. Assume a set of R random fields which can represent randomness in the geometry and material parameters as yield stress, friction forces, etc., as well as time-in variant randomness in the external load $\mathbf{b}(x_k) = \{b_1(x_k)\ b_2(x_k)\ \ldots\ b_R(x_k)\}$, $k = 1, 2, 3$. Define the first two statistical moments for the random fields $b_r(x_k)$, $r = 1, 2, \ldots, R$,

$$E(b_r) = b_r^0 = \int_{-\infty}^{+\infty} b_r p_1(b_r)\, db_r \tag{14.1}$$

$$Cov(b_r, b_s) = S_b^{rs} = \int_{-\infty}^{+\infty}\int_{-\infty}^{+\infty} (b_r - b_r^0)(b_s - b_s^0) p_2(b_r, b_s)\, db_r db_s \tag{14.2}$$

$r, s = 1, 2, \ldots, R$, where $E(b_r)$ is the expected value (mean), of the random variable b_r, $Cov\ (b_r, b_s)$ is the covariance between b_r and b_s, and p_1 and p_2 is a one-dimensional and a two-dimensional probability density functions respectively. The definition (14.2) can be replaced by

$$S_b^{rs} = \alpha_{b_r} \alpha_{b_s} b_r^0 b_s^0 \mu_{b_r b_s} \tag{14.3}$$

with

$$\alpha_{b_r} = \left[\frac{Var(b_r)}{b_r^0}\right]^{\frac{1}{2}}$$

$$\mu_{b_r b_s} = \int_{-\infty}^{+\infty}\int_{-\infty}^{+\infty} b_r b_s p_2(b_r, b_s)\, db_r db_s \tag{14.4}$$

where $Var(b_r)$, $\mu_{b_r b_s}$ and α_{b_r} denote variance, correlation functions and the coefficients of variation, respectively.

The second-order perturbation technique applied to the stochastic approach of the virtual energy principle in metal forming of rigid-viscoplastic material involves expanding all the random field variables in the problem, i.e. proportionality moduli $\mathbf{P}(\mathbf{b}(\mathbf{x}), \mathbf{x})$ given by Eq. (12.12), boundary tractions $\mathbf{t}\ (\mathbf{b}(\mathbf{x}), \mathbf{x})$ and velocity $\mathbf{V}(\mathbf{b}(\mathbf{x}),\mathbf{x})$ about the spatial expectations of the random field variables $\mathbf{b}(\mathbf{x}) = \{b_r(\mathbf{x})\}$, denoted by $\mathbf{b}^0(\mathbf{x}) = \{b_r^0(\mathbf{x})\}$, via Taylor series with a given small parameter γ and retaining terms up to second order. These expansions are expressed as

$$\begin{aligned}
\mathbf{P}(\mathbf{b}(\mathbf{x}),\mathbf{x}) &= \mathbf{P}^0(\mathbf{b}^0(\mathbf{x}),\mathbf{x}) + \gamma\,\mathbf{P}^r(\mathbf{b}^0(\mathbf{x}),\mathbf{x})\,\Delta b_r(\mathbf{x}) \\
&\quad + \frac{1}{2}\gamma^2\,\mathbf{P}^{rs}(\mathbf{b}^0(\mathbf{x}),\mathbf{x})\Delta b_r(\mathbf{x})\,\Delta b_s(\mathbf{x}) \\
\mathbf{t}(\mathbf{b}(\mathbf{x}),\mathbf{x}) &= \mathbf{t}^0(\mathbf{b}^0(\mathbf{x}),\mathbf{x}) + \gamma\,\mathbf{t}^{,r}(\mathbf{b}^0(\mathbf{x}),\mathbf{x})\Delta b_r(\mathbf{x}) \\
&\quad + \frac{1}{2}\gamma^2\,\mathbf{t}^{,rs}(\mathbf{b}^0(\mathbf{x}),\mathbf{x})\Delta b_r(\mathbf{x})\,\Delta b_s(\mathbf{x}) \\
\mathbf{V}(\mathbf{b}(\mathbf{x}),\mathbf{x}) &= \mathbf{V}^0(\mathbf{b}^0(\mathbf{x}),\mathbf{x}) + \gamma\,\mathbf{V}^{,r}(\mathbf{b}^0(\mathbf{x}),\mathbf{x})\Delta b_r(\mathbf{x}) \\
&\quad + \frac{1}{2}\gamma^2\,\mathbf{V}^{,rs}(\mathbf{b}^0(\mathbf{x}),\mathbf{x})\Delta b_r(\mathbf{x})\,\Delta b_s(\mathbf{x})
\end{aligned} \tag{14.5}$$

where

$$\gamma \Delta b_r(\mathbf{x}) = \delta b_r(\mathbf{x}) = \gamma\left(b_r(\mathbf{x}) - b_r^0(\mathbf{x})\right) \tag{14.6}$$

is the first-order variation of $b_r(\mathbf{x})$ about $b_r^0(\mathbf{x})$, and

$$\begin{aligned}
\gamma^2 \Delta b_r(\mathbf{x})\Delta b_s(\mathbf{x}) &= \delta b_r(\mathbf{x})\ \delta b_s(\mathbf{x}) \\
&= \gamma^2\left(b_r(\mathbf{x}) - b_r^0(\mathbf{x})\right)\left(b_s(\mathbf{x}) - b_{sr}^0(\mathbf{x})\right)
\end{aligned} \tag{14.7}$$

is the second-order variation of $b_r(\mathbf{x})$ and $b_s(\mathbf{x})$ about $b_r^0(\mathbf{x})$ and $b_s^0(\mathbf{x})$, respectively. The notation $(.)^0$ represents the value of the functions taken at b_r^0 while $(.)^{,r}$ and $(.)^{,rs}$ stand for the first and second (mixed) partial derivatives with respect to the random field variables $b_r(\mathbf{x})$ evaluated at their expectations, respectively. By using the perturbation approach the expansions (13.5) are substituted into the virtual work theorem (10.15)

$$\int_V \mathbf{P}\,\mathbf{A}^m \mathbf{V} \mathbf{A}^n\,\delta\mathbf{V}\,dV = \int_V \mathbf{t}\,\delta\mathbf{V}\,d(\partial V) \tag{14.8}$$

where the following notations are adopted

$$\dot{\boldsymbol{\varepsilon}} = \mathbf{A}^k\mathbf{V} \qquad \dot{\varepsilon}_{ij} = A_{ijk}V_k \tag{14.9}$$

Collecting terms of equal orders the following zeroth-, first- and second-order virtual work principles are derived:

Zeroth-order (γ^0 terms, one equation)

$$\int_V \mathbf{P}^0\mathbf{A}^m\mathbf{V}^0\mathbf{A}^n\,\delta\mathbf{V}\,dV = \int_V \mathbf{t}^0\,\delta\mathbf{V}\,d(\partial V) \tag{14.10}$$

First-order (γ^1 terms, R equations)

$$\int_V \mathbf{P}^0 \mathbf{A}^m \mathbf{V}^{,r} \mathbf{A}^n \, \delta \mathbf{V} \, dV = \int_{\partial V} \mathbf{t}^{,r} \, \delta \mathbf{V} \, d(\partial V) - \int_V \mathbf{P}^{,r} \mathbf{A}^m \mathbf{V}^0_m \mathbf{A}^n \, \delta \mathbf{V} \, dV \tag{14.11}$$

Second-order (γ^2 terms, one equation)

$$\begin{aligned} \int_V \mathbf{P}^0 \mathbf{A}^m \mathbf{V}^{,rs} S_b^{rs} \mathbf{A}^n \, \delta \mathbf{V}_n \, dV &= \int_{\partial V} t^{,rs} \delta \mathbf{V} \, d(\partial V) \\ &- \int_V \left(2\,\mathbf{P}^{,r} \mathbf{A}^m \mathbf{V}^{,s} + \mathbf{P}^{,rs} \mathbf{A}^m \mathbf{V}^0 \right) S_b^{rs} \mathbf{A}^n \, \delta \mathbf{V} \, dV \end{aligned} \tag{14.12}$$

The function $\delta\mathbf{V}$ should satisfy the kinematic boundary conditions. The second-order equation is obtained by multiplying the R-variate probability density function $p_R(b_1, b_2, ..., b_R)$ by the γ^2-terms and integrating over the domain of the random field variables $\mathbf{b}(\mathbf{x})$. For instance, the γ^2-term involving $\mathbf{t}^{,rs}\left(\mathbf{b}^0(\mathbf{x}), \mathbf{x}\right)$ reads

$$\begin{aligned} &\int_{-\infty}^{+\infty} \left[\int_{\partial V_\sigma} \gamma^2 \, \mathbf{t}^{,rs}\left(\mathbf{b}^0(\mathbf{x}), \mathbf{x}\right) \Delta b_r(\mathbf{x}) \, \Delta b_s(\mathbf{x}) \, \delta \mathbf{V} \, d(\partial V) \right] p_R\left(\mathbf{b}(\mathbf{x})\right) d\mathbf{b} \\ &= \gamma^2 \int_{\partial \Omega_\sigma} \mathbf{t}^{,rs} S_b^{rs} \delta \mathbf{V} \, d(\partial V) \end{aligned} \tag{14.13}$$

The occurrence of the double sums $(.)^{,rs} S_b^{rs}$ and $(.)^{,r} (.)^{,s} S_b^{rs}$ in the formulation enables us to deal with only one equation (14.12), instead of R(R+1)/2 second-order equations apparently required because Eq. (14.5) is symmetric with respect to r and s, which is particularly important in a numerical approach. Equation (14.10) being identical to the deterministic virtual work theorem for the rigid-viscoplastic material model can serve as the basis to obtain the zeroth-order velocities $\mathbf{V}^0(\mathbf{x})$. The higher-order terms $\mathbf{V}^{,r}(\mathbf{x})$ and $\mathbf{V}^{,rs}(\mathbf{x})$ can be evaluated from Eqs. (14.11) and (14.12). All terms involved on the left-hand side of Eqs. (14.10) - (14.12) are identical except for the unknown functions. The probabilistic characteristics of the problem can be translated into the right-hand sides of the equation. The expressions (14.10) – (14.12) solved for $\mathbf{V}^0(\mathbf{x})$, $\mathbf{V}^{,r}(\mathbf{x})$ and $\mathbf{V}^{,rs}(\mathbf{x})$ of the random velocity field $\mathbf{V}\left(\mathbf{b}(\mathbf{x}), \mathbf{x}\right)$ may be calculated, for a given γ. Setting $\gamma = 0$ yields the deterministic solution. The solution in our case is obtained by setting $\gamma = 1$ which, stipulates that the random field variables $\mathbf{b}(\mathbf{x})$ is small. By introducing the expanded equation $(14.5)_3$ into the expression for the mean value of the random velocity field $\mathbf{V}\left(\mathbf{b}(\mathbf{x}), \mathbf{x}\right)$

$$E\left(\mathbf{V}\left(\mathbf{b}(\mathbf{x}), \mathbf{x}\right)\right) = \int_{-\infty}^{+\infty} \mathbf{V}\left(\mathbf{b}(\mathbf{x}), \mathbf{x}\right) p_R\left(\mathbf{b}(\mathbf{x})\right) d\mathbf{b} \tag{14.14}$$

The second-order estimate of velocity field **V**(**b**(**x**), **x**) is

$$\mathrm{E}\big(\mathbf{V}(\mathbf{b}(\mathbf{x}),\mathbf{x})\big)=\mathbf{V}^{0}(\mathbf{x})+\frac{1}{2}\mathbf{V}^{,rs}(\mathbf{x})\,\mathrm{S}_{\mathrm{b}}^{\mathrm{rs}} \tag{14.15}$$

since

$$\begin{aligned}\mathrm{E}(\mathbf{V})&=\int_{-\infty}^{+\infty}\Big\{\mathbf{V}^{0}\big(\mathbf{b}^{0}(\mathbf{x}),\mathbf{x}\big)+\mathbf{V}^{,r}\big(\mathbf{b}^{0}(\mathbf{x}),\mathbf{x}\big)\Delta \mathrm{b}_{r}(\mathbf{x})\\&\quad+\frac{1}{2}\mathbf{V}^{,rs}\big(\mathbf{b}^{0}(\mathbf{x}),\mathbf{x}\big)\Delta \mathrm{b}_{r}(\mathbf{x})\Delta \mathrm{b}_{s}(\mathbf{x})\Big\}\mathrm{p}_{\mathrm{R}}\big(\mathbf{b}(\mathbf{x})\big)\mathrm{d}\,\mathbf{b}\\&=\mathbf{V}^{0}\big(\mathbf{b}^{0}(\mathbf{x}),\mathbf{x}\big)\underbrace{\int_{-\infty}^{+\infty}\mathrm{p}_{\mathrm{R}}\big(\mathbf{b}(\mathbf{x})\big)\mathrm{d}\,\mathbf{b}}_{=1}+\mathbf{V}^{,r}\big(\mathbf{b}^{0}(\mathbf{x}),\mathbf{x}\big)\underbrace{\int_{-\infty}^{+\infty}\Delta \mathrm{b}_{r}(\mathbf{x})\mathrm{p}_{\mathrm{R}}\big(\mathbf{b}(\mathbf{x})\big)\mathrm{d}\,\mathbf{b}}_{=0}\\&\quad+\frac{1}{2}\mathbf{V}^{,rs}\big(\mathbf{b}^{0}(\mathbf{x}),\mathbf{x}\big)\int_{-\infty}^{+\infty}\Delta \mathrm{b}_{r}(\mathbf{x})\Delta \mathrm{b}_{s}(\mathbf{x})\mathrm{p}_{\mathrm{R}}\big(\mathbf{b}(\mathbf{x})\big)\mathrm{d}\,\mathbf{b}\end{aligned} \tag{14.16}$$

If the first-order accuracy of the velocity estimation is required, then Eq. (14.15) reduces to

$$\mathrm{E}\big(\mathbf{V}^{,rs}\big(\mathbf{b}^{0}(\mathbf{x}),\mathbf{x}\big)\big)=\mathbf{V}^{0}(\mathbf{x}) \tag{14.17}$$

The first-order cross-covariances of $\mathbf{V}\big(\mathbf{b}(\mathbf{x}^{1}),\mathbf{x}^{1}\big)$ and $\mathbf{V}\big(\mathbf{b}(\mathbf{x}^{2}),\mathbf{x}^{2}\big)$ are determined by substituting the second-order expansion of the random velocity field $\mathbf{V}\big(\mathbf{b}(\mathbf{x}),\mathbf{x}\big)$ into the expression for the cross-covariance

$$\begin{aligned}&\mathrm{Cov}\big(\mathbf{V}\big(\mathbf{b}(\mathbf{x}^{1}),\mathbf{x}^{1}\big),\mathbf{V}\big(\mathbf{b}(\mathbf{x}^{2}),\mathbf{x}^{2}\big)\big)=\mathrm{S}_{\mathrm{v}}^{\mathrm{ij}}(\mathbf{x}^{1},\mathbf{x}^{2})\\&=\int_{-\infty}^{+\infty}\Big\{\mathbf{V}\big(\mathbf{b}(\mathbf{x}^{1}),\mathbf{x}^{1}\big)-\mathrm{E}\big(\mathbf{V}\big(\mathbf{b}(\mathbf{x}^{1}),\mathbf{x}^{1}\big)\big)\Big\}\\&\times\Big\{\mathbf{V}\big(\mathbf{b}(\mathbf{x}^{2}),\mathbf{x}^{2}\big)-\mathrm{E}\big(\mathbf{V}^{*}\big(\mathbf{b}(\mathbf{x}^{2}),\mathbf{x}^{2}\big)\big)\Big\}\mathrm{p}_{\mathrm{R}}\big(\mathbf{b}(\mathbf{x})\big)\,\mathrm{d}\,\mathbf{b}\end{aligned} \tag{14.18}$$

to get

$$\mathrm{S}_{\mathrm{v}}^{\mathrm{ij}}\big(\mathbf{x}^{1},\mathbf{x}^{2}\big)=\mathbf{V}^{,r}\big(\mathbf{x}^{1}\big)\,\mathbf{V}^{,s}\big(\mathbf{x}^{2}\big)\,\mathrm{S}_{\mathrm{b}}^{\mathrm{rs}} \tag{14.19}$$

The strain rate probabilistic characteristics are

- second-order accurate mean value

$$\begin{aligned}&\mathrm{E}\big(\dot{\boldsymbol{\varepsilon}}(\mathbf{b}(\mathbf{x}),\mathbf{x})\big)=\mathrm{E}\Big[\frac{1}{2}\big(\mathrm{grad}\,\mathbf{V}(\mathbf{b}(\mathbf{x}),\mathbf{x})+{}^{\mathrm{T}}\mathrm{grad}\,\mathbf{V}(\mathbf{b}(\mathbf{x}),\mathbf{x})\big)\Big]\\&=\frac{1}{2}\Big[\mathrm{grad}\,\mathbf{V}^{0}(\mathbf{x})+{}^{\mathrm{T}}\mathrm{grad}\,\mathbf{V}^{0}(\mathbf{x})+\frac{1}{2}\big(\mathrm{grad}\,\mathbf{V}^{,rs}(\mathbf{x})+{}^{\mathrm{T}}\mathrm{grad}\,\mathbf{V}^{,rs}(\mathbf{x})\big)\mathrm{S}_{\mathrm{b}}^{\mathrm{rs}}\Big]\\&=\dot{\boldsymbol{\varepsilon}}^{0}(\mathbf{x})+\frac{1}{2}\dot{\boldsymbol{\varepsilon}}^{,rs}(\mathbf{x})\,\mathrm{S}_{\mathrm{b}}^{\mathrm{rs}}\end{aligned} \tag{14.20}$$

where

$$\begin{aligned}\dot{\boldsymbol{\varepsilon}}\big(\mathbf{b}(\mathbf{x}),\mathbf{x}\big) &= \dot{\boldsymbol{\varepsilon}}^{0}\big(\mathbf{b}^{0}(\mathbf{x}),\mathbf{x}\big)+\dot{\boldsymbol{\varepsilon}}^{,r}\big(\mathbf{b}^{0}(\mathbf{x}),\mathbf{x}\big)\Delta b_{r}(\mathbf{x}) \\ &\quad +\frac{1}{2}\dot{\boldsymbol{\varepsilon}}^{,rs}\big(\mathbf{b}^{0}(\mathbf{x}),\mathbf{x}\big)\Delta b_{r}(\mathbf{x})\,\Delta b_{s}(\mathbf{x})\end{aligned} \tag{14.21}$$

– first-order accurate cross-covariance

$$\begin{aligned}&\mathrm{Cov}\big(\dot{\boldsymbol{\varepsilon}}\big(\mathbf{b}(\mathbf{x}^{1}),\mathbf{x}^{1}\big),\dot{\boldsymbol{\varepsilon}}\big(\mathbf{b}^{0}(\mathbf{x}^{2}),\mathbf{x}^{2}\big)\big)=S_{\varepsilon}^{ijkl}(\mathbf{x}^{1},\mathbf{x}^{2}) \\ &=\frac{1}{4}\big(\mathrm{grad}\,\mathbf{V}^{,r}(\mathbf{x}^{1})+{}^{T}\mathrm{grad}\,\mathbf{V}^{,r}(\mathbf{x}^{1})\big)\big(\mathrm{grad}\,\mathbf{V}^{,s}(\mathbf{x}^{2})+{}^{T}\mathrm{grad}\,\mathbf{V}^{,s}(\mathbf{x}^{2})\big)S_{b}^{rs} \\ &=\boldsymbol{\varepsilon}^{,r}\big(\mathbf{x}^{1}\big)\boldsymbol{\varepsilon}^{,s}\big(\mathbf{x}^{2}\big)S_{b}^{rs}\end{aligned} \tag{14.22}$$

The first two moments for the stresses are

$$\boldsymbol{\sigma}=\mathbf{P}\dot{\boldsymbol{\varepsilon}}=\left(\mathbf{P}^{0}+\mathbf{P}^{,r}\,\Delta b_{r}+\frac{1}{2}\mathbf{P}^{,rs}\,\Delta b_{r}\Delta b_{s}\right)\left(\dot{\boldsymbol{\varepsilon}}^{0}+\dot{\boldsymbol{\varepsilon}}^{,u}\Delta b_{v}+\frac{1}{2}\dot{\boldsymbol{\varepsilon}}^{,uv}\Delta b_{u}\Delta b_{v}\right) \tag{14.23}$$

By employing Eq. (14.23) in stress equations similar to Eqs. (14.14) and (14.18) and neglecting the variations of an order higher than two we arrive at second-order accurate mean value

$$\begin{aligned}\mathrm{E}\big(\boldsymbol{\sigma}\big(\mathbf{b}(\mathbf{x}),\mathbf{x}\big)\big) &= \mathbf{P}^{0}(\mathbf{x})\dot{\boldsymbol{\varepsilon}}^{0}(\mathbf{x})+\frac{1}{2}\big(\mathbf{P}^{,rs}(\mathbf{x})\dot{\boldsymbol{\varepsilon}}^{0}(\mathbf{x}) \\ &\quad +2\,\mathbf{P}^{,r}(\mathbf{x})\dot{\boldsymbol{\varepsilon}}^{,s}(\mathbf{x})+\mathbf{P}^{0}(\mathbf{x})\dot{\boldsymbol{\varepsilon}}^{,rs}(\mathbf{x})\big)\,S_{b}^{rs}\end{aligned} \tag{14.24}$$

and the first-order accurate cross-covariance

$$\begin{aligned}&\mathrm{Cov}\big(\boldsymbol{\sigma}\big[\mathbf{b}(\mathbf{x}^{1}),\mathbf{x}^{1}\big],\boldsymbol{\sigma}^{*}\big[\mathbf{b}(\mathbf{x}^{2}),\mathbf{x}^{2}\big]\big)=S_{\sigma}^{ijkl}\big(\mathbf{x}^{1},\mathbf{x}^{2}\big) \\ &=\Big[P_{ijmn}^{,r}\big(x_{k}^{1}\big)P_{kl\tilde{m}\tilde{n}}^{,s}\big(x_{k}^{2}\big)\dot{\varepsilon}_{mn}^{0}\big(x_{k}^{1}\big)\dot{\varepsilon}_{\tilde{m}\tilde{n}}^{0}\big(x_{k}^{2}\big) \\ &\quad+P_{ijmn}^{,r}\big(x_{k}^{1}\big)P_{kl\tilde{m}\tilde{n}}^{0}\big(x_{k}^{2}\big)\dot{\varepsilon}_{mn}^{0}\big(x_{k}^{1}\big)\dot{\varepsilon}_{\tilde{m}\tilde{n}}^{,s}\big(x_{k}^{2}\big) \\ &\quad+P_{ijmn}^{0}\big(x_{k}^{1}\big)P_{kl\tilde{m}\tilde{n}}^{,r}\big(x_{k}^{2}\big)\dot{\varepsilon}_{mn}^{,s}\big(x_{k}^{1}\big)\dot{\varepsilon}_{\tilde{m}\tilde{n}}^{0}\big(x_{k}^{2}\big) \\ &\quad+P_{ijmn}^{0}\big(x_{k}^{1}\big)P_{kl\tilde{m}\tilde{n}}^{0}\big(x_{k}^{2}\big)\dot{\varepsilon}_{mn}^{,r}\big(x_{k}^{1}\big)\dot{\varepsilon}_{\tilde{m}\tilde{n}}^{,s}\big(x_{k}^{2}\big)\Big]S_{b}^{rs} \\ &\quad+P_{ijmn}^{0}\big(x_{k}^{1}\big)P_{kl\tilde{m}\tilde{n}}^{0}\big(x_{k}^{2}\big)\dot{\varepsilon}_{mn}^{,r}\big(x_{k}^{1}\big)\dot{\varepsilon}_{\tilde{m}\tilde{n}}^{,s}\big(x_{k}^{2}\big)\big)\,S_{b}^{rs}\end{aligned} \tag{14.25}$$

14.3 Discretized Random Variable

The sequence of variational statements given in section 14.2 may serve as the basis for a spatially discretized formulation. Assume that the domain of interest V is discretized. The basic idea of the mean-based, second-order, second-moment analysis in a stochastic discretized problem is to expand via Taylor series all the vector and matrix stochastic field variables about the mean values of random variables $\mathbf{b}(\mathbf{x})$, to retain only up to second-order terms and to use in the analysis only the first two statistical moments. The expressions for the expectations and cross-covariances (autocovariances) of the nodal velocities can be obtained in terms of the nodal velocity derivatives with respect to the random variables. In the stochastic numerical approach the fields $\mathbf{b}(\mathbf{x})$ have to be represented by a set of basic random variables. To discretize $\mathbf{b}(\mathbf{x})$ by expressing them in terms of point values the following approximation is used

$$\mathbf{b}(\mathbf{x}) = \mathbf{N}_{\bar{\alpha}}(\mathbf{x})\,\mathbf{b}_{\bar{\alpha}} \tag{14.26}$$

where $\mathbf{N}_{\bar{\alpha}}$ are shape functions and $\mathbf{b}_{\bar{\alpha}}$ is the matrix of random parameter nodal values.

The same shape functions as in Eq. (14.26) as used for velocity approximation.

$$\mathbf{V}(\mathbf{x}) = \mathbf{N}_{\bar{\alpha}}(\mathbf{x})\,\mathbf{V}_{\bar{\alpha}} \tag{14.27}$$

where $\mathbf{V}_{\bar{\alpha}}$ is the matrix of nodal velocities. The matrix $\mathbf{V}_{\bar{\alpha}}$ can be related to the nodal velocity vector $\mathbf{V}_{\alpha}$ by the transformation

$$\mathbf{V}_{\bar{\alpha}} = \mathbf{B}_{\bar{\alpha}\alpha}\,\mathbf{V}_{\alpha} \tag{14.28}$$

which substituted into Eq. (14.27) gives

$$\mathbf{V}(\mathbf{x}) = \mathbf{N}_{\alpha}(\mathbf{x})\,\mathbf{V}_{\alpha} \tag{14.29}$$

provided we denote

$$\mathbf{N}_{\alpha}(\mathbf{x}) = \mathbf{N}_{\bar{\alpha}}(\mathbf{x})\,\mathbf{B}_{\bar{\alpha}\alpha} \tag{14.30}$$

A vector of nodal random variables $\mathbf{b}_{\rho}$ related to the matrix $\mathbf{b}_{\bar{\alpha}}$ is introduced by an appropriate transformation

$$\mathbf{b}_{\bar{\alpha}} = \mathbf{B}_{\bar{\alpha}\rho}\,\mathbf{b}_{\rho} \tag{14.31}$$

Then Eq. (14.26) is

$$\mathbf{b}(\mathbf{x}) = \mathbf{N}_{\bar{\alpha}}(\mathbf{x})\,\mathbf{B}_{\bar{\alpha}\rho}\,\mathbf{b}_{\rho} = \mathbf{N}_{\rho}(\mathbf{x})\,\mathbf{b}_{\rho} \tag{14.32}$$

which may be regarded as the random variable counterpart of the velocity expansion Eq. (14.27). By Eq. (14.26)

$$\mathrm{E}\big(\mathbf{b}(\mathbf{x})\big) = \mathbf{b}^{0}(\mathbf{x}) = \mathbf{N}_{\rho}(\mathbf{x})\,\mathbf{b}^{0} \tag{14.33}$$

$$\mathrm{Cov}\big(\mathrm{b}_{r}(\mathbf{x}),\mathrm{b}_{s}(\mathbf{x})\big) = \mathrm{S}_{\mathrm{b}}^{rs} = \mathbf{N}_{r\rho}(\mathbf{x})\,\mathbf{N}_{s\sigma}(\mathbf{x})\mathrm{S}_{\mathrm{b}}^{\rho\sigma} \tag{14.34}$$

and

$$\Delta\,\mathbf{b}(\mathbf{x}) = \mathbf{N}_{\rho}(\mathbf{x})\Delta\,\mathbf{b}_{\rho} \tag{14.35}$$

where

$$\Delta \mathbf{b}_\rho = \mathbf{b}_\rho - \mathbf{b}_\rho^0 \tag{14.36}$$

and $\mathbf{b}_\rho^0$ and $S_b^{\rho\sigma}$ stand for the mean value vector and the covariance matrix of the nodal random variable vector $\mathbf{b}_\rho$, respectively.

14.4 The Numerical Stochastic Rigid - Viscoplastic Model

Consider the finite element equation (12.12) in which the matrix $\mathbf{K}_V$ and the vectors $\mathbf{Q}$ and $\mathbf{V}$ are functions of the discretized random variable $\mathbf{b} = \mathbf{b}(\mathbf{x})$

$$\mathbf{K}_V(\mathbf{b})\mathbf{V}(\mathbf{b}) = \mathbf{Q}(\mathbf{b}) \tag{14.37}$$

All the random functions are expanded about the mean value E($\mathbf{b}$) via a Taylor series and only up to second order terms are retained. For any small parameter γ we have

$$\begin{aligned} \mathbf{V}(\mathbf{b},t) &= E(\mathbf{V}(t)) + \gamma \sum_{i=1}^{q} E(\mathbf{V}_{b_i}(t)\Delta b_i) \\ &+ \frac{1}{2}\gamma^2 \sum_{i,j=1}^{q} E(\mathbf{V}_{b_i b_j}(t))\Delta b_i \Delta b_j \end{aligned} \tag{14.38}$$

where Δb_i represents the first-order variation of b_i about $E(b_i)$ where b_i are the nodal values of b($\mathbf{x}$), that is the values of b at x_i, i = 1, …, R and for any function g the following notations are used $E(g(x)) = g(x, E(b))$, $E(g_{bi}) = \frac{\partial g}{\partial b_i}$, $E(g_{b_i b_j}) = \frac{\partial^2 g}{\partial b_i \partial b_j}$. In a similar way as it is in (14.38) we can express $\mathbf{K}_V(\mathbf{b})$ and $\mathbf{Q}(\mathbf{b})$

$$\begin{aligned} \mathbf{K}_V(\mathbf{b}) &= E(\mathbf{K}_V) + \gamma \sum_{i=1}^{q} E(\mathbf{K}_{Vb_i})\Delta b_i + \\ &+ \frac{1}{2}\gamma^2 \sum_{i,j=1}^{2} E(\mathbf{K}_{Vb_i b_j})\Delta b_i \Delta b_j \end{aligned} \tag{14.39}$$

$$\begin{aligned} \mathbf{Q}(\mathbf{b}) &= E(\mathbf{Q}) + \gamma \sum_{i=1}^{q} E(\mathbf{Q}_{b_i})\Delta b_i + \\ &+ \frac{1}{2}\gamma^2 \sum_{i,j=1}^{2} E(\mathbf{Q}_{b_i b_j})\Delta b_i b_j \end{aligned} \tag{14.40}$$

Substitution of equations (14.38), (14.39) and (14.40) into equation (14.37) and collecting terms of order γ^0, γ^1 and γ^2 the following equations for $E(\mathbf{V})$, $E(\mathbf{V}_{b_i})$ and $E(\mathbf{V}_{b_i b_j})$ are derived:

Zero order (γ^0 terms)

$$E(\mathbf{K}_V)E(\mathbf{V}) = E(\mathbf{Q}) \tag{14.41}$$

First order (γ^1 terms)

$$E(\mathbf{K}_V)E(\mathbf{V}_{b_i}) = E(\mathbf{Q}_{1b_i}) \tag{14.42}$$

where

$$E(\mathbf{Q}_{1b_i}) = E(\mathbf{Q}_{b_i}) - \left(E(\mathbf{K}_{Vb_i})E(\mathbf{V})\right)$$

Second order (γ^2 terms)

$$E(\mathbf{K}_V)\hat{\mathbf{V}}_2 = \hat{\mathbf{Q}}_2 \tag{14.43}$$

where

$$\begin{aligned}\hat{\mathbf{Q}}_2 &= \sum_{i,j=1}^{q}\left\{\left[\frac{1}{2}E(\mathbf{Q}_{b_i b_j})\mathrm{Cov}(b_i, b_j)\right]\right\} + \\ &- \sum_{i,j=1}^{q}\left\{\left[\frac{1}{2}E(\mathbf{K}_{Vb_i b_j})E(\mathbf{V}) + E(\mathbf{K}_{Vb_i})E(\mathbf{V}_{b_i})\right]\mathrm{Cov}(b_i, b_j)\right\}\end{aligned} \tag{14.44}$$

$$\hat{\mathbf{V}}_2 = \frac{1}{2}\sum_{i,j=1}^{q} E(\mathbf{V}_{b_i b_j})\mathrm{Cov}(b_i, b_j)$$

$$\mathrm{Cov}(b_i, b_j) = \left[\mathrm{Var}(b(\mathbf{x}_i)), \mathrm{Var}(b(\mathbf{x}_j))\right]^{\frac{1}{2}} R(b(\mathbf{x}_i), b(\mathbf{x}_j)) \tag{14.45}$$

and $R(b(\mathbf{x}_i), b(\mathbf{x}_j))$ is the autocorrelation. The definition for the expectation and cross-covariance of the velocity matrix **V** are given by

$$E(\mathbf{V}) = \int_{-\infty}^{+\infty} \mathbf{V}(\mathbf{b})p(\mathbf{b})d\,\mathbf{b} \tag{14.46}$$

and

$$\mathrm{Cov}(\mathbf{V}^i, \mathbf{V}^j) = \int_{-\infty}^{+\infty} \left(\mathbf{V}^i - E(\mathbf{V}^i)\right)\left(\mathbf{V}^j - E(\mathbf{V}^j)\right) p(\mathbf{b})\,d\,\mathbf{b} \tag{14.47}$$

where $p(\mathbf{b})$ is the joint probability density function. The second-order estimate of the mean value of **V** is obtained from Eq. (14.38) to give

$$E(\mathbf{V}) = \mathbf{V}(E(\mathbf{b})) + \frac{1}{2}\left\{\sum_{i,j=1}^{R} E\left(\mathbf{V}_{b_i b_j}\right) \text{Cov}\left(b_i, b_j\right)\right\} \qquad (14.48)$$

If the element strain-rate vector $\dot{\boldsymbol{\varepsilon}}$ is expressed in the form $\dot{\boldsymbol{\varepsilon}} = \mathbf{B}\mathbf{V}$, then the mean value and the cross-covariance of $\dot{\boldsymbol{\varepsilon}}$ can be expressed as

$$E(\dot{\boldsymbol{\varepsilon}}) = \mathbf{B}\, E(\mathbf{V}) + \frac{1}{2}\left\{\sum_{i,j=1}^{R} \mathbf{B}\, E\, \mathbf{V}_{b_i b_j}\, \text{Cov}\left(b_i, b_j\right)\right\} \qquad (14.49)$$

and

$$\text{Cov}\left(\dot{\boldsymbol{\varepsilon}}^1, \dot{\boldsymbol{\varepsilon}}^2\right) = \left\{\sum_{i,j=1}^{R} \left(\mathbf{B}^1\, E\left(\mathbf{V}_{b_i}^1\right)\right)\left(\mathbf{B}^2\, E\left(\mathbf{V}_{b_j}^2\right)\right)^{\mathrm{T}} \text{Cov}\left(b_i, b_j\right)\right\} \qquad (14.50)$$

The mean value and cross-covariance of $\boldsymbol{\sigma}$ can be shown to be

$$E(\boldsymbol{\sigma}) = E(\mathbf{P})\, E(\dot{\boldsymbol{\varepsilon}}) + \left\{\sum_{i,j=1}^{R}\left[E\left(\mathbf{P}_{b_i}\right)\mathbf{B}\, E\left(\mathbf{V}_{b_j}\right)\right.\right.$$
$$\left.\left. + \frac{1}{2}E\left(\mathbf{P}_{b_i b_j}\right)\mathbf{B}\, E(\mathbf{V})\right]\text{Cov}\left(b_i, b_j\right)\right\} \qquad (14.51)$$

and

$$\text{Cov}\left(\boldsymbol{\sigma}^1, \boldsymbol{\sigma}^2\right) = \left\{\sum_{i,j=1}^{R}\left[\left(E\left(\mathbf{P}^1\right)\mathbf{B}^1\, E\left(\mathbf{V}_{b_i}^1\right)\right)\left(E\left(\mathbf{P}^2\right)\mathbf{B}^2\, E\left(\mathbf{V}_{b_i}^2\right)\right)^{\mathrm{T}}\right.\right.$$
$$+ \left(E\left(\mathbf{P}_{b_i}^1\right)\mathbf{B}^1\, E\left(\mathbf{V}^1\right)\right)\left(E\left(\mathbf{P}_{b_j}^2\right)\mathbf{B}^2\, E\left(\mathbf{V}^2\right)\right)^{\mathrm{T}}$$
$$+ \left(E\left(\mathbf{P}^1\right)\mathbf{B}^1\, E\left(\mathbf{V}_{b_i}^1\right)\right)\left(E\left(\mathbf{P}_{b_j}^2\right)\mathbf{B}^2\, E\left(\mathbf{V}^2\right)\right)^{\mathrm{T}}$$
$$\left.\left. + \left(E\left(\mathbf{P}_{b_i}^1\right)\mathbf{B}^1\, E\left(\mathbf{V}^1\right)\right)\left(E\left(\mathbf{P}^2\right)\mathbf{B}^2\, E\left(\mathbf{V}_{b_i}^2\right)\right)^{\mathrm{T}}\right]\text{Cov}\left(b_i, b_j\right)\right\} \qquad (14.52)$$

respectively.

Part X

Contact and Friction

15 Contact and Friction

15.1 Introduction

In metalworking processes the workpiece is deformed by the contact with the die. The pressure required for deformation generates a stress normal to the die surface. The movement of the workpiece relative to the die surface generates a shear stress at the interface. The friction between the die and the workpiece arises with potential for wear of both die and workpiece materials.

15.2 Boundary Conditions

Consider the body B in the state of deformation of volume V and the surface ∂V at time t (Fig. 15.1). The surface ∂V of the body B consists of the free surface and the contact surface with the die. The contact surface can be divided into the slipping surface denoted by ∂V_V and the sticking surface denoted by ∂V_o. In the metal forming process the free surface, contact surface and slipping surface can change in the time of deformation.

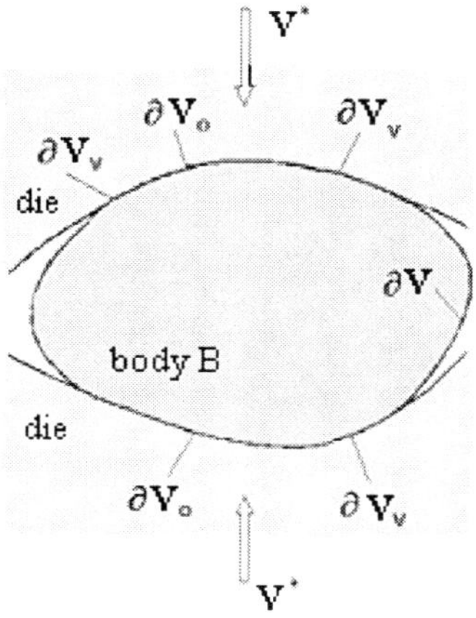

Fig. 15.1. Slipping and sticking surfaces during plastic deformation of body B in rigid dies

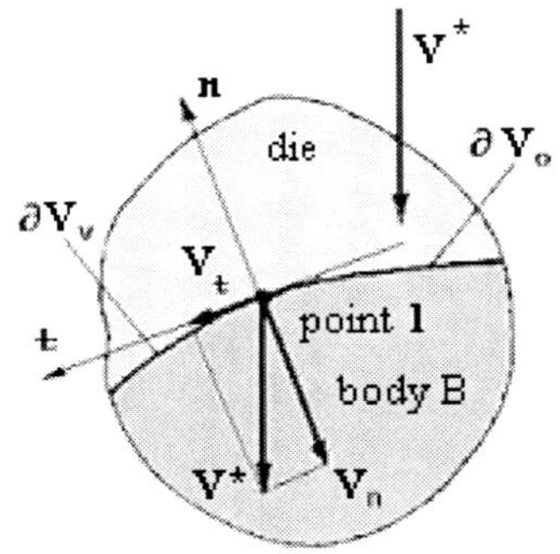

Fig. 15.2. Distribution of velocities on the contact surface with the die

Consider kinematic boundary conditions. Assume that the rigid die moves with the velocity

$$\mathbf{V} = \mathbf{V}^* (\mathbf{X}, t) \tag{15.1}$$

By the assumption of the die rigidity $\dot{\boldsymbol{\varepsilon}}^* = 0$. Moreover the rigid rotation of the die implies

$$\dot{\boldsymbol{\omega}}^* = \frac{1}{2}(\text{grad}\,\mathbf{V} - {}^{T}\text{grad}\,\mathbf{V}) \tag{15.2}$$

On the sticking surface ∂V_o

$$\mathbf{V} = \mathbf{V}^* \tag{15.3}$$

where $\mathbf{V}^*$ is a given velocity field. On the slipping surface ∂V_V (Fig. 15.2) velocity $\mathbf{V}^*$ can be decomposed on normal $\mathbf{V}_n$ and tangent $\mathbf{V}_t$ parts according to the relation

$$\mathbf{V}^* = \mathbf{V}_n + \mathbf{V}_t \tag{15.4}$$

where $\mathbf{V}_n = \mathbf{V}^* \cdot \mathbf{n}$ and $\mathbf{n}$ is the unit normal vector to the surface ∂V_V.

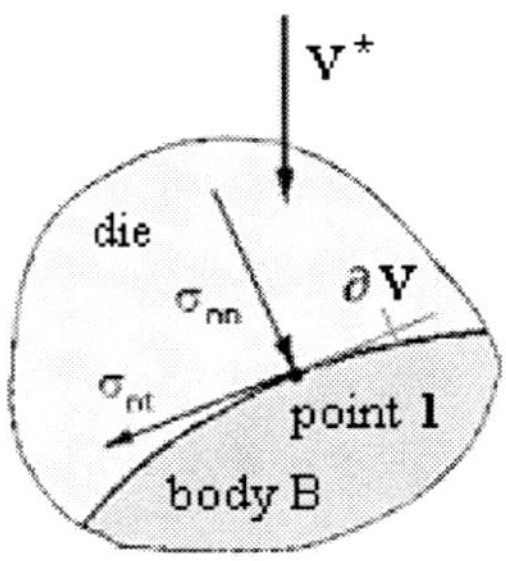

Fig. 15.3. Components of stresses on the contact surface with the die

For point **X** belonging to the free surface we have

$$\boldsymbol{\sigma} \cdot \mathbf{n} = 0 \tag{15.5}$$

The slipping on the surface ∂V_V is associated with the tangent stress σ_{nt} caused by friction. This stress depends on such parameters as the normal stress to the surface σ_{nn}, the slipping velocity $\mathbf{V}_t$, the friction coefficient f. Applying the Coulomb law we have

$$\sigma_{n\tau} = f\,\sigma_{nn} \quad \text{for} \quad \left| f\,\sigma_{nn} \right| \le \bar{\tau} \tag{15.6}$$

$$\sigma_{n\tau} = \tau \quad \text{for} \quad \left| f\,\sigma_{nn} \right| > \bar{\tau} \tag{15.7}$$

where $\bar{\tau}$ is the shearing stress.

15.3 Thermal Boundary Conditions

On the contact surface between the die and the workpiece the thermal boundary condition of the fourth kind is applied. If the contact is ideal then

$$\theta_B = \theta_D \tag{15.8}$$

where the subscripts 'B' and 'D' denote the body and the die temperature respectively.

Considering the heat generated by the friction $\mathbf{t}\,\mathbf{V}_t$ as the result of action of the fiction a force $\mathbf{t}$ on the contact surface

$$\mathbf{t}\,\mathbf{V}_t = k_B \frac{\partial \theta_B}{\partial n} - k_D \frac{\partial \theta_D}{\partial n} \tag{15.9}$$

If both bodies are cold and the contact surface is heated only during the friction process, then the heat fluxes to the die and to the material are summarized, taking into account the expression $\dfrac{\partial \theta_D}{\partial n}$ (Fig. 15.4).

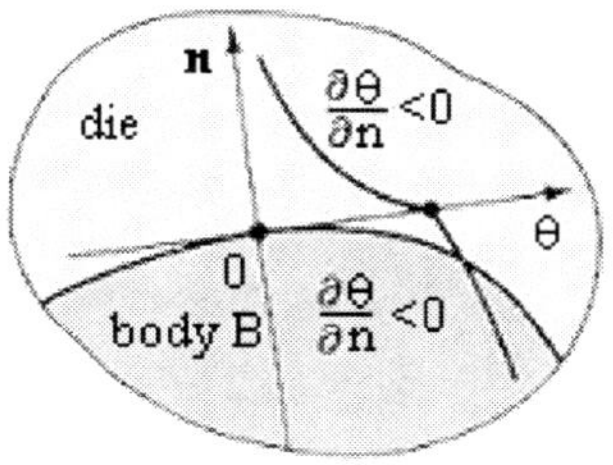

Fig. 15.4. Die and body B heated by friction on the contact surface

If the temperature of the body is higher than that of the die temperature, then the heat from hot material and generated by friction heats the die (Fig. 15.5). On the sticking surface $\mathbf{t}\,\mathbf{V}_t = 0$, because $\mathbf{V}_t = 0$.

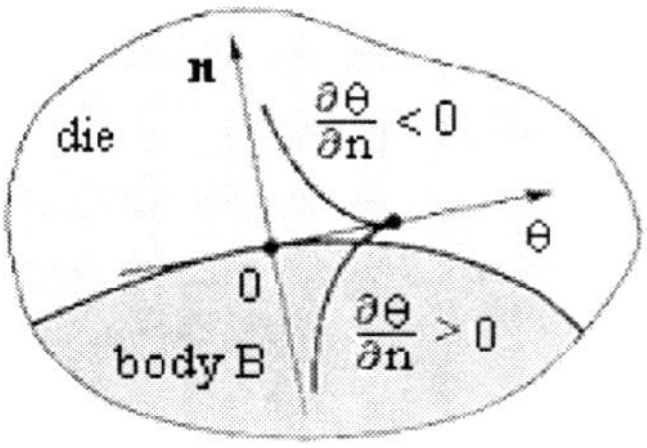

Fig. 15.5. Temperature distribution in the die and body B in hot forming

15.4 Friction

Friction is the resistance to motion encountered when one body slides over another. In metal-working processes it arises from the sliding of the workpiece against the die. This phenomenon is described by the Coulomb law. The Coulomb law for friction stress σ_{nt} is given by Eqs. (15.6) - (15.7). The friction force stress σ_{nt} depends on the normal pressure σ_{nn}, which the value implies from the traction forces **t.**

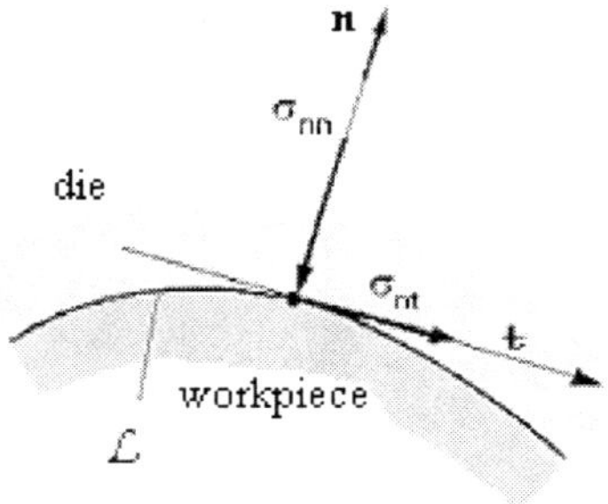

Fig. 15.6. Coulomb friction stress on the contact surface die-workpiece

The surface of contact $\mathcal{L}_{1,2}$ (Fig.15.7) can be treated as a discontinuous surface with the slipping velocity

$$[[\mathbf{V}]] = \mathbf{V}^{1,2} = \mathbf{V}^1 - \mathbf{V}^2 \tag{15.10}$$

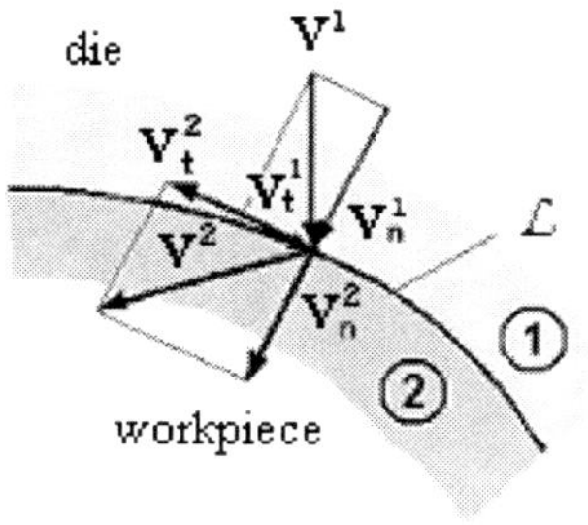

Fig. 15.7. Slipping velocity on the contact surface die-workpiece

The work dissipated through friction can be treated as the work dissipated on the surface of discontinuity. This work is expressed as

$$\dot{W} = \int_{L_{(1,2)}} m\bar{\tau}[[\mathbf{V}]]dL \qquad (15.11)$$

where m ($0 \leq m \leq 1$) is treated as a degree of sticking to the die surface. In this way the Coulomb law is substituted by shearing stress law, which for a constant coefficient m gives a constant value of friction stress $m\bar{\tau}$ on the surface considered.

15.5 Lubrication

In some cases friction is helpful. High friction on the punch surface helps increase reductions in deep drawing and ironing. Sometimes friction has to have at least some small value, as in rolling to assume entry of the workpiece into the roll gap and helps maintain rolling without skidding of the workpiece. In most cases friction is preferably reduced to zero by the introduction of a lubricant. Lubricants respond to interface conditions in different ways leading to a variety of lubricating mechanisms. The analysis of lubrication mechanisms is out the scope of the book.

Part XI

Simplified Equations

16 Simplified Equations of Metal Forming

In this section standard analytical solutions evaluating plastic deformation have been included. These equations describe the deformation in a simple way, which sometimes is very helpful. It is a present practice to formulate constitutive equations using assumptions, which are not universally accepted, but in most cases they are necessary to obtain analytical solutions.

16.1 Upsetting

Consider an elementary volume separated from the deformed body to determine normal and tangent stresses on the contact surface between a cylindrical workpiece and the die (Fig. 16.1). Denote stresses on the boundary of the elementary volume by normal stress σ_z, friction stress τ_{rz}, radial stress σ_r and circumferential stress σ_ϑ.

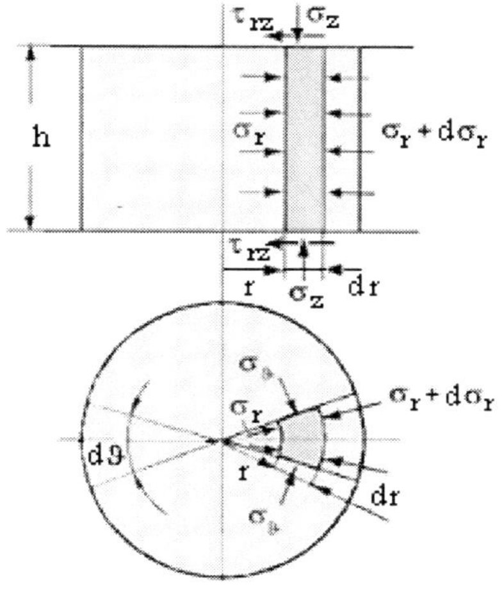

Fig. 16.1. Stresses in an elementary volume in the upsetting of a cylindrical workpiece

The equilibrium of forces on an elementary volume is

$$\sigma_r r d\vartheta h + 2\sigma_\vartheta \sin\frac{d\vartheta}{2} drh - (\sigma_r + d\sigma_r)(r + dr)d\vartheta h - 2\tau_{rz} r d\vartheta dr = 0 \qquad (16.1)$$

For small angles we have

$$\sin\frac{d\vartheta}{2} \cong \frac{d\vartheta}{2} \qquad (16.2)$$

The problem considered is axisymmetrical so radial stress σ_r and circumferential stress σ_ϑ are equal. By Eqs. (16.1) and (16.2), omitting the terms of small values of higher orders

$$d\sigma_r h + 2\tau_{rz} dr = 0 \qquad (16.3)$$

Assume the plasticity criterion

$$d\sigma_r - d\sigma_z = 0 \qquad (16.4)$$

The expression (16.3) can be solved if the tangent stress τ_{rz} on the contact surface is known. Assume that the following relation on the contact surface holds

$$\tau_{rz} = f\sigma_z \qquad (16.5)$$

where f is the friction coefficient. By Eqs. (16.3), (16.4) and (16.5) we get

$$d\sigma_z h + 2f\sigma_z dr = 0 \qquad (16.6)$$

After separation of variables, Eq. (16.6) takes the form

$$\frac{d\sigma_z}{\sigma_z} = -\frac{2f dr}{h} \qquad (16.7)$$

By integrating Eq. (16.7) we get

$$\ln\sigma_z\Big|_{-\sigma_o}^{\sigma_z} = -\frac{2fr}{h}\Big|_{r_o}^{r} \qquad (16.8)$$

Assume that the lower limit of integration is the yield limit σ_o. By (16.8) we get

$$\ln\frac{\sigma_z}{-\sigma_o} = -\frac{2f}{h}\left(r - r_o\right) \qquad (16.9)$$

By Eq. (16.9) we get the following expression for normal stress σ_z

$$\sigma_z = -\sigma_o e^{\frac{2f}{h}(r_o - r)} \qquad (16.10)$$

The tangent stress τ_{rz} is

$$\tau_{rz} = -f\sigma_o e^{\frac{2f}{h}(r_o - r)} \qquad (16.11)$$

By Eq. (16.11) the tangent stress τ_{rz} for a high friction coefficient and for a high ratio of workpiece diameter to its height can pass the theoretical values which results from the plasticity criterion of maximal tangent stress or the energy condition of pure non-dilatational deformation. In the upsetting of low workpieces

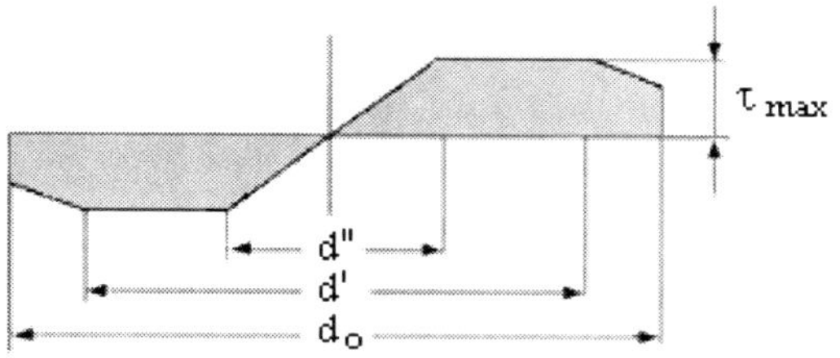

Fig. 16.2. Friction stresses in the upsetting of a cylindrical workpiece

assume the existence of a zone where the friction stress reaches the maximal value equal to half of the yield limit

$$\tau_{rz} = \tau_{max} = -0.5\,\sigma_o \tag{16.12}$$

The possibility of occurrence of such a zone increases if the friction coefficient increases. The equilibrium equation in this zone (Fig. 16.2) is the same as for the first one

$$d\sigma_z h + 2\tau_{rz} dr = 0 \tag{16.13}$$

By Eq. (16.12), after reparation of variables we get

$$d\sigma_z = \frac{\sigma_o}{h} dr \tag{16.14}$$

By integrating we have

$$\sigma_z \Big|_{\sigma_z'}^{\sigma_z} = \frac{\sigma_o}{h} r \Big|_{r'}^{r} \tag{16.15}$$

where the notation $(\cdot)'$ is used to express the respective values between the first and the second zones. By substituting the integration limits we get

$$\sigma_z - \sigma_z' = \frac{\sigma_o}{h}(r - r') \tag{16.16}$$

The radius r is determined from the condition that the tangent stress on the boundary of the first zone reaches the maximal value equal to half of the yield limit

$$f\sigma_o e^{\frac{2f}{h}(r_o - r')} = 0.5\sigma_o \tag{16.17}$$

Hence

$$r' = \frac{h}{2f}\ln 2f + r_o \tag{16.18}$$

Then the normal stress σ_z is

$$\sigma_z' = -\sigma_o e^{-\ln 2f} = -\frac{\sigma_o}{2f} \tag{16.19}$$

Finally we get the expression for normal stress in the second zone

$$\sigma_z = -\frac{\sigma_o}{2f} - \frac{\sigma_o}{h}\left(\frac{h}{2f}\ln 2f + r_o - r\right) \tag{16.20}$$

By Eq. (16.20) the normal stress σ_z in the second zone is a linear function of radius r. The second zone ends at the point in which the radius r is equal to the height h of the workpiece. Then the normal stress σ_z is

$$\sigma_z'' = -\frac{\sigma_o}{2f} - \frac{\sigma_o}{h}\left(\frac{h}{2f}\ln 2f + r_o - h\right) \tag{16.21}$$

Based on experiments, assume that the tangent stress on the contact surface between the die and the workpiece in the third zone decreases linearly with respect to the radius reaching at the axis of the workpiece the value equal to

$$\tau_{rz} = -\frac{\sigma_o}{2}\frac{r}{h} \tag{16.22}$$

The equilibrium condition in the third zone is the same as for the remaining two zones

$$d\sigma_z h + 2\tau_{rz} dr = 0 \tag{16.23}$$

By Eqs. (16.22) and (16.23) we get

$$d\sigma_z = \frac{\sigma_o}{h^2} r dr \tag{16.24}$$

After integration we have

$$\sigma_z \Big|_{\sigma_z''}^{\sigma_z} = \frac{\sigma_o}{h^2}\frac{r^2}{2}\Big|_h^r \tag{16.25}$$

Substituting the integration limits

$$\sigma_z - \sigma_z'' = \frac{\sigma_o}{2h^2}\left(r^2 - h^2\right) \tag{16.26}$$

Finally we get the expression for normal stress in the third zone as

$$\sigma_z = -\sigma_o \left(\frac{1}{2f} + \frac{\ln 2f}{2f} + \frac{r_o}{h} - \frac{h^2 - r^2}{2h^2} - 1 \right) \tag{16.27}$$

The highest pressure per unit area appears at the axis of the workpiece decreasing in the direction of the free outer surface (Fig. 16.3). The unit pressure is dependent

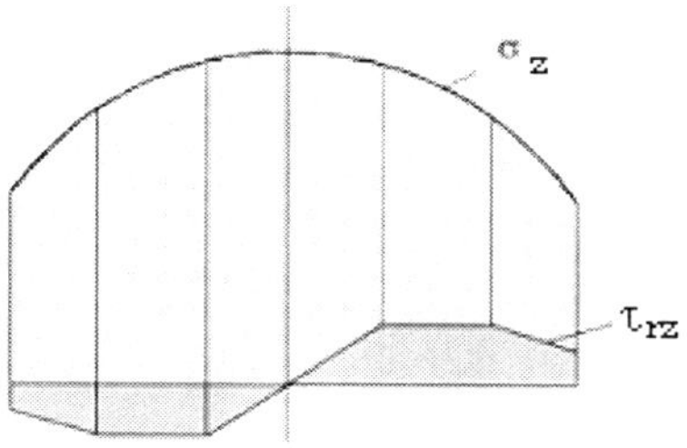

Fig. 16.3. Pressure and tangent stresses in the upsetting of a cylindrical workpiece

on the workpiece shape characterized by the ratio of the diameter to its height and friction coefficient. If this ratio and friction coefficient is higher, then the unit and total pressures are higher. For this reason high unit pressures are observed in the upsetting of thin workpieces. In order to reduce the unit pressure lubrication is applied.

16.2 Rolling

The rolling pressures and friction stresses on the contact surface between the material and the rolls are fundamental parameters of the rolling process. Consider the equilibrium of forces on an elementary volume separated from the domain of plastic deformation (Fig. 16.4.) bounded by two cylindrical surfaces of radii r and r + dr and the rolls surfaces.

Assume that the radial stress σ_r is uniformly distributed on the cylindrical surface. Consider radial stress σ_r, circumferential stress σ_ϑ and stress in z direction σ_z on an elementary volume, all of positive signs. The equilibrium equation in the rolling direction is

$$\sigma_r r\varphi - (\sigma_r + d\sigma_r)(r + dr)\varphi - 2\tau_{r\vartheta} dr \cos\frac{\varphi}{2} + 2\sigma_\vartheta dr \sin\frac{\varphi}{2} = 0 \tag{16.28}$$

Assuming $\cos\frac{\varphi}{2} \cong 1$ and $\sin\frac{\varphi}{2} \cong \frac{\varphi}{2}$ we get the equilibrium equation in the form

$$\frac{d\sigma_r}{dr} + \frac{\sigma_r - \sigma_\vartheta}{r} + \frac{2\tau_{r\vartheta}}{r\vartheta} = 0 \tag{16.29}$$

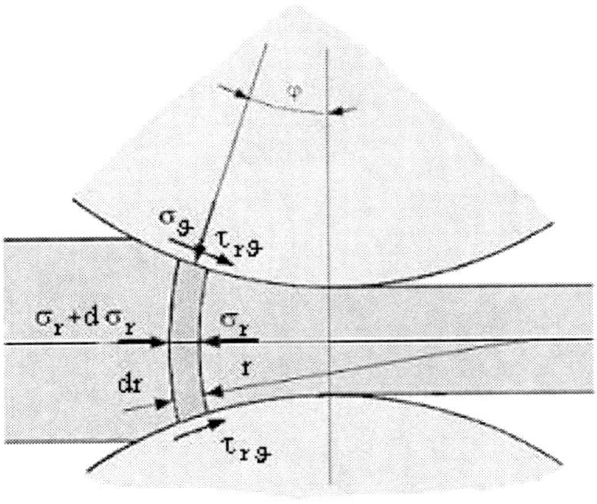

Fig. 16.4. Stresses in an elementary volume in rolling

The friction stresses in the forward slip zone are opposite to the direction of material flow, then the equilibrium equation in this zone is

$$\frac{d\sigma_r}{dr} + \frac{\sigma_r - \sigma_\vartheta}{r} + \frac{2\tau_{r\vartheta}}{2\varphi} = 0 \tag{16.30}$$

The height of the rolled material at the cross-section of the plastic deformation domain is a function of the radius of the location point and the roll bite angle

$$h \cong r\varphi \tag{16.31}$$

The equilibrium equation in forward and backward slip zones is

$$\frac{d\sigma_r}{dh} + \frac{\sigma_r - \sigma_\vartheta}{h} \pm \frac{2\tau_{r\vartheta}}{\varphi h} = 0 \tag{16.32}$$

Three characteristic subzones in backward and forward slip zones are assumed based on experimental results (Fig. 16.5). In slipping subzones

$$\tau_{r\vartheta} = f\sigma_\vartheta \tag{16.33}$$

In sticking subzones the tangent stress has a maximal value equal to half of the yield limit. In dead subzones the tangent stress decreases linearly to zero value and then changes the sign and growths to maximal value. In the slipping subzone the equilibrium equation is

$$\frac{d\sigma_r}{dh} + \frac{\sigma_r - \sigma_\vartheta}{h} \pm \frac{2\tau_{r\vartheta}}{\varphi h} = 0 \tag{16.34}$$

Assume the plasticity criterion

$$\sigma_r - \sigma_\vartheta = \sigma_o \tag{16.35}$$

which gives

$$d\sigma_r - d\sigma_\vartheta = 0$$

and the friction law

$$\tau_{r\vartheta} = f\sigma_\vartheta \tag{16.36}$$

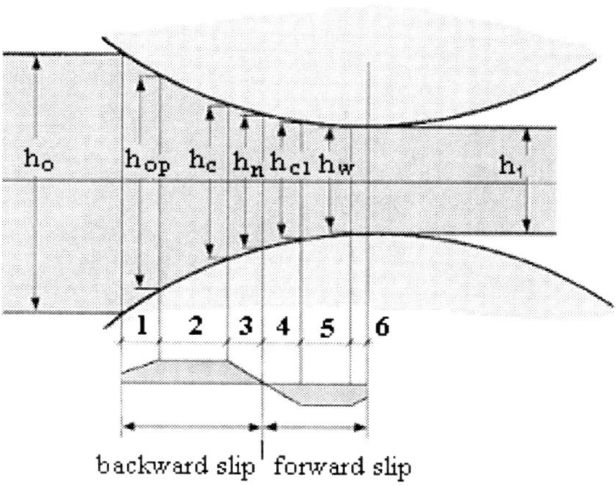

Fig. 16.5. Friction stresses in the domain of plastic deformation
1, 6 – slipping subzone, 2, 5 – sticking subzone, 3, 4 – dead subzone

The equilibrium equation takes the form

$$\frac{d\sigma_\vartheta}{dh} + \frac{\sigma_o}{h} + \frac{2f}{\vartheta}\frac{\sigma_\vartheta}{h} = 0 \tag{16.37}$$

Denoting

$$\frac{2f}{\varphi} = \delta \tag{16.38}$$

and by separation of variables we get

$$\frac{d\sigma_\vartheta}{\delta\sigma_\vartheta + \sigma_o} = -\frac{dh}{h} \tag{16.39}$$

By integrating we have

$$\frac{1}{\delta}\ln\left(\delta\sigma_\vartheta + \sigma_o\right) = -\ln h + \ln c \tag{16.40}$$

Hence

$$\left(\delta\sigma_\vartheta + \sigma_o\right)^{\frac{1}{\delta}} = \frac{C}{h} \tag{16.41}$$

The constant C is obtained from the boundary condition. If $h = h_o$, then $\sigma_r = 0$, and the circumferential stress reaches the yield limit. Taking into account that the circumferential stress is negative we have

$$\sigma_\vartheta = -\sigma_o \tag{16.42}$$

The constant C takes the form

$$C = h_o\left(-\delta\sigma_o - \sigma_o\right)^{\frac{1}{\delta}} \tag{16.43}$$

By Eqs. (16.43) and (16.41) we get

$$\sigma_\vartheta = -\frac{\sigma_o}{\delta}\left[(\delta-1)\left(\frac{h_o}{h}\right)^{\delta} + 1\right] \tag{16.44}$$

In a forward slip zone where friction stresses are determined by the friction law the same description is applied. For the cross-section where the material leaves the deformation zone we have (Fig. 16.5)

$$h = h_1 \qquad \sigma_r = 0 \qquad \sigma_\vartheta = -\sigma_o \tag{16.45}$$

As the result we get

$$\sigma_\vartheta = -\frac{\sigma_o}{\delta}\left[(\delta+1)\left(\frac{h}{h_1}\right)^{\delta} - 1\right] \tag{16.46}$$

The friction stresses in the slipping zone can be determined by the friction law. In a backward slip zone the friction stresses are given by

$$\tau_{r\vartheta} = \frac{f\sigma_o}{\delta}\left[(\delta-1)\left(\frac{h_o}{h}\right)^{\delta} + 1\right] \tag{16.47}$$

and in a forward slipping subzone of the forward slip zone

$$\tau_{r\vartheta} = -\frac{f\sigma_o}{\delta}\left[(\delta+1)\left(\frac{h}{h_1}\right)^{\delta} - 1\right] \tag{16.48}$$

The height of rolled material at the plane between the slipping and sticking subzones is determined by the assumption that the friction stresses have a maximal value equal to the yield limit

$$h = h_{op} \quad \tau_{r\vartheta} = -f\sigma_\vartheta = \frac{\sigma_o f}{\delta}\left[(\delta-1)\left(\frac{h}{h_{op}}\right)^{\delta} + 1\right] = \frac{\sigma_o}{2} \tag{16.49}$$

$$h = h_w \quad \tau_{r\vartheta} = f\sigma_\vartheta = -\frac{\sigma_o f}{\delta}\left[(\delta+1)\left(\frac{h_w}{h_1}\right)^{\delta} - 1\right] = -\frac{\sigma_o}{2} \tag{16.50}$$

Finally we get

$$\frac{h_{op}}{h_o} = \left(\frac{\delta - 1}{\frac{\delta}{2f} - 1} \right)^{\frac{1}{\delta}} \qquad \frac{h_w}{h_1} = \left(\frac{\frac{\delta}{2f} + 1}{\delta + 1} \right)^{\frac{1}{\delta}} \tag{16.51}$$

In a sticking subzone the friction stresses reach a maximal value equal to half of the yield limit by the hypothesis of maximal tangent stresses. The yield criterion has the form

$$(\sigma_r - \sigma_\vartheta)^2 + 4\tau_{r\vartheta}^2 = 4k^2 \tag{16.52}$$

where $k = \frac{2}{\sqrt{3}}\sigma_o$ by the Huber-Mises yield criterion. The radial and circumferential stresses are equal. Then the yield criterion of maximal tangent stress has the form

$$\tau_{r\vartheta} = \pm\, 0.5\, \sigma_o \tag{16.53}$$

The equilibrium equation

$$\frac{d\sigma_r}{dh} + \frac{\sigma_r - \sigma_\vartheta}{h} \pm \frac{2\tau_{r\vartheta}}{\varphi h} = 0 \tag{16.54}$$

takes the form

$$\frac{d\sigma_r}{dh} \pm \frac{\sigma_o}{\varphi h} = 0 \tag{16.55}$$

By the equilibrium equation for a sticking subzone in a backward slip zone we have

$$\frac{d\sigma_\vartheta}{dh} - \frac{\sigma_o}{\varphi h} = 0 \tag{16.56}$$

By (16.56) we get

$$\sigma_\vartheta = \frac{\sigma_o}{\varphi} \ln h + C \tag{16.57}$$

The constant C is determined by the boundary conditions. In the plane separating the slipping and sticking subzones

$$\text{if } h = h_{op}, \qquad \text{then } f\sigma_\vartheta = -\,0.5\,\sigma_o \qquad \text{and } \sigma_\vartheta = -\frac{\sigma_o}{2f} \tag{16.58}$$

The constant C is

$$C = -\frac{\sigma_o}{2f} - \frac{\sigma_o}{\vartheta} \ln h_{op} \tag{16.59}$$

Finally we get the following expression describing the distribution of pressures in sticking subzone from the side of entry of material to the plastic deformation zone

$$\sigma_\vartheta = -\frac{\sigma_o}{2f}\left[\frac{2f}{\varphi}\ln\left(\frac{h_{op}}{h}\right)+1\right] \tag{16.60}$$

Similarly the unit pressures are determined from the side where the material comes out from the plastic deformation zone. The boundary conditions for this zone are

$$\text{if } h = h_w, \qquad \text{then } f\,\sigma_\vartheta = -0.5\,\sigma_o \qquad \text{and } \sigma_\vartheta = -\frac{\sigma_o}{2f} \tag{16.61}$$

By the boundary conditions for a sticking subzone we get the equation describing the distribution of unit pressures

$$\sigma_\vartheta = -\frac{\sigma_o}{2f}\left[1+\delta\ln\frac{h}{h_w}\right] \tag{16.62}$$

In dead subzones the friction stresses decrease from maximal values to zero at a neutral plane, and then change the direction and growth to a maximal value. In a dead subzone of a backward slip subzone the unit friction forces are

$$\tau_{r\vartheta} = -\frac{\sigma_o}{2}\frac{x}{h_c} \tag{16.63}$$

and in a dead subzone of a forward slip zone

$$\tau_{r\vartheta} = \frac{\sigma_o}{2}\frac{x}{h_{cl}} \tag{16.64}$$

By the expression (16.63) we get the equation for the determination of unit pressures in a dead zone. In a dead subzone of a backward slip zone

$$\sigma_\vartheta = \sigma_c - \frac{\sigma_o}{\varphi^2}\left[\frac{h_c - h}{h_c} - \frac{h_n}{h_c}\ln\frac{h_c}{h}\right] \tag{16.65}$$

and in a dead subzone of a forward slip zone

$$\sigma_\vartheta = \sigma_{cl} - \frac{\sigma_o}{\varphi^2}\left[\frac{h_{cl} - h}{h_{cl}} - \frac{h_n}{h_{cl}}\ln\frac{h_{cl}}{h}\right]$$

The pressures have the maximal value at the plane of separation between backward and forward slip zones decreasing next in both directions (Fig. 16.6). The change of pressure is dependent on the shape of the rolls and the friction conditions. If the ratio of length to the height of rolled material is higher, then the unit pressures are higher. An increase of the friction coefficient leads to an increase of unit pressures.

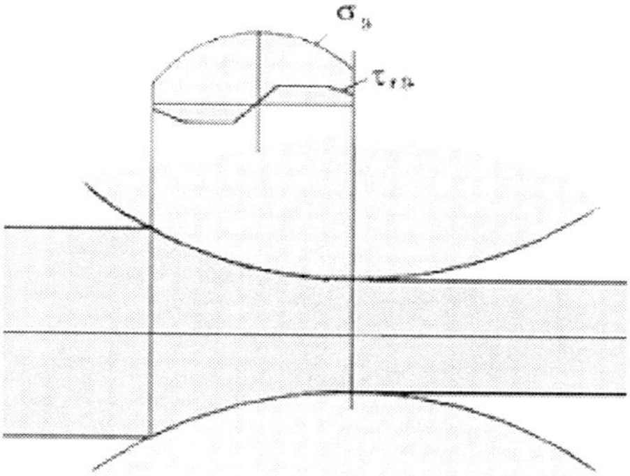

Fig. 16.6. Pressures and stresses forces in rolling

16.3 Bending

16.3.1 The Bending of Narrow Elements

Consider the bending of a rectangular bar of small curvature whose width is considerably smaller than its thickness. For such a ratio of dimensions the deformation of material in the direction of width is free. The stress state corresponds to simple tension for layers lying outside the neutral axis and to uniaxial compression for layers lying on inner side. The width of tension layers decreases and of the compressed ones increases.

The analysis of stresses and strains is carried out in a cylindrical coordinate system. Denote by σ_r the stress in thickness direction, σ_ϑ in circumferential direction and σ_z in width direction.

In the first stage of bending the material is elastic and the strain in circumferential direction is determined by (Fig. 16.7)

$$\varepsilon_\vartheta = \frac{x}{\rho} \tag{16.66}$$

where ρ is the element radius and x is the distance from the neutral axis. By Hook's law the stress σ_ϑ in circumferential direction is

$$\sigma_\vartheta = E\frac{x}{\rho} \tag{16.67}$$

where E is the Young's modulus.

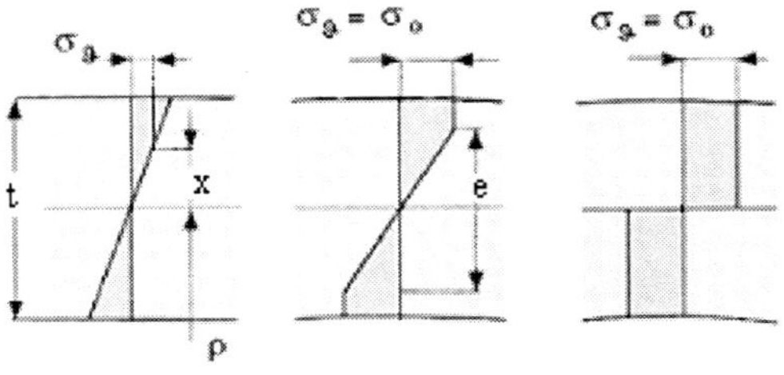

Fig. 16.7. Stresses at the cross-section of a bended band

By (16.67) the circumferential stresses σ_ϑ are a linear function of x. The maximal strain occurs at the outer and inner layers of the bending element

$$\varepsilon_\vartheta = \pm\frac{t}{2\rho} \tag{16.68}$$

The circumferential stress σ_ϑ reaches the maximal value in these places. The distribution of stresses for elastic, elasto-plastic and plastic states is given in Fig. 16.7. The bending moment M is

$$M = \int_{-\frac{t}{2}}^{\frac{t}{2}} \sigma_\vartheta x\, b\, dx \tag{16.69}$$

where t is the thickness and b is the second dimension of the element. By (16.67) and (16.69) we get

$$M = \int_{-\frac{t}{2}}^{\frac{t}{2}} \frac{E\, b\, x^2 dx}{\rho} \tag{16.70}$$

and finally

$$M = E\frac{bt^3}{12}\frac{1}{\rho} \tag{16.71}$$

Hence

$$\frac{1}{\rho} = \frac{M}{EJ} \tag{16.72}$$

where $J = \dfrac{bt^3}{12}$ is the moment of inertia of the bended cross-section related to the neutral axis. The expression (16.71) represents the straight line in coordinates 1/ρ - M (Fig. 16.8).

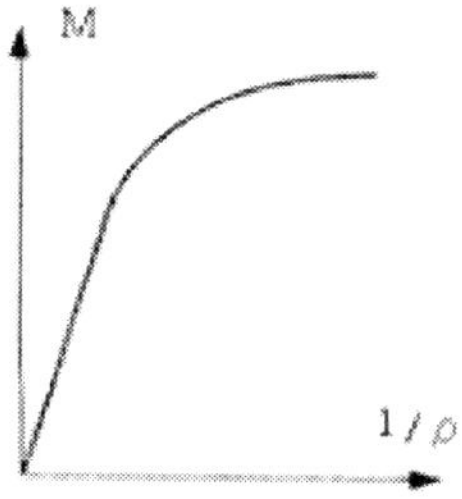

Fig. 16.8. Bending curve

If the boundary of an element is in a plastic range then the bending moment is obtained from (16.70). We get

$$M = \frac{bt^2}{\sigma}\sigma_o \tag{16.73}$$

If the moment increases, then the thickness of plastic layers increase and the moment is

$$M = 2\int_0^{\frac{e}{2}} E\frac{x}{\rho}x\,b\,dx + 2\int_{-\frac{e}{2}}^{\frac{e}{2}} \sigma_o x\,b\,dx \tag{16.74}$$

Hence

$$M = \frac{bt^2}{4}\sigma_o\left(1 - \frac{e^2}{3t^3}\right) \tag{16.75}$$

If $e \to 0$, then the state corresponds to a completely plastic one. In this case the moment reaches the maximal value

$$M = \frac{bt^2}{4}\sigma_o \tag{16.76}$$

16.3.2 The Bending of Wide Bands

Assume $\varepsilon_z = 0$ in the process of the bending of wide bands. Hence

$$\varepsilon_r + \varepsilon_\vartheta = 0 \tag{16.77}$$

If $\varepsilon_z = 0$, then neglecting the radial stress we get the relation between σ_z and σ_ϑ

$$\sigma_z = \nu\, \sigma_\vartheta \tag{16.78}$$

and finally

$$\sigma_\vartheta = \frac{E}{1-\nu^2}\,\varepsilon_\vartheta \tag{16.79}$$

16.3.3 The Bending of Wide and Thick Bands

In the analysis presented, radial stress in the material element was not taken into account, which is permitted in the bending of wide bands.

Consider the equilibrium of forces on an elementary volume separated by two cylindrical surfaces of radii r and r + dr and two planes composing the angle $d\vartheta$ (Fig. 16.9)

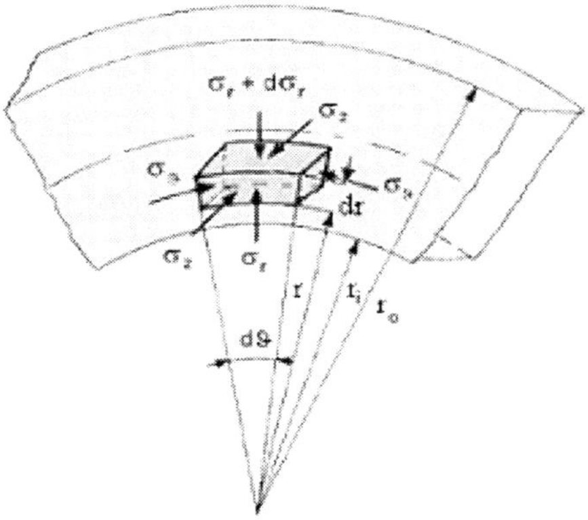

Fig. 16.9. Stresses in an elementary volume

The equilibrium of forces on an elementary volume in radial direction is

$$(\sigma_r + d\sigma_r)\,b\,(r + dr)\,d\vartheta - \sigma_r\, b\, r\, d\vartheta - 2\sigma_\vartheta \sin\frac{d\vartheta}{2}\, b\, dr = 0 \tag{16.80}$$

By (16.80) we get

$$\frac{d\sigma_r}{dr} + \frac{\sigma_r - \sigma_\vartheta}{r} = 0 \tag{16.81}$$

For the plane state of strain the stress in z direction is

$$\sigma_z = \frac{\sigma_r + \sigma_\vartheta}{2} \tag{16.82}$$

Assume the Huber-Mises yield criterion

$$\sigma_r - \sigma_\vartheta = \frac{2}{\sqrt{3}}\sigma_o = \sigma_o' \tag{16.83}$$

By (16.81) and (16.83) we get

$$\frac{d\sigma_r}{dr} + \frac{\sigma_o'}{r} = 0 \tag{16.84}$$

By the boundary conditions the radial stresses on inner and outer surfaces are equal to zero. By solving the equilibrium equation with the boundary conditions we get (Fig. 16.9)

$$\sigma_r = -\sigma_o' \ln\frac{r}{r_i} \tag{16.85}$$

where r_i is the inner radius. The circumferential stresses are

$$\sigma_\vartheta = -\sigma_o'\left(1 + \ln\frac{r}{r_i}\right) \tag{16.86}$$

The stress in z direction is

$$\sigma_z = -\frac{\sigma_o'\left(1 + 2\ln\frac{r}{r_i}\right)}{2} \tag{16.87}$$

From the above radial and circumferential stresses in r, ϑ and z directions in the inner zone increase when approaching the neutral axis. Assume the Huber-Mises yield criterion

$$\sigma_\vartheta - \sigma_r = \sigma_o' \tag{16.88}$$

The equilibrium equation on the outer surface is

$$\frac{d\sigma_r}{dr} + \frac{\sigma_r - \sigma_\vartheta}{r} = 0 \tag{16.89}$$

The yield criterion and boundary conditions give

$$\sigma_r = -\sigma_o' \ln\frac{r_o}{r} \tag{16.90}$$

where r_o is the outer radius.

Circumferential stress in the tension zone is

$$\sigma_\vartheta = -\sigma_o^{'}\left(1 - \ln\frac{r_o}{r}\right) \qquad (16.91)$$

and in direction of generator z

$$\sigma_z = -\frac{\sigma_o^{'}\left(1 - 2\ln\frac{r_o}{r}\right)}{2} \qquad (16.92)$$

By Eqs. (16.90), (16.91) and (16.92) the radial stresses in the outer zone increase when approaching the neutral axis, and the circumferential stresses decrease, and in the direction of generator z are positive and increase when approaching the neutral axis.

The radial stresses cannot jump on the boundary between the tension and compression zones. By this condition the radius of neutral surface r_n is calculated from the expression

$$\ln\frac{r_o}{r_n} = \ln\frac{r_n}{r_i} \qquad (16.93)$$

Hence

$$r_n = \sqrt{r_o\, r_i} \qquad (16.94)$$

By (16.94) the radius of the neutral surface is equal to the geometric mean of the inner and outer radii. Since the geometrical mean is smaller then the arithmetic one, the thickness of the outer layer is higher than the thickness of the compressed layer. If the inner and outer radii tend to infinity, then the geometric and arithmetic means are equal and the neutral layer is in the middle of the thickness. If the element is bended, the differences of the layers thickness in tension and compression increase.

The bending moment is

$$M = b\int_{r_i}^{r_o} \sigma_\vartheta\, r\, dr \qquad (16.95)$$

which gives

$$M = \frac{bt^2}{4}\sigma_o \qquad (16.96)$$

16.3.4 Deformation

In the bending of wide bands, assume that the strain in z-direction is equal to zero. The outer layers, i.e. between outer and neutral surfaces, are elongated in a circumferential direction. Depending on the location of the neutral surface, the growth of material thickness occurs if the increase of the thickness of the inner layer exceeds the decrease of the thickness of the outer layer or the decrease of the thickness if the decrease of the thickness of the outer layer exceeds the increase of the thickness of the inner layer.

The changes of the bended band are determined by considering the deformation in an elementary volume separated by the central angle ϑ, the outer radius r_o and the inner radius r_i (Fig. 16.10). Denote the changes of the central angle by ϑ, the outer radius by dr_o and the inner radius dr_i. If the plane state of strain is considered, then the area of elementary volume is constant in deformation. The area of domain **1234** (Fig. 16.10) is

$$A_o = \frac{\vartheta}{2}\left(r_o^2 - r_n^2\right) = \text{const} \tag{16.97}$$

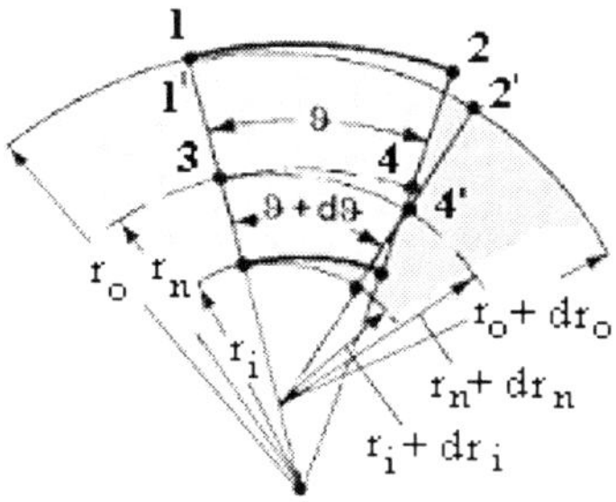

Fig. 16.10. Element of a bended band

By differentiating the above equation we get

$$d\vartheta r_o^2 + 2\vartheta r_o dr - d\vartheta r_n^2 + 2\vartheta r_n dr_n = 0 \tag{16.98}$$

Since the length of a neutral layer does not change $\vartheta\ r_n = \text{const}$, differentiation of Eq. (16.98) gives

$$d\vartheta\ r_n + \vartheta\ dr_n = 0 \tag{16.99}$$

By (16.97) and (16.99) we get

$$d\vartheta r_o^2 + 2\vartheta r_o dr_o + d\vartheta r_n^2 = 0 \tag{16.100}$$

Hence

$$dr_o = -\frac{r_o^2 + r_n^2}{2r_o}\frac{d\vartheta}{\vartheta} \quad (16.101)$$

In a similar way the changes of the inner radius are determined

$$dr_i = -\frac{r_i^2 + r_n^2}{2r_i}\frac{d\vartheta}{\vartheta} \quad (16.102)$$

The change of thickness of a bended band is equal to the differences of changes of the inner and outer radii

$$dg = dr_z - dr_w \quad (16.103)$$

By Eqs. (16.101), (16.102) and (16.103) we get

$$dt = \left[\frac{r_i^2 + r_n^2}{r_i} - \frac{r_o^2 + r_n^2}{r_o}\right]\frac{d\vartheta}{2\vartheta} \quad (16.104)$$

By the expression $t = r_o - r_i$ we get the final expression describing the changes of the thickness of a bended band

$$dt = t\left(\frac{r_n^2}{r_o r_i} - 1\right)\frac{d\vartheta}{2\vartheta} \quad (16.105)$$

We see that in bending without circumferential external forces applied the radius of a neutral surface is the geometric mean of the inner and outer radii

$$r_n = \sqrt{r_o r_i} \quad (16.106)$$

Without external forces applied by Eq. (16.105) the material thickness does not change. In bending however with external forces applied the material thickness changes. Bending with positive forces applied leads to a decrease of the thickness and bending with negative forces leads to an increase of the thickness. In the case of bending without external forces applied, however, the thickness of layers can change. In the bending of wide bands the inner layers are elongated in a circumferential direction, which leads to a decrease of its thickness, and the inner layers are reduced in a circumferential direction, which leads to an increase of the thickness at this layer. Such changes of thickness cause the flow of material in direction to the outer surface. The neutral surface flows in the opposite direction.

Consider the deformation of a material layer closed to a neutral surface in its negative side. If the curvature increases the layer is reduced in the circumferential direction and approaches the neutral layer. For some curvature the layer becomes the neutral layer and when the curvature increases this layer is on the positive side, which leads to changes of stress and strain signs. At a specific moment of time this layer can reach the initial length. The movement of the neutral surface denoted by a is equal to the differences of arithmetic and geometric means of the inner and outer radii

$$a = \frac{r_o + r_i}{2} - \sqrt{r_o r_i} \tag{16.107}$$

16.3.5 Spring-Back

In the beginning of bending material is in an elastic state. The relation between the bending moment and the curvature of the neutral surface is

$$\frac{1}{\rho} = \frac{M}{EJ} \tag{16.108}$$

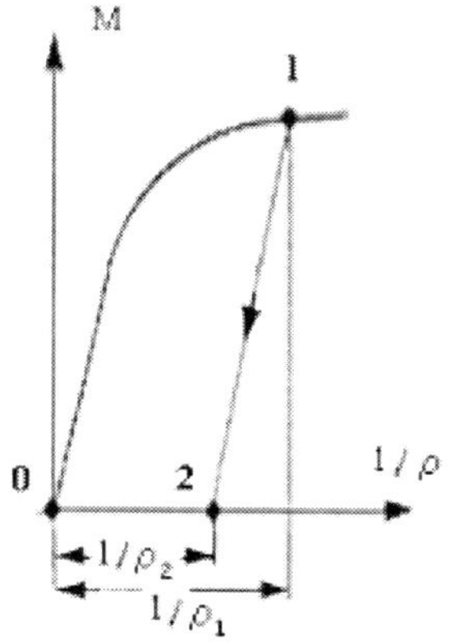

Fig. 16.11. Scheme of the unloading process

If the stress exceeds the yield limit, then the increase of bending moment leads to a fast increase of curvature. In the unloading process elastic strain gives the decrease of curvature, which is illustrated by line **12** in Fig 16.11. The changes of curvature in the loading process in the case of narrow bands and given by Eq. (16.108)

$$\frac{1}{\rho_1} - \frac{1}{\rho_2} = \frac{M}{EJ} = \frac{2}{t}\frac{\sigma_g}{E} \tag{16.109}$$

where σ_g is the bending stress.

In the case of the bending of wide bands

$$\frac{1}{\rho_1}-\frac{1}{\rho_2}=\frac{M\left(1-\nu^2\right)}{EJ}=\frac{2\left(1-\nu^2\right)\sigma_g}{t\,E} \tag{16.110}$$

The unloading process is characterized by the spring-back angle (Fig. 16.12).

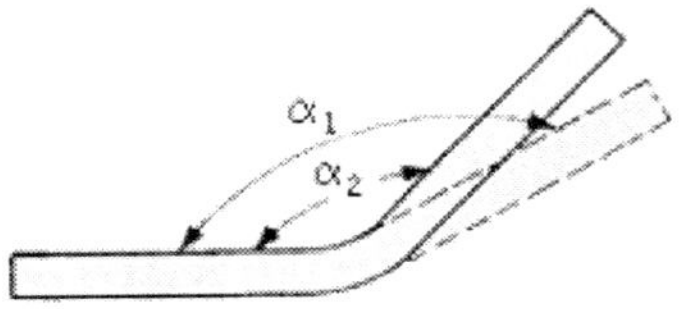

Fig. 16.12. The spring-back of material

By the assumption that the length l of a bended element does not change

$$\alpha_1\rho_1=\alpha_2\rho_2=l \tag{16.111}$$

we get the following expression describing the change of the spring-back angle

$$\alpha_1-\alpha_2=\frac{Ml}{EJ}=\frac{2\sigma_g l}{t\,E} \tag{16.112}$$

and for wide bands

$$\alpha_1-\alpha_2=\frac{Ml}{EJ}=\frac{2l\left(1-\nu^2\right)\sigma_g}{t\,E} \tag{16.113}$$

16.3.6 Residual Stresses

The distribution of stresses for an elastic-plastic state in the loading process is given in Fig. 16.13.

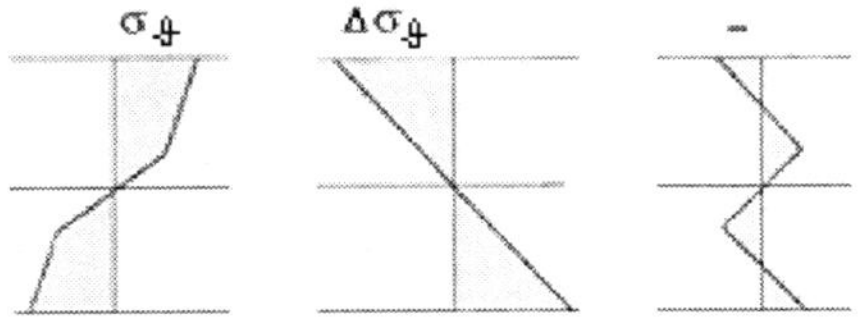

Fig. 16.13. Stresses in the loading and unloading processes

In the unloading process the increment of circumferential stresses is $\Delta\sigma_\vartheta$. The loaded layers are shortened. The incremental stress $\Delta\sigma_\vartheta$ has an opposite sign and its distribution is linear, because in the unloading process deformation is elastic. The stresses in the loading and unloading processes have to give the same value of bending moment related to a neutral axis. The sum of stresses σ_9 and $\Delta\sigma_9$ gives the so-called residual stresses. In the loading process the residual stresses remain negative in outer tension layers and in inner layers positive (Fig. 16.13). Consider the loading process in the same direction. The material undergoes elastic and plastic deformations. The next loading process leads to a decrease of stresses at boundary layers. For some value of the bending moment stress on the outer surface is zero and further loading gives stress of the opposite sign. If the bending moment has the opposite sign, then the initial plastic strain appears for a smaller value of the bending moment (Fig. 16.14). In this case stresses in the boundary

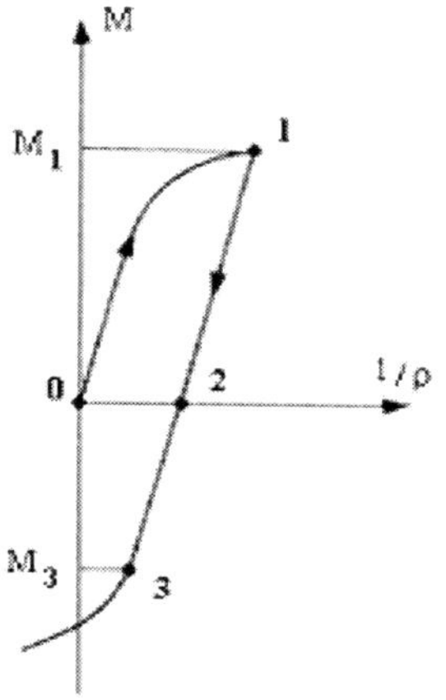

Fig. 16.14. The influence of residual stresses on the shape of a bending curvature

layer have the same sign as residual stresses. The real stresses in the boundary layer are the sum of residual stresses and stresses from the bending moment which leads to early plasticization, than described in the first case.

16.4 Wire Drawing

Consider the equilibrium equation in an elementary volume separated in such a way to ensure the uniform state of stresses described by the principal stresses on the surfaces of separation.

Elementary volume is separated by two spherical surfaces whose distance is $dx/\cos\alpha$ (Fig. 16.15). Consider the unit pressure σ_n and the friction stress τ acting on the surface of contact of material with the conical die. Assume for simplification that the unit pressure σ_n and the friction stress do not influence its sign. The friction law is

$$\tau = f\sigma_n \tag{16.114}$$

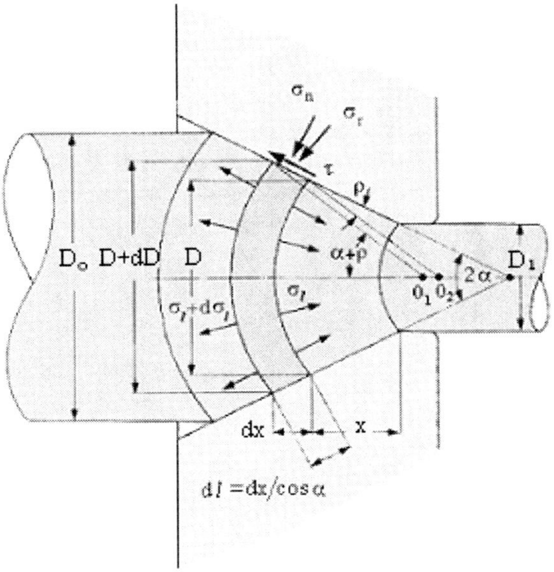

Fig. 16.15. Stresses in an elementary volume of the drawing zone

The equilibrium of forces on an elementary volume in the drawing direction is

$$(\sigma_l + d\sigma_l)(A + dA) - \sigma_l A + \frac{dx}{\cos\alpha}\pi D\sigma_n(\sin\alpha + f\cos\alpha) = 0 \tag{16.115}$$

where

$$A = \frac{\pi D^2}{4} \quad dx = \frac{dD}{2\,tg\alpha} \quad dA = \frac{\pi}{2} D\,dD \tag{16.116}$$

and D is the diameter of the wire and α is the angle of the cone.

The expression on a normal force N acting on an elementary surface dA is (Fig. 16.16)

$$dN = \sigma_n\,dA \tag{16.117}$$

The elementary friction force is

$$dT = dNf \tag{16.118}$$

The resultant force dR is (Fig. 16.16)

$$dR = \sqrt{dN^2 + dT^2} = \sigma_n dF\sqrt{1 + f^2} \tag{16.119}$$

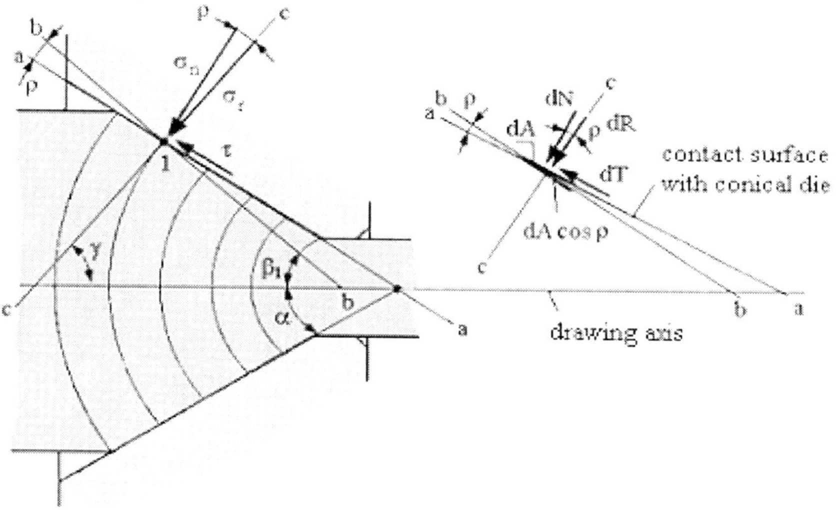

Fig. 16.16. The relation between principal and normal stresses on the surface of contact with the die

The direction of force dR is the same as the direction of principal stress σ_r at the point of contact with the die. The principal stress σ_r is determined from the expression

$$\sigma_r = \frac{dR}{dA\cos\alpha} = \frac{\sigma_n\sqrt{1+f^2}}{\cos\alpha} = \frac{\sigma_n}{\cos^2\rho} \tag{16.120}$$

where ρ is the friction angle i.e. $f = \text{tg}\,\rho$. The trajectory of principal radial stresses distributed in the material is shown in (Fig. 16.17).

By the Huber-Mises yield criterion

$$\sigma_l - \sigma_r = \sigma_o \tag{16.121}$$

Since the radial stresses σ_r are negative, the yield criterion can be rewritten as

$$\sigma_l + \sigma_r = \sigma_o$$

By (16.120) we get

$$\sigma_l + \frac{\sigma_n}{\cos^2\rho} = \sigma_o \tag{16.122}$$

Assume the hardening function of the form

$$\sigma_o = \sigma_{oo} + \varphi\frac{D_o^2 - D^2}{D_o^2} \tag{16.123}$$

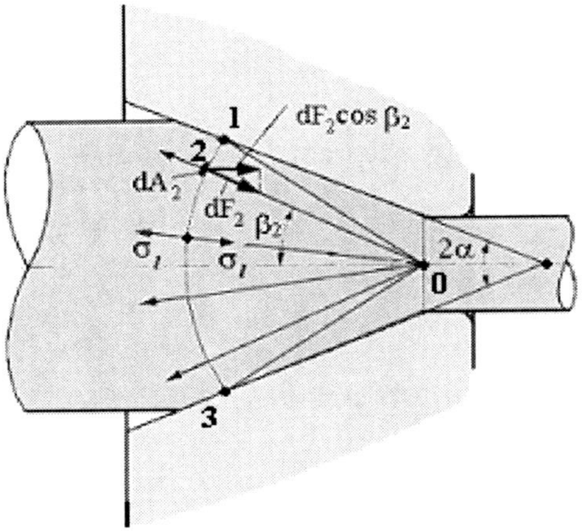

Fig. 16.17. Forces along arch **13** being the trajectory of principal radial stresses

By (16.122) we get

$$\sigma_n = \left(\sigma_{oo} + \varphi\frac{D_o^2 - D^2}{D_o^2} - \sigma_l\right)\cos^2\rho \tag{16.124}$$

By (16.115) and (16.124) we have

$$\frac{d\sigma_l}{dD} + \sigma_l\frac{b}{D} + \frac{2a}{D}\left(\sigma_{oo} + \varphi\frac{D_o^2 - D^2}{D_o^2}\right) = 0 \tag{16.125}$$

where $a = \cos^2\rho\ (1 + f\mathrm{ctg}\alpha) - 1$ and $b = \dfrac{a+1}{a}$ are the constant parameters of the drawing process. After the solution we get

$$\sigma_l = D^{-b}\left\{C - \left[(\sigma_{oo} - \varphi)\frac{2a}{b}D^b - \varphi\frac{2a}{b+2}\frac{D^{b+2}}{D_o^2}\right]\right\} \tag{16.126}$$

The constant C is determined by the boundary conditions on the surface separating the elastic and plastic parts. For $D = D_o$ the stress σ_l is equal to σ_{lo} and comes from the external stress σ_o or, if is zero, then stress σ_l results from a reaction of the elastic part of the material

$$\sigma_l = \sigma_{lo} = D_o^{-b}\left\{C - \left[(\sigma_{oo} + \varphi)\frac{2a}{b}D_o^b - \varphi\frac{2a}{b+2}D_o^b\right]\right\} \tag{16.127}$$

Hence

$$C = D_o^b\left[\sigma_{lo} + 2a\left(\frac{\sigma_{oo} + \varphi}{b} - \frac{\varphi}{b+2}\right)\right]$$

By (16.126) and (16.127) we get

$$\sigma_l = \left(\frac{D_o}{D}\right)^b \left[\sigma_{lo} + 2a\left(\frac{\sigma_{oo} - \varphi}{b} - \frac{\varphi}{b+2}\right)\right] + 2a\left[\frac{\varphi}{b+2}\left(\frac{D}{D_o}\right)^2 - \frac{\sigma_{oo} + \varphi}{b}\right] \tag{16.128}$$

If b → -2 some terms in Eq. (16.128) are undetermined, so

$$\sigma_l = \lim_{b\to -2}\left\{\left(\frac{D_o}{D}\right)^b \left[\sigma_{lo} + 2a\left(\frac{\sigma_{oo} + \varphi}{b} - \frac{\varphi}{b+2}\right)\right] + 2a\left[\frac{\varphi}{b+2}\left(\frac{D}{D_o}\right)^2 - \frac{\sigma_{oo} + \varphi}{b}\right]\right\} = \lim_{b\to -2}\left\{\frac{\left(\frac{D}{D_o}\right)^2 - \left(\frac{D}{D_o}\right)^{-b}}{b+2}\varphi 2a + \left(\frac{D_o}{D}\right)^b\left[\sigma_{lo} + 2a\frac{\sigma_{oo} + \varphi}{b}\right] - 2a\frac{\sigma_{oo} + \varphi}{b}\right\} \tag{16.129}$$

If $b = -2$ the expression

$$\frac{\left(\frac{D}{D_o}\right)^2 - \left(\frac{D}{D_o}\right)^{-b}}{b+2}\varphi 2a$$

is undetermined. By the l'Hospital rule

$$\lim_{b\to -2}\frac{\left(\frac{D}{D_o}\right)^2 - \left(\frac{D}{D_o}\right)^{-b}}{b+2}\varphi 2a = \lim_{b\to -2}\frac{\frac{d}{db}\left[\left(\frac{D}{D_o}\right)^2 - \left(\frac{D}{D_o}\right)^{-b}\right]}{\frac{d}{db}(b+2)}\varphi 2a = 2a\varphi \ln\frac{D}{D_o}\left(\frac{D}{D_o}\right)^{-b} \tag{16.130}$$

By (16.130) we get for b = - 2

$$\sigma_l = \left(\frac{D_o}{D}\right)^b \left[\sigma_{lo} + 2a\left(\frac{\sigma_{oo} - \varphi}{-2} - \varphi \ln\frac{D_o}{D}\right)\right] - 2a\frac{\sigma_{oo} + \varphi}{-2} \tag{16.131}$$

By the yield criterion (16.124) and Eq. (16.120), the stress σ_r for $b \neq -2$ is

$$\sigma_r = \left(\frac{D_o}{D}\right)^b \left[\sigma_{lo} + 2a\left(\frac{\sigma_{oo} - \varphi}{b} - \frac{\varphi}{b+2}\right)\right] + 2a\left[\frac{\varphi}{b+2}\left(\frac{D}{D_o}\right)^2 - \frac{\sigma_{oo} + \varphi}{b}\right] + \\ - \left[\sigma_{po} + \varphi\left(1 - \frac{D^2}{D_o^2}\right)\right] \tag{16.132}$$

for $b = -2$

$$\sigma_r = \left(\frac{D_o}{D}\right)^b \left[\sigma_{lo} + 2a\left(\frac{\sigma_{oo} + \varphi}{b} - \varphi \ln \frac{D_o}{D}\right)\right] \\ - 2a\frac{\sigma_{oo} + \varphi}{b} - \left[\sigma_{oo} + \varphi\left(1 + \frac{D^2}{D_o^2}\right)\right] \tag{16.133}$$

The term $\left(\frac{D_o}{D}\right)^b$ can be expanded in the series

$$\left(\frac{D_o}{D}\right)^b = 1 + b \ln \frac{D_o}{D} + \frac{1}{2!}\left(b \ln \frac{D_o}{D}\right)^2 + \frac{1}{3!}\left(b \ln \frac{D_o}{D}\right)^3 + \ldots \tag{16.134}$$

and taking two terms from Eq. (16.134) we get
for $b \neq -2$

$$\sigma_l = 2a\left[\left(\sigma_{oo} + \varphi - b\frac{\varphi}{b+2}\right)\ln\frac{D_o}{D} - \frac{\varphi}{b+2}\frac{D_o^2 - D^2}{D_o^2} + \left(1 + b \ln \frac{D_o}{D}\right)\sigma_{lo}\right] \tag{16.135}$$

for $b = -2$

$$\sigma_l = 2a\left[\left(\sigma_{oo} - b\,\varphi \ln \frac{D_o}{D}\right)\ln\frac{D_o}{D} + \left(1 + b \ln \frac{D_o}{D}\right)\sigma_{lo}\right] \tag{16.136}$$

for $b \neq -2$

$$\sigma_r = 2a\left[\left(\sigma_{oo} + \varphi - \frac{\varphi b}{b+2}\right)\ln\frac{D_o}{D} - \frac{\varphi}{b+2}\frac{D_o^2 - D^2}{D_o^2}\right] \\ + \left(1 + b \ln \frac{D_o}{D}\right)\sigma_{lo} - \left(\sigma_{oo} - \varphi\frac{D_o^2 - D^2}{D_o^2}\right) \tag{16.137}$$

for b = -2

$$\sigma_r = 2a\left[\left(\sigma_{oo} - b\,\varphi \ln\frac{D_o}{D}\right)\ln\frac{D_o}{D} + \left(1 + b\ln\frac{D_o}{D}\right)\sigma_{lo} - \left(\sigma_{oo} - \varphi\frac{D_o^2 - D^2}{D_o^2}\right)\right] \tag{16.138}$$

In the region between the conical die and the sizing part the increase of stress σ_l is proportional to the ratio cap surface – wire cross-section (Fig. 16.18)

$$\frac{A_{cap}}{A_1} = \frac{\pi\left(R_1^2 + h^2\right)}{\pi R_1^2} = 1 + tg^2\frac{\beta}{2} = \frac{1}{\cos^2\frac{\beta}{2}} \tag{16.139}$$

where h is the cap height, R_1 is the radius and $\beta = \alpha + \rho$.

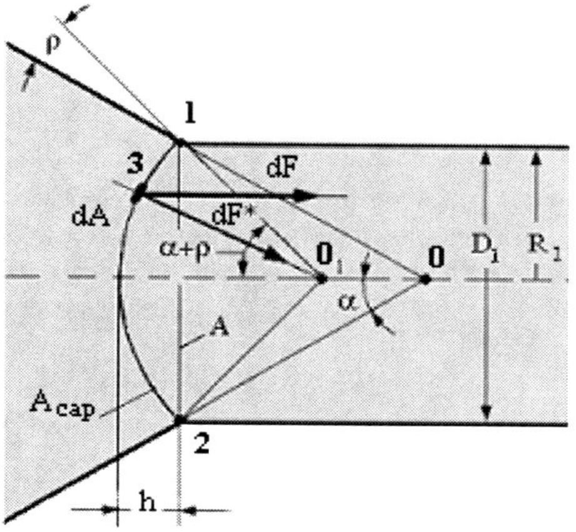

Fig. 16.18. Forces in exit from the plastic deformation zone

The mean stress σ_{lm} in the entrance of material to the sizing part is for $b \neq -2$

$$\sigma_{lm} = \frac{2a}{\cos^2\frac{\alpha+\rho}{2}}\left[\left(\sigma_{oo} + \varphi - b\frac{\varphi}{b+2}\right)\ln\frac{D_o}{D_1} - \frac{\varphi}{b+2}\,\frac{D_o^2 - D_1^2}{D_o^2}\right] + \left(1 + b\ln\frac{D_o}{D_1}\right)\frac{\sigma_{lo}}{\cos^2\frac{\alpha+\rho}{2}} \tag{16.140}$$

for b = -2

$$\sigma_{lm} = \frac{2a}{\cos^2\frac{\alpha+\rho}{2}}\left(\sigma_{oo} - b\,\varphi \ln\frac{D_o}{D_1}\right)\ln\frac{D_o}{D_1} + \left(1 + b\ln\frac{D_o}{D_1}\right)\frac{\sigma_{lo}}{\cos^2\frac{\alpha+\rho}{2}} \tag{16.141}$$

The above expressions determine the mean value of stress σ_l. The real stress σ_l is distributed non-uniformly in the cross-section of the wire. The highest is on the surface and the lowest is at the wire axis.

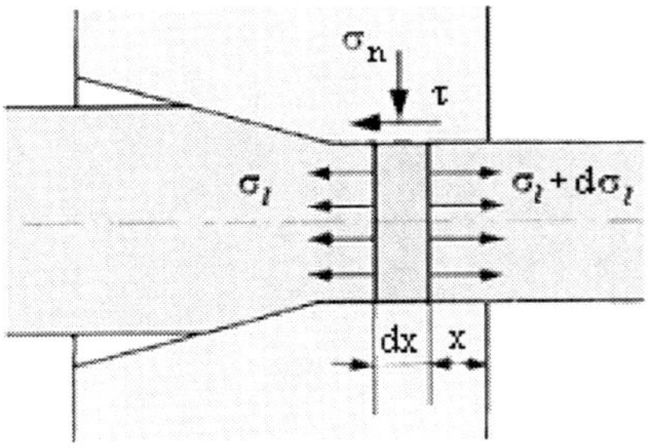

Fig. 16.19. Stresses in an elementary volume in the sizing zone

The further increase of stress σ_l takes place in the sizing part by the action of friction forces on the contact surface. In the sizing zone the wire does not undergo plastic deformation. In order to determine the influence of the sizing zone on the drawing stresses, we write the equilibrium of forces on an elementary volume separated from the sizing region (Fig. 16.19)

$$(\sigma_l + d\sigma_l)\frac{\pi}{4}D_1^2 - \sigma_l\frac{\pi}{4}D_1^2 - f\sigma_n \pi D_1 dx = 0 \tag{16.142}$$

which gives

$$d\sigma_l D_1 - 4f\sigma_n dx = 0 \tag{16.143}$$

Assume in the sizing part the plasticity criterion

$$\sigma_l + \sigma_n = \sigma_{o1} \tag{16.144}$$

By Eqs. (16.144) and (16.143) we get

$$\frac{d\sigma_l}{\sigma_l - \sigma_{o1}} = -\frac{4f}{D_1}dx \tag{16.145}$$

and after integration

$$\sigma_l - \sigma_{o1} = C_1 e^{\frac{-4fx}{D_1}} \tag{16.146}$$

The constant C_1 is determined by the boundary condition. The stress σ_l changes in a continuous way, so σ_l at the end of the cross-section is equal to the stress σ_l in the initial sizing zone. If x = 0, then $\sigma_l = \sigma_{l mean}$, and $C_1 = \sigma_{l mean} - \sigma_{ol}$. By (16.146) we get

$$\sigma_l = \sigma_{ol} + \left(\sigma_{l\,mean} - \sigma_{ol}\right) e^{\frac{-4f}{D_l}x} \tag{16.147}$$

The term $e^{\frac{-4f}{D_l}x}$ can be expanded in the series

$$e^{\frac{-4f}{D_l}x} = 1 - \frac{4f}{D_l}x + \frac{1}{2!}\left(-\frac{4f}{D_l}x\right)^2 + \frac{1}{3!}\left(-\frac{4f}{D_l}x\right)^3 \tag{16.148}$$

Taking the first two terms of the expression (16.148)

$$\sigma_l = \sigma_{l\,mean}\left(1 - \frac{4f}{D_l}x\right) + \frac{4f}{D_l} \cdot \sigma_{ol} \tag{16.149}$$

The final expression for drawing stress σ_d is
for $b \neq -2$

$$\sigma_d = \left\{ \frac{2a}{\cos^2\frac{\alpha+\rho}{2}} \left[\left(\sigma_{oo} + \varphi - b\frac{\varphi}{b+2} \right) \ln\frac{D_o}{D_l} - \frac{\varphi}{b+2}\frac{D_o^2 - D_l^2}{D_o^2} \right] + \left(1 + b\ln\frac{D_o}{D_l}\right)\sigma_o \right\} \left(1 - \frac{4fl_d}{D_l}\right) + \frac{4f}{D_l} l_d \sigma_{ol} \tag{16.150}$$

for $b = -2$

$$\sigma_d = \left[\frac{2a}{\cos^2\frac{\alpha+\rho}{2}} \left(\sigma_{oo} - b\,\varphi\ln\frac{D_o}{D_l} \right) \ln\frac{D_o}{D_l} + \left(1 + b\ln\frac{D_o}{D_l}\right) \frac{\sigma_{lo}}{\cos^2\frac{\alpha+\rho}{2}} \right] \cdot \left(1 - \frac{4fl_d}{D_l}\right) + \frac{4fl_d}{D_l}\sigma_{oo} \tag{16.151}$$

The above equations do not take into account in a sufficient way the influence of the bending of layers in the entrance zone to the die and in the zone between the sizing part and the conical die as well as deformations resulting from friction forces. The influence of the above parameters increases with the increase of the friction coefficient as well as the angle of the conical die and in such cases the above expressions underrate the value of drawing stress.

16.5 Sink Drawing

Sink drawing is used to reduce the outer radius of the tube with a small change of wall thickness. Consider in the drawing zone the elementary volume (Fig. 16.20) separated by two planes passing by the axis of symmetry, two conical surfaces whose generating lines are perpendicular to the conical die and the conical surface of the tube. Consider stress σ_l, circumferential stress σ_ϑ, radial stress σ_r and the tangent stress on the surface of contact between the die and the tube (Fig. 16.21). The equilibrium equation written in direction of σ_l is

$$\begin{aligned}&(\sigma_l + d\sigma_l)(r + dr)\,t\,d\vartheta + \tau r_z \frac{dr}{\sin\alpha} d\vartheta \\ &+ 2\sigma_\vartheta \frac{dr}{\sin\alpha} t \frac{d\vartheta}{2} \sin\alpha - \sigma_l r g d\vartheta = 0\end{aligned} \tag{16.152}$$

where r_z is the outer radius. By omitting the terms of the small values

$$\sigma_l + d\sigma_l \frac{r}{dr} + \sigma_\vartheta + \tau \frac{r_z}{t \sin\alpha} = 0 \tag{16.153}$$

The equilibrium equation in the direction of radial stress σ_r is

$$2\sigma_\vartheta t \frac{dr}{\sin\alpha} \frac{d\vartheta}{2} \cos\alpha - \sigma_r r_z \frac{dr}{\sin\alpha} d\vartheta = 0 \tag{16.154}$$

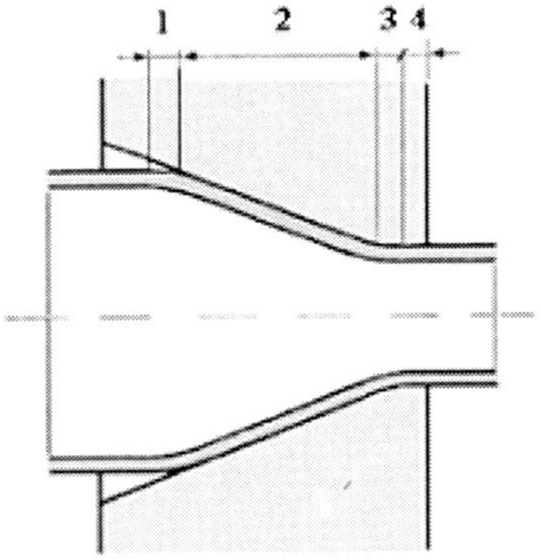

Fig. 16.20. Sink drawing **1** - deformation without contact; **2** – the drawing zone; **3** – the transient zone; **4** – the sizing zone

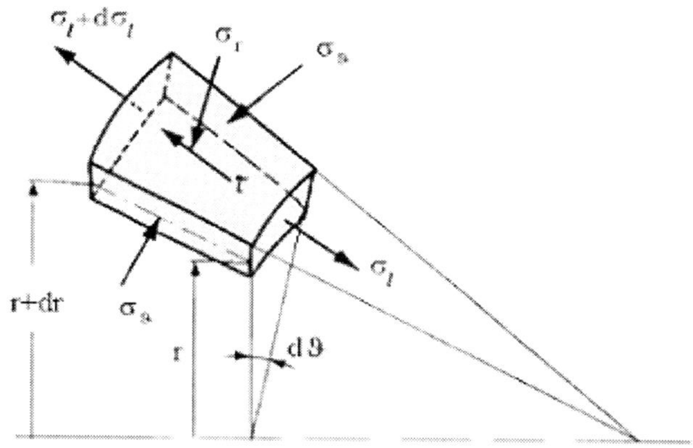

Fig. 16.21. An elementary volume in the drawing zone

By (16.154) we get

$$\frac{\sigma_r}{t} - \frac{\sigma_\vartheta}{r_z}\cos\alpha = 0 \tag{16.155}$$

Assume on the contact surface

$$\tau = f\sigma_r \tag{16.156}$$

By (16.153), (16.155) and (16.156) we get

$$\sigma_l + d\sigma_l \frac{r}{dr} + \sigma_\vartheta + \sigma_\vartheta f \operatorname{ctg}\alpha = 0 \tag{16.157}$$

In order to solve Eq. (16.157) the relation between σ_l and σ_ϑ should be known. Since material is in a plastic state, stresses fulfill the yield criterion. Assume

$$\sigma_l - \sigma_\vartheta = \beta\sigma_o \tag{16.158}$$

where β is a coefficient taking into account the influence of mean stress on the plastic state. Since circumferential stresses are positive the yield criterion is written in the form

$$\sigma_l - (-\sigma_\vartheta) = \beta\sigma_o \tag{16.159}$$

Assume

$$\sigma_o = \sigma_{oo} + B_1\varepsilon_i^n \tag{16.160}$$

where

$$\varepsilon_i = \frac{\sqrt{2}}{3}\sqrt{(\varepsilon_l - \varepsilon_r)^2 + (\varepsilon_r - \varepsilon_\vartheta)^2 + (\varepsilon_\vartheta - \varepsilon_l)^2} \tag{16.161}$$

The change of the wall thickness is not significant in the drawing of tubes, so assume

$$\varepsilon_r = 0$$

By the condition of constant volume

$$\varepsilon_l = -\varepsilon_\vartheta \tag{16.162}$$

By Eq. (16.161)

$$\varepsilon_i = \frac{2}{\sqrt{3}}\varepsilon_l \tag{16.163}$$

where

$$\varepsilon_l = \ln\frac{r_o}{r} \tag{16.164}$$

Finally we get from Eq. (16.160)

$$\sigma_o = \sigma_{oo} + B\ln^n\frac{r_o}{r} \tag{16.165}$$

$$B = \left(\frac{2}{\sqrt{3}}\right)^n B_1 \tag{16.166}$$

By the yield criterion (16.159) and expression (16.166), the equilibrium condition (16.157) takes the form

$$f\,\mathrm{ctg}\alpha\,\sigma_l - d\sigma_l\frac{r}{dr} - \left(\sigma_{oo} + B\ln^n\frac{r_o}{r}\right)\left(1 + f\,\mathrm{ctg}\alpha\right) = 0 \tag{16.167}$$

Solving Eq. (16.167) with the boundary conditions if $r = r_o$, then $\sigma_l = 0$ we get

$$\begin{aligned}\sigma_l &= \beta\left(1 + f\,\mathrm{ctg}\alpha\right)\sigma_{oo}\left(1 - \frac{1}{2}f\,\mathrm{ctg}\alpha\ln\frac{r_o}{r}\right)\ln\frac{r_o}{r} \\ &+ \frac{\beta(1 + f\,\mathrm{ctg}\alpha)B}{1+n}\left(1 - \frac{f\,\mathrm{ctg}\alpha}{2+n}\ln\frac{r_o}{r}\right)\ln^{1+n}\frac{r_o}{r}\end{aligned} \tag{16.168}$$

The increase of stress in the transient zone is

$$\sigma_l = \frac{\sigma_{ol}t}{4R_1} \tag{16.169}$$

where σ_{ol} is the stress in the axis direction.

In the beginning of the sizing zone the stress along the drawing axis is

$$\sigma_l = \beta\left(1 + f\ \mathrm{ctg}\alpha\right)\sigma_{oo}\left(1 - \frac{1}{2} f\ \mathrm{ctg}\alpha \ln\frac{r_o}{r}\right)\ln\frac{r_o}{r} + \frac{\beta\left(1 + f\ \mathrm{ctg}\alpha\right)B}{1+n}\left(1 - \frac{f\ \mathrm{ctg}\alpha}{2+n}\ln\frac{r_o}{r}\right)\ln^{1+r}\frac{r_o}{r} + \frac{\sigma_{ol}t}{4R_1} \tag{16.170}$$

The equilibrium of forces on an elementary volume (Fig. 16.22) in the sizing zone is

$$\left(\sigma_l + d\sigma_l\right)r_1 t d\vartheta - \sigma_l r_1 t d\vartheta - f\sigma_r dx\, r_z d\vartheta = 0 \tag{16.171}$$

where r_1 is the radius of the sizing part. Hence

$$d\sigma_l r_1 t - f\sigma_r r_z dx = 0 \tag{16.172}$$

The equilibrium equation in the radial direction is

$$\sigma_r r_z dx\, d\vartheta - 2\sigma_\vartheta t dx \frac{d\vartheta}{2} = 0 \tag{16.173}$$

Hence

$$\sigma_r = \frac{\sigma_\vartheta t}{r_z} \tag{16.174}$$

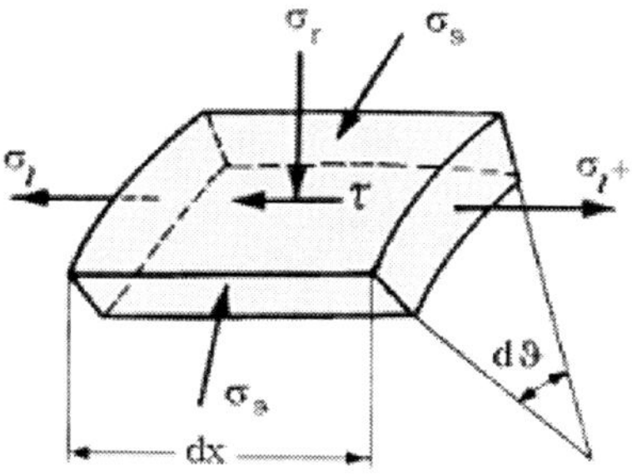

Fig. 16.22. An elementary volume in the sizing zone

Assume that in the sizing zone the material is in a plastic state and fulfills the yield criterion

$$\sigma_l + \sigma_\vartheta = \beta\,\sigma_{ol} \tag{16.175}$$

Hence

$$\sigma_\vartheta = \beta\,\sigma_{ol} - \sigma_l \tag{16.176}$$

By (16.172), (16.174) and (16.175) we get

$$\frac{d\sigma_l}{\sigma_l - \beta\sigma_{ol}} = -f\frac{dx}{r_l} \tag{16.177}$$

By integrating of equation (16.177) and taking into account the boundary conditions if x = 0, then $\sigma_l = \sigma_{l1}$, we get

$$\sigma_l = \sigma_{l1} e^{\frac{-fx}{r_l}} + \beta\sigma_{ol}\left(1 - e^{\frac{-fx}{r_l}}\right) \tag{16.178}$$

At the end of the sizing zone for x = *l* we get by (16.169) the following expression for the drawing stress

$$\begin{aligned}\sigma_{ld} = &\left[\beta(1 + f\,\mathrm{ctg}\alpha)\sigma_{oo}\left(1 - \frac{1}{2}f\,\mathrm{ctg}\alpha\ln\frac{r_o}{r_l}\right)\ln\frac{r_o}{r_l}\right. \\ &\left. + \frac{\beta(1 + f\,\mathrm{ctg}\alpha)B}{1+n}\left(1 - \frac{f\,\mathrm{ctg}\alpha}{2+n}\ln\frac{r_o}{r_l}\right)\ln^{1+n}\frac{r_o}{r_l} + \frac{\sigma_{ol}t}{4R_l}\right] \\ &\times e^{\frac{-fl}{r_l}} + \beta\sigma_{ol}\left(1 - e^{\frac{-fl}{r_l}}\right)\end{aligned} \tag{16.179}$$

Assume that the following relations hold

$$\frac{\varepsilon_l}{\varepsilon_r} = \frac{\sigma_l - \sigma_m}{\sigma_r - \sigma_m} \qquad \frac{\varepsilon_r}{\varepsilon_\vartheta} = \frac{\sigma_r - \sigma_m}{\sigma_\vartheta - \sigma_m} \tag{16.180}$$

where

$$\sigma_m = \frac{\sigma_l + \sigma_r + \sigma_\vartheta}{3} \tag{16.181}$$

The relation between radial and circumferential stresses is determined from the equilibrium equation in the ring cut from the tube of a thickness equal to one

$$\sigma_{rm} D = \sigma_\vartheta 2t \tag{16.182}$$

Hence

$$\sigma_{rm} = \frac{\sigma_\vartheta\, 2t}{D} \tag{16.183}$$

The mean radial stresses are determined by the expression

$$\sigma_{rmean} = \frac{\int_0^t \sigma_r dt}{t} = a\sigma_{rm} \tag{16.184}$$

where a is the coefficient characterizing the distribution of σ_r along the thickness. Assume the yield criterion

$$\sigma_l - \sigma_\vartheta = \beta\sigma_o \tag{16.185}$$

By (16.183), (16.184) and (16.185) we get

$$\ln\frac{t_o}{t_1} = \frac{4\frac{at}{D}(\beta\sigma_o - \sigma_l) - \beta\sigma_o + 2\sigma_l}{2\beta\sigma_o - \sigma_l - \frac{2at}{D}(\beta\sigma_o - \sigma_l)}\ln\frac{d_o}{d_1} \tag{16.186}$$

Introduce the coefficient $\omega = \sigma_l/\sigma_o$. Then the expression (16.186) takes the form

$$\ln\frac{t_o}{t_1} = \frac{4\frac{at}{D}(\beta - \omega) - \beta + 2\omega}{2\beta - \omega - \frac{2at}{D}(\beta - \omega)}\ln\frac{d_o}{d_1} \tag{16.187}$$

The above expression serves as a basis in the analysis of the wall thickness related to the parameters of the process. If the strain components ε_l, ε_r, ε_ϑ do not satisfy

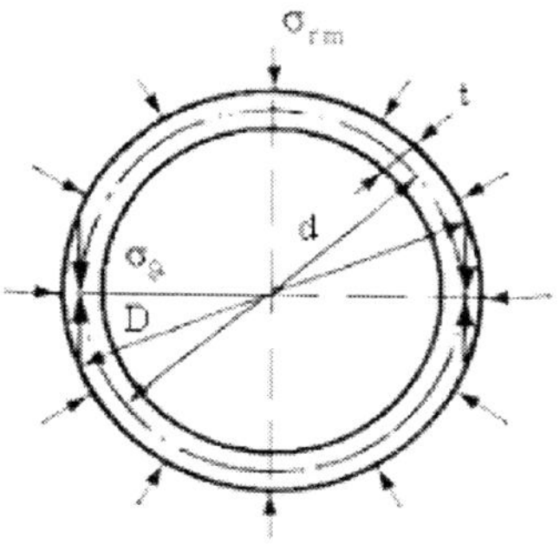

Fig. 16.23. Forces acting in the cross-section of the ring

the proportionality condition (16.180) the incremental expression is introduced

$$\frac{d\varepsilon_r}{d\varepsilon_\vartheta} = \frac{\sigma_r - \sigma_m}{\sigma_\vartheta - \sigma_m} \tag{16.188}$$

where

$$d\varepsilon_r = \frac{dt}{t} \qquad d\varepsilon_\vartheta = \frac{dr}{r} \tag{16.189}$$

By the yield criterion (16.185) and relations between the circumferential and radial stresses (16.183), (16.184) and the relation (16.188) we get

$$\frac{dt}{t} = \frac{4\frac{at}{D}\left(\beta\sigma_o - \sigma_l\right) - \beta\sigma_o + 2\sigma_l}{2\beta\sigma_o - \sigma_l - \frac{2at}{D}\left(\beta\sigma_o - \sigma_l\right)}\frac{dr}{r} \tag{16.190}$$

16.6 The Sinking of Tubes in Sinking Mills

The reduction of tube diameter is carried out in hot rolling. The process is carried out in sinking mills. The scheme of sinking is illustrated in Fig. (16.24). The most frequently system used in the sizing processes is a circle-oval or by applying three sinking mills. In rolling the outer radius as well as the wall thickness are changed.

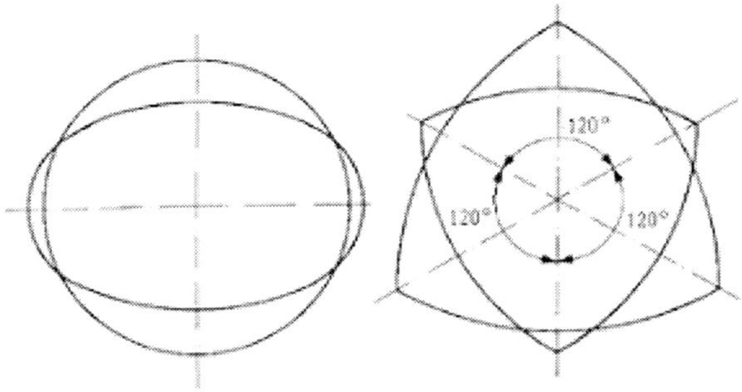

Fig. 16.24. Sinking in sinking mills

Assume that the longitudinal σ_z and radial σ_r stresses are neglected and the circumferential stresses σ_ϑ reach the value of yield limit σ_o

$$\sigma_\vartheta = -\sigma_o \tag{16.191}$$

The changes in wall thickness of the tube are determined by the relation

$$\frac{d\varepsilon_r}{d\varepsilon_\vartheta} = \frac{\sigma_r - \sigma_m}{\sigma_\vartheta - \sigma_m} \tag{16.192}$$

where σ_m is the mean stress

$$\sigma_m = \frac{\sigma_r + \sigma_\vartheta + \sigma_z}{3} \tag{16.193}$$

and

$$d\varepsilon_r = \frac{dt}{t} \qquad d\varepsilon_\vartheta = \frac{dD}{D} \tag{16.194}$$

By (16.192) and (16.194) we get

$$\frac{dt}{t} = -\frac{1}{2}\frac{dD}{D} \tag{16.195}$$

Integrating Eq. (16.195) we get

$$\ln\frac{t_1}{t_o} = -\frac{1}{2}\ln\frac{D_1}{D_o} \tag{16.196}$$

where subscript o denotes the reference and 1 the current value. Hence

$$t_1 = t_o\sqrt{\frac{D_o}{D_1}} \tag{16.197}$$

By Eq. (16.196) the wall thickness in rolling increases which is the fault of the process. However we can influence the wall thickness by applying tension in rolling. The tension in rolling is characterized by the ratio of longitudinal stress σ_z to yield stress σ_o

$$n = \frac{\sigma_z}{\sigma_o} \tag{16.198}$$

The circumferential stress σ_ϑ is determined by the yield criterion

$$\sigma_z - \sigma_\vartheta = \sigma_o \tag{16.199}$$

By Eqs. (16.198) and (16.199) we get

$$\sigma_\vartheta = (n-1)\sigma_o \tag{16.200}$$

The mean stress is

$$\sigma_m = \frac{\sigma_r + \sigma_\vartheta + \sigma_z}{3} = \frac{(2n-1)}{3}\sigma_o \tag{16.201}$$

By Eq. (16.192) we get

$$\frac{dt}{t} = \frac{dD}{D}\frac{1-2n}{n-2} \tag{16.202}$$

Consider the case for which the wall thickness does not change. Since dD/D is different from zero, then the wall thickness does not change

$$\frac{1-2n}{n-2} = 0 \tag{16.203}$$

Hence

$$n = \frac{1}{2}$$

This means for a positive stress σ_z equal to the half of the yield limit the wall thickness does not change. If $n > \frac{1}{2}$, then $\frac{1-2n}{n-2}$ is positive, which means that the wall thickness changes. By integrating Eq. (16.202) we get

$$\ln\frac{t_1}{t_o} = \frac{1-2n}{n-2}\ln\frac{D_1}{D_o} \tag{16.204}$$

By (16.204) the wall thickness after reduction can be determined. Theoretically n can reach a value equal to one. Then

$$\ln\frac{t_1}{t_o} = \ln\frac{D_1}{D_o} \tag{16.205}$$

This means the radial strain ε_r and circumferential strain ε_ϑ are equal. The tension in rolling is assumed in practice as 0.7 – 0.8.

16.7 The Fullering of Round Bars

In the fullering of round bars made of materials of low yield stress the shape fullers are applied (Fig. 16.25). Consider the case denoted by **4** in Fig. 16.25 where the fullers surround the bars entirely. Let the bar axis be the z axis. Because of axial symmetry the circumferential and radial strains are equal and negative, and the strain along the z axis is positive. The analysis of stresses is carried out in

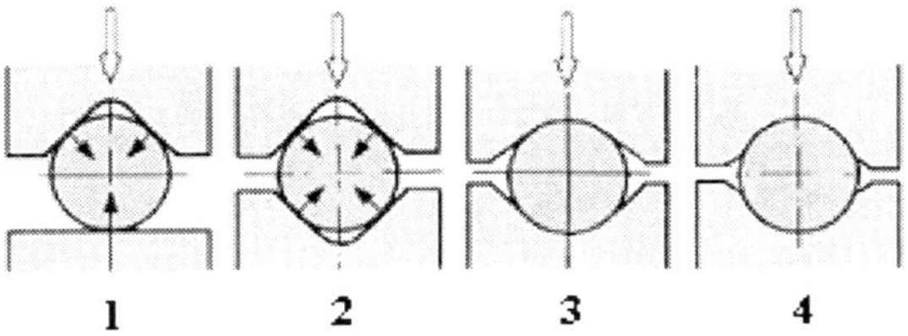

Fig. 16.25. Fullers in the elongation of round bars.

an elementary volume (Fig. 16.26) separated by two planes perpendicular to the z axis whose distance is dz.

Consider the radial σ_r and friction stress τ_{zr} on the contact surface between the die and the workpiece. On the bottom of the surface consider stresses σ_z and $\sigma_z + d\sigma_z$.

In the fullering of round bars in shape fullers the ratio of width of the fuller to the diameter of the bar is less than 2. Assume that the friction forces increase proportionally when moving from the fuller axis.

$$\tau_{zr} = -2f\sigma_o \frac{z}{l_o} \tag{16.206}$$

where l_o is the length of the fuller.

The equilibrium of forces on an elementary volume in the direction of the bar axis is

$$(\sigma_z + d\sigma_z)\frac{\pi D^2}{4} - \sigma_z \frac{\pi D^2}{4} + \tau_{zr}\pi D dz = 0 \tag{16.207}$$

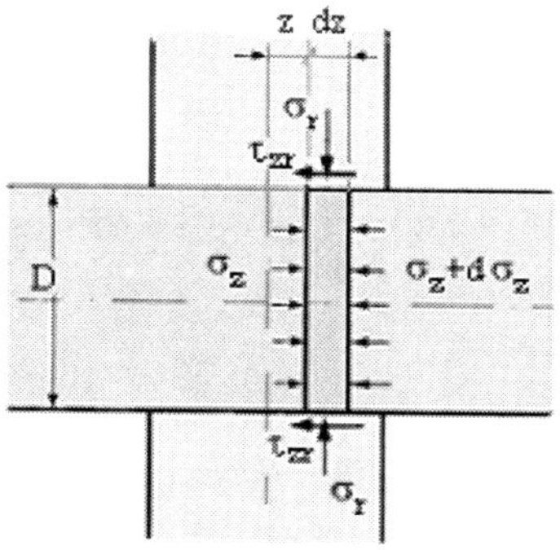

Fig. 16.26. An elementary volume in the fullering of round bars.

Hence

$$\frac{d\sigma_z}{dz} + \frac{4\tau_{zr}}{D} = 0 \tag{16.208}$$

By axisymmetry after differentiating of the yield criterion we get

$$d\sigma_z = d\sigma_r \tag{16.209}$$

By (16.206), (16.209) and (16.208)

$$\frac{d\sigma_r}{dz} - \frac{8f\sigma_o z}{Dl_o} = 0 \tag{16.210}$$

After integrating we have

$$\sigma_r = \frac{4f\sigma_o}{Dl_o} z^2 + C \tag{16.211}$$

The constant C is determined by the boundary condition; if $z = 0.5l_o$, then $\sigma_r = -\sigma_o$. Then we get

$$C = -\sigma_o\left(1+\frac{fl_o}{D}\right) \tag{16.212}$$

By (16.212) and (16.211) we get

$$\sigma_r = -\sigma_o\left[1+\frac{4f}{Dl_o}\left(\frac{l_o^2}{4}-z^2\right)\right] \tag{16.213}$$

The fullering force is

$$P = 2\int_0^{0.5l_o}|\sigma_r|D\,dz = \sigma_o D l_o\left(1+\frac{2}{3}f\frac{l_o}{D}\right) \tag{16.214}$$

and the mean pressure

$$p_{mean} = \sigma_o\left(1+\frac{2}{3}f\frac{l_o}{D}\right) \tag{16.215}$$

By Eq. (16.215) the mean unit pressures increase with the friction coefficient and with the ratio of the length of fuller l_o to its diameter D.

16.8 Punching

Consider the domain of deformation divided into the subdomains denoted by **1** and **2** (Fig. 16.27). Assume the radial stresses between subdomains **1** and **2**

$$\sigma_r = -1.1\ \sigma_o \ln\frac{D}{d} \tag{16.216}$$

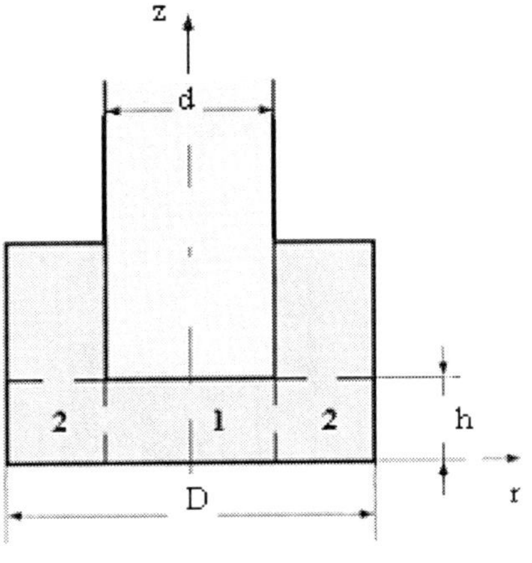

Fig. 16.27. Punching

and the stress along the z axis

$$\sigma_z = -\sigma_o\left(1 + 1.1\ln\frac{D}{d}\right) \tag{16.217}$$

In the first domain the equilibrium equation has the form

$$d\sigma_z = \frac{\sigma_o}{h}dr \tag{16.218}$$

After integration we get

$$\sigma_z = \frac{\sigma_o}{h}r + C \tag{16.219}$$

The constant C is determined by the boundary conditions on the surface between subdomains **1** and **2**

$$\frac{\sigma_o}{h}\frac{d}{2} + C = -\sigma_o\left(1 + 1.1\ln\frac{D}{d}\right) \tag{16.220}$$

Hence

$$C = -\sigma_o\left(1 + 1.1\ln\frac{D}{d} + \frac{d}{2h}\right) \tag{16.221}$$

Finally we get the following expression describing the change of stress σ_z

$$\sigma_z = -\sigma_o\left(1 + 1.1\ln\frac{D}{d} + \frac{0.5d - r}{h}\right) \tag{16.222}$$

The value of the punching force is

$$P = 2\pi\sigma_o\int_0^{0.5d}\left(1 + 1.1\ln\frac{D}{d} + \frac{0.5d - r}{h}\right)rdr \tag{16.223}$$

After integration we get

$$P = \sigma_o\left(1 + 1.1\ln\frac{D}{d} + \frac{1}{6}\frac{d}{h}\right)\frac{\pi d^2}{4} \tag{16.224}$$

The mean unit pressure p_m is

$$p_m = \sigma_o\left(1 + 1.1\ln\frac{D}{d} + \frac{1}{6}\frac{d}{h}\right) \tag{16.225}$$

16.9 Sheet Drawing

The cupping operation is the first operation in forming of a cylindrical drawpiece with a sheet bottom (Fig. 16.28). The drawing carried out at room temperatures

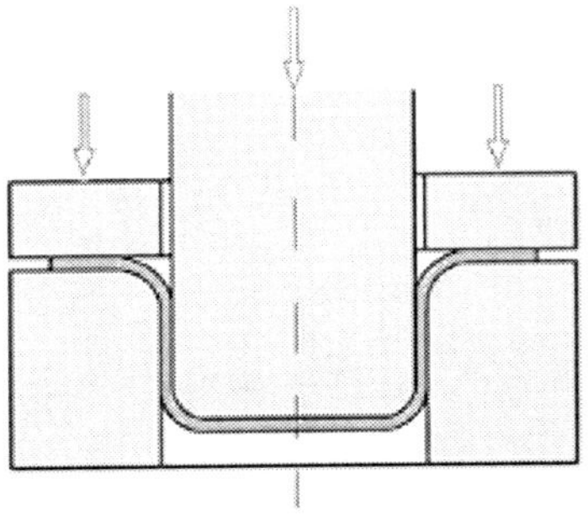

Fig. 16.28. Sheet forming

leads to a hardening of the material. The highest strains are on the border of the drawpiece, and moving to the bottom of the drawpiece the material undergoes less strain. Fig. 16.29 illustrates the state of stresses in the deformed region. Consider the state of strain in two points of the drawpiece. Point **1** is on the border of the workpiece and point **2** is in some distance from the border. At point **1** the radial stress is equal to zero. If we neglect the stresses from the blank holder, then assuming the yield criterion of maximal tangent stresses, the circumferential stress reaches the values of the yield limit. In cylindrical coordinates at point **1** we have $\sigma_r = 0$, $\sigma_\vartheta = -\sigma_o$ and $\sigma_z = 0$. Assume that the radial stress in the beginning of the sizing zone is less than the yield limit of the deformed material.

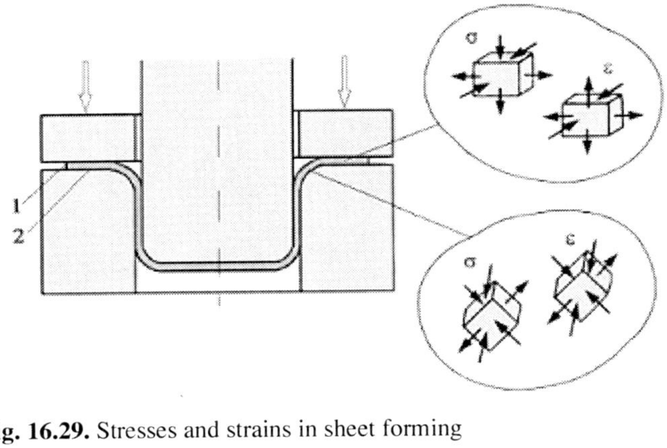

Fig. 16.29. Stresses and strains in sheet forming

Assume

$$\frac{\sigma_r - \sigma_m}{\sigma_\vartheta - \sigma_m} = \frac{d\varepsilon_r}{d\varepsilon_\vartheta} \qquad (16.226)$$

Taking into account the values of stresses at point **1** we have

$$\frac{\frac{1}{3}\sigma_o}{-\frac{2}{3}\sigma_o} = \frac{d\varepsilon_r}{d\varepsilon_\vartheta} \qquad (16.227)$$

By the condition of constant volume

$$d\varepsilon_r + d\varepsilon_\vartheta + d\varepsilon_z = 0 \qquad (16.228)$$

we get

$$d\varepsilon_z = -\frac{1}{2} d\varepsilon_\vartheta \qquad (16.229)$$

From the above the highest as to absolute value is circumferential strain, and the radial strain as well as the strain in thickness direction is half less. Since the circumferential strain ε_ϑ is negative, the strains ε_r and ε_z are positive. This means the circumferential reduction of material is associated with the increase of thickness and elongation along the radius. At some distance from free surface stresses are assumed to satisfy the expression

$$\sigma_r = n\sigma_o \qquad (16.230)$$

where

$$0 < n < 1$$

By the yield criterion

$$\sigma_\vartheta = (n - 1)\,\sigma_o \qquad (16.231)$$

The mean stress is

$$\sigma_m = \frac{(2n - 1)\sigma_o}{3} \qquad (16.232)$$

By (16.226) we get

$$d\varepsilon_r = d\varepsilon_\vartheta \frac{n + 1}{n - 2} \qquad (16.233)$$

The incremental strain in the thickness direction by the condition of constant volume is

$$d\varepsilon_z = -\varepsilon_\vartheta \left(\frac{2n - 1}{n - 2} \right) \qquad (16.234)$$

By Eq. (16.234) we can determine the parameter n for which the material thickness does not change $d\varepsilon_z = 0$. By Eq. (16.234) we get

$$n = \frac{1}{2} \tag{16.235}$$

By the equilibrium equation we can determine the radius, for which condition (16.236) is satisfied. Neglecting friction forces in the equilibrium equation

$$r = 0.61R \tag{16.236}$$

The increase of thickness occurs in the domain where the condition $r = 0.61R$ is satisfied. In the domain where $r < 0.61R$, a decrease of thickness takes place. Depending on the dimensions of the material and the die diameter the different zones of decrease or increase of thickness or only increase can appear. The thickness of the wall near the bottom for high drawpieces is reduced, and at a further distance the thickness increases. For low drawpieces we observe the increase of wall thickness along the entire height. The friction forces cause an additional increase of stress σ_r which decreases the thickness.

Drawing is the cold forming process where hardening takes place. The highest deformations are near the free surface. Moving in the direction to the bottom of the drawpiece the hardening is lower. In the drawing process the material hardening is a negative phenomenon. It causes the increase of force and decrease of the admitted degree of deformation. The ideal case would be to obtain along the generating line when we remove from the drawpiece bottom the decrease of plastic stress. This can be obtained by applying first the heating process of the drawpiece and next cooling the already deformed part. In such a case the greater degree of deformation in one operation can be applied.

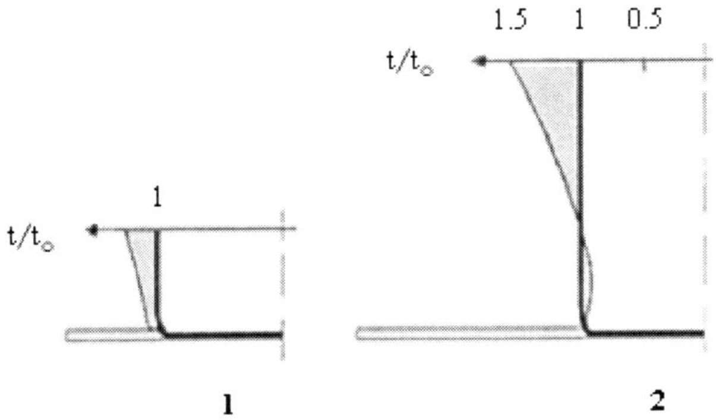

Fig. 16.30. Distribution of the drawpiece thickness
1 – low drawpiece, **2** – high drawpiece

16.10 Backward Extrusion

Consider the domain of deformation divided in to three subdomains (Fig. 16.31). Assume that the strains in subdomains are uniform and assume the discontinuity of tangent velocity on subdomains borders. Assume in subdomain **1**

$$u_r = \frac{1}{2}\varepsilon_r r \tag{16.237}$$

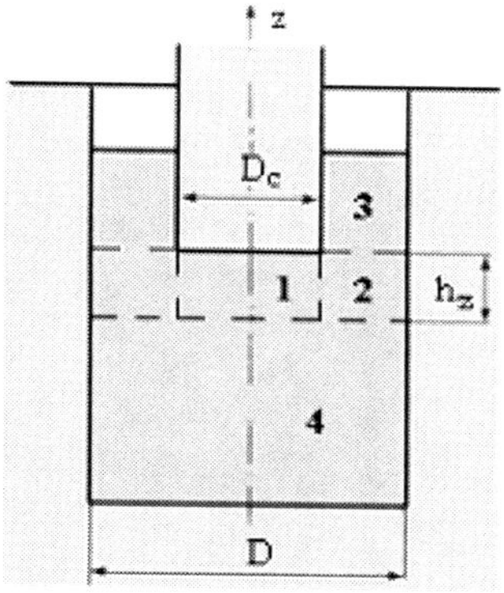

Fig. 16.31. Deformation zones in backward extrusion

The state of strain is axisymmetric

$$\varepsilon_r = \varepsilon_\vartheta = -\frac{1}{2}\varepsilon_z = \frac{1}{2}\varepsilon \tag{16.238}$$

In an axisymmetrical case

$$\varepsilon_i = |\varepsilon_z| = \varepsilon \tag{16.239}$$

By the virtual work theorem, the work of internal forces in subdomain **1** is

$$w_1 = \sigma_o \int_0^{2\pi}\int_0^{h_z}\int_0^{R_c} \varepsilon_r d\vartheta dr dz = \pi\varepsilon h_z \sigma_o R_c^2 \tag{16.240}$$

The work of friction forces is

$$w_2 = \psi k \int_0^{2\pi}\int_0^{R_c} \frac{1}{2}\varepsilon r^2 d\vartheta dr = \frac{1}{2}\pi\psi k \varepsilon R_c^3 \tag{16.241}$$

where ψ is the parameter characterizing the ratio of tangent stress τ to maximal tangent stress which results from the Huber-Mises yield criterion

$$\tau = \psi \frac{\sigma_o}{\sqrt{3}} \tag{16.242}$$

and k is the parameter characterizing the work of the internal forces on the shearing surface $w_s = \int k[[V]]dA$.

The work of the shearing forces on the surface z = 0 between subdomains **1** and **4**

$$w_3 = k \int_0^{2\pi} \int_0^{R_c} \frac{1}{2} \varepsilon r^2 d\vartheta dr = \frac{1}{3} \pi k \varepsilon \tag{16.243}$$

In subdomain **2**

$$u_r = \frac{1}{2} \varepsilon' \frac{R^2 - r^2}{r} \tag{16.244}$$

where ε' is the relative change of height h_z in subdomain **2**

$$\varepsilon' = \varepsilon \frac{R_c^2}{R^2 - R_c^2} \tag{16.245}$$

The remaining components of strain for subdomain **2** are

$$\varepsilon_r = -\frac{1}{2} \varepsilon' \left(1 + \frac{R^2}{r^2} \right) \tag{16.246}$$

$$\varepsilon_\vartheta = -\frac{1}{2} \varepsilon' \left(1 - \frac{R^2}{r^2} \right) \tag{16.247}$$

The intensivity is

$$\varepsilon_i = \frac{\varepsilon'}{\sqrt{3}} \sqrt{3 + \frac{R^4}{r^4}} \tag{16.248}$$

The work of the internal forces is

$$\begin{aligned} w_4 &= \frac{\sigma_o}{\sqrt{3}} \int_0^{2\pi} \int_0^{h_z} \int_{R_c}^{R} \varepsilon' \sqrt{3 + \frac{R^4}{r^4}} \, r d\vartheta dz dr = \\ &= \frac{\pi}{\sqrt{3}} \sigma_o \varepsilon h_z R_c B \end{aligned} \tag{16.249}$$

where

$$B = \frac{\ln\left[\frac{1}{3}\left(\frac{R^2}{R_c^2} + \sqrt{3 + \frac{R^4}{r^4}}\right)\right] - \sqrt{1 + 3\frac{R_c^4}{R^4}} + 2}{1 - \left(\frac{R_c}{R}\right)^2} \tag{16.250}$$

The work of the friction forces on the vertical surface

$$w_5 = \psi k \int_0^{2\pi}\int_0^{h} \varepsilon' z R d\vartheta dz = \psi\pi k R \varepsilon' h_z^2 \tag{16.251}$$

The work of the shearing forces on the vertical surface between subdomains **1** and **2**

$$w_6 = k \int_0^{2\pi}\int_0^{h_z} (\varepsilon' + \varepsilon) z R_c d\vartheta dz = \pi k R_c (\varepsilon' + \varepsilon) h_z^2 \tag{16.252}$$

The shearing work between subdomains **2** and **3**, and **2** and **4** is

$$\begin{aligned} w_7 &= 2k \int_0^{2\pi}\int_{R_c}^{R} \frac{1}{2}\varepsilon' \frac{R^2 - r^2}{r} r d\vartheta dr = \\ &= 2\pi k \varepsilon' \left(\frac{2}{3}R^3 + \frac{2}{3}R_c^3 + \frac{2}{3}R^2 R_c\right) \end{aligned} \tag{16.253}$$

The total work is

$$\begin{aligned} w = \pi R_c^2 \varepsilon h_z \sigma_o \Bigg[1 + \frac{1+\psi}{3\sqrt{3}} \frac{R_c}{h_z} + \frac{B}{3} + 1.15 \frac{R_c}{h_z} \times \\ \times \frac{\frac{2}{3}\frac{R}{R_c} + \frac{1}{3}\frac{R_c^2}{R^2} - 1 + \frac{1}{2}\frac{h_z^2}{R_c^2}\left(\psi\frac{R_c}{R} + 1\right)}{1 + \left(\frac{R_c}{R}\right)^2} \end{aligned} \tag{16.254}$$

Since

$$w = P\Delta h = p\pi R_c^2 \Delta h \tag{16.255}$$

we get the expression describing the ratio of the absolute value of the mean pressure to the yield stress

$$\frac{p}{\sigma_o} = 1 + \frac{B}{3} + \frac{1+\psi}{3\sqrt{3}}\frac{R_c}{h_z} + 1.15\frac{R_c}{h_z} \times \\ \times \frac{\frac{2}{3}\frac{R}{R_c} + \frac{1}{3}\frac{R_c^2}{R^2} - 1 + \frac{1}{2}\frac{h_z^2}{R_c^2}\left(1 + \psi\frac{R_c}{R}\right)}{1 - \left(\frac{R_c}{R}\right)^2} \tag{16.256}$$

The total work on kinematically admissible displacements satisfies

$$\frac{\partial w}{\partial h_z} = 0 \tag{16.257}$$

By Eq. (16.257)

$$\frac{h_z}{D_c} = \sqrt{\frac{\frac{1+\psi}{2}\left(1 - \frac{D_c^2}{D^2}\right) + \frac{1}{2}\left(\frac{2}{3}\frac{D}{D_c} + \frac{1}{3}\frac{D_c^2}{D^2} - 1\right)}{1 + \psi\frac{D_c}{D}}} \tag{16.258}$$

By Eq. (16.258) the friction forces characterized by parameter ψ do not have significant influence on the value of ratio h_z/D_c. The ratio D_c/D is of great importance, and influences mean normal pressure.

16.11 The Upsetting of Cylindrical Elements Between Rings

In upsetting the domain of deformation can be divided in to three subdomains denoted by **1**, **2** and **3** (Fig. 16.32).

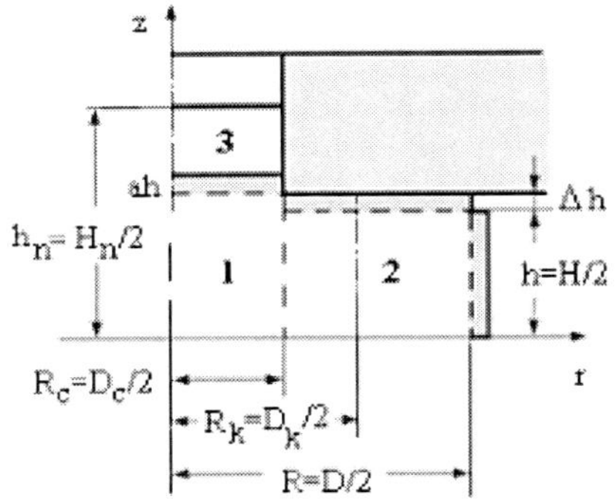

Fig. 16.32. Subdomains of the deformation in upsetting between rings

Assume the components of displacement and strain

$$u_z = az, \qquad u_r = \frac{1}{2}ar \qquad \varepsilon_z = -a, \qquad \varepsilon_r = \frac{1}{2}a, \qquad \varepsilon_\vartheta = \frac{1}{2}a \tag{16.259}$$

where parameter a takes into account the change of height in subdomain **1**. The work of internal forces in subdomain **1** is

$$w_1 = \sigma_o \int_0^{2\pi}\int_0^{h}\int_0^{R_c} \varepsilon_i r d\vartheta dr dz \tag{16.260}$$

Assume

$$\varepsilon_i = |a| \tag{16.261}$$

After integrating

$$w_1 = \pi\sigma_o |a| h R_c^2 \tag{16.262}$$

Assume that the flat cross-sections before deformation remain flat after deformation, and on the boundary between subdomains **1** and **3** the shearing of material occurs. The work of the shearing forces is

$$w_2 = k \int_0^{2\pi}\int_0^{R_c} u r dr d\vartheta = \frac{1}{3}\pi k |a| R_c^3 \tag{16.263}$$

For subdomain **2** $\varepsilon_z = -\varepsilon$, where $\varepsilon = \frac{\Delta h}{h}$. The condition of constant volume is

$$-\varepsilon + \frac{u_r}{r} + \frac{\partial u_r}{\partial r} = 0 \tag{16.264}$$

By solving Eq. (16.264) with the boundary conditions saying that the radial displacements on the boundary of subdomains **1** and **2** are equal, we get

$$\begin{aligned} u_r &= \frac{a}{2}\frac{R_c^2}{r} + \frac{\varepsilon}{2}\frac{r^2 - R_c^2}{r} \\ \varepsilon_r &= -\frac{a}{2}\frac{R_c^2}{r^2} + \frac{\varepsilon}{2}\left(1 + \frac{R_c^2}{r^2}\right) \\ \varepsilon_\vartheta &= \frac{a}{2}\frac{R_c^2}{r^2} + \frac{\varepsilon}{2}\left(1 - \frac{R_c^2}{r^2}\right) \end{aligned} \tag{16.265}$$

The strain work in subdomain **2** is

$$w_3 = \sigma_o \int_0^{2\pi} \int_0^{h} \int_{R_c}^{R} \varepsilon_i r d\vartheta dz dr \tag{16.266}$$

where

$$\varepsilon_i = \frac{1}{\sqrt{3}} \sqrt{3\varepsilon^2 + (a - \varepsilon)^2 \frac{R_c^4}{r^4}} \tag{16.267}$$

After integrating we get

$$w_3 = \pi\sigma_o \varepsilon h \left[a_1 \ln \frac{\left(a_1 + \sqrt{R_c^4 + a_1^2}\right) R^2}{\left(a_1 + \sqrt{R^4 + a_1^2}\right) R_c^2} + \sqrt{R^4 + a_1^2} - \sqrt{R_c^4 + a_1^2} \right. \tag{16.268}$$

where

$$a_1^2 = \frac{1}{3} R_c^4 \left(\frac{a}{\varepsilon} - 1 \right)^2 \tag{16.269}$$

The work of the shearing forces between subdomains **1** and **2** is

$$w_4 = 2\pi R_c k \int_0^h \left| u_{z1} - u_{z2} \right| dz = \pi R_c k h^2 (\varepsilon - a) \tag{16.270}$$

where u_{z1} and u_{z2} are displacements related to subdomains **1** and **2**, respectively. The surface for which the radial displacement is equal to zero can be determined as follows

$$R_k = R_c \sqrt{1 - \frac{a}{\varepsilon}} \tag{16.271}$$

The work of the friction forces is

$$w_5 = \psi k \int_0^{2\pi} \int_{R_c}^{R_k} \left| u_r \right| r d\vartheta dr + \int_0^{2\pi} \int_{R_c}^{R_k} u_r r d\vartheta dr + \tag{16.272}$$

After integration we get

$$w_5 = \pi\psi k \varepsilon \left[\frac{4}{3} R_k - R_k^2 (R + R_c) + \frac{1}{3} \left(R^3 + R_c^3\right) \right. \tag{16.273}$$

Assume that in the subdomain **3** the work of plastic strain is neglected because this work is small in comparison to the subdomain **2**. The deformation of material should assure that the total strain work reaches the minimal value. Consider parameter a characterizing the deformation. If a < 0, then a decrease of height occurs in subdomain **1**, and an increase if a > 0. The strain work reaches the lowest value, if the first derivative with respect to a is zero. Then

$$\frac{h^2}{R_c^2} - \left(\beta - \sqrt{3}\right)\frac{h}{R_c} - \psi\left(\frac{R}{R_c} + 1 - 2\sqrt{1 - \frac{a}{\varepsilon}}\right) + \frac{1}{3} = 0 \tag{16.274}$$

where

$$\beta = \ln \frac{\frac{R_c^2}{R}\left(1 - \frac{a}{\varepsilon}\right) + \sqrt{3 + \frac{R_c^4}{R^4}\left(1 - \frac{a}{\varepsilon}\right)^2}}{1 - \frac{a}{\varepsilon} + \sqrt{3 + \left(1 - \frac{a}{\varepsilon}\right)^2}} \tag{16.275}$$

For the first stage

$$\frac{a}{\varepsilon} = 1 - \frac{1}{\varphi}\left[1 - \sqrt{1 - \frac{B_4}{B_6}\varphi^2}\right] \tag{16.276}$$

where

$$\varphi = \frac{D_c}{D}$$

$$B_6 = \varphi^2 B_4 + B_5$$

$$B_5 = \varphi k^2 \psi\left(1 - \varphi^2\right)$$

$$B_4 = \frac{\varphi}{1 - \varphi^2}\left(B_3 + \Psi B_2\right) + \frac{\varphi^2}{3}\left(\varphi^3 + \psi\right) + 3.465k + 2kB_1$$

$$B_3 = \frac{1}{3}\varphi^2\left(\varphi^3 - 3\varphi + 2\right) \tag{16.277}$$

$$B_2 = \frac{1}{3}\left(1 - 3\varphi^2 - 2\varphi^3\right)$$

$$B_1 = \sqrt{3 + \varphi^4} - 3\varphi^2 - \varphi^2 \ln\frac{\varphi^2 + \sqrt{3 + \varphi^4}}{3}$$

$$k = \frac{H}{D}$$

By the condition of constant volume

$$\frac{dh}{dR} = \frac{2Rh}{R_c^2\left(1-\frac{a}{\varepsilon}\right)-R^2} \tag{16.278}$$

The expression (16.278) should be integrated with boundary conditions

$$\frac{D}{D_o}=1, \qquad \frac{H_c}{H_o}=1, \qquad \frac{H}{H_o}=1$$

The most often used relations are

$$2 \le \frac{D_o}{H_o} \le 4, \quad 0.2 \le \frac{D_c}{D_o} \le 0.8, \quad 1 \ge \frac{H}{H_o} \ge 0.1 \tag{16.279}$$

16.12 Forward Extrusion of Cylindrical Workpieces

In forward extrusion we apply flat and conical dies (Fig. 16.33). Assume that in

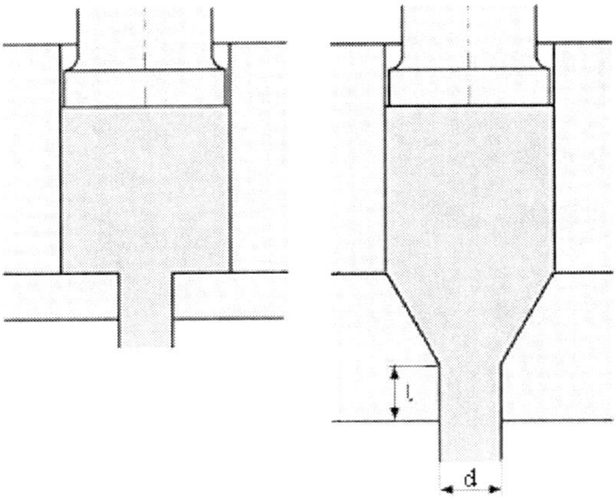

Fig. 16.33. Forward extrusion with flat and conical dies

the sizing zone the radial stress σ_{r1} is equal to the yield limit σ_o

$$|\sigma_{r1}| = \sigma_o \tag{16.280}$$

The force P_1 necessary for extrusion in the sizing part is expressed as

$$P_1 = f_1|\sigma_{r1}|\pi d l = f_1\sigma_o \pi d\, l \tag{16.281}$$

and the stress

$$p_1 = \frac{P_1}{A_1} = \frac{f_1 \sigma_o \pi d\, l}{\frac{\pi d^2}{4}} = \sigma_o \frac{4 f_1 l}{d} \tag{16.282}$$

The total work in the conical part consists of three terms, i.e. the pure plastic deformation, the work of friction forces on the contact surface and the work resulting from the reaction of the sizing zone. Consider the domain of plastic deformation (Fig. 16.34). Consider a spherical coordinate system. Then

$$\varepsilon_R = \frac{\partial u_R}{\partial R}, \qquad \varepsilon_\varphi = \varepsilon_\vartheta = \frac{u_R}{R} \tag{16.283}$$

By the condition of constant volume

$$\frac{\partial u_R}{\partial R} + 2\frac{u_R}{R} = 0 \tag{16.284}$$

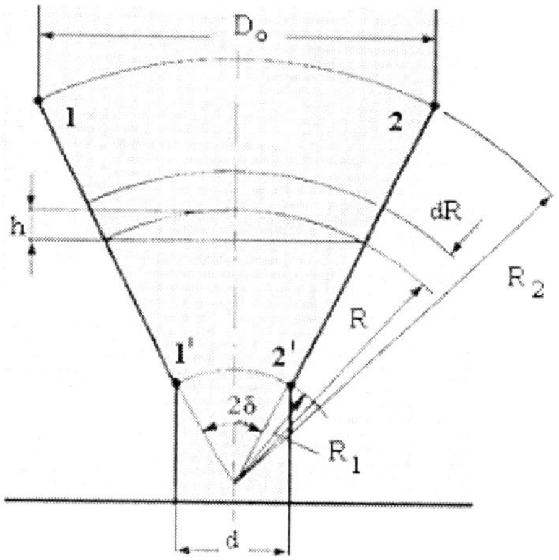

Fig. 16.34. The domain of plastic deformation

The expression (16.284) can be written as

$$\frac{\partial\left(u_R R^2\right)}{\partial R} = 0 \tag{16.285}$$

After integrating

$$u_R R^2 = f(\varphi) \tag{16.286}$$

For the point with the coordinate $\varphi = 0$ and $R = R_2$ we have

$$u_R = w$$

where w is the displacement in the z direction. Assume for simplicity, that the function $f(\varphi)$ does not depend on φ and takes the value wR_2^2

$$u_R R^2 = wR_2^2 \tag{16.287}$$

Hence

$$u_R = \frac{wR_2^2}{R^2} \tag{16.288}$$

and

$$\varepsilon_R = \frac{2wR_z^2}{R^3} \tag{16.289}$$

In this case, if $\varepsilon_\varphi = \varepsilon_\vartheta$ we have

$$\varepsilon_i = \varepsilon_R \tag{16.290}$$

The plastic strain work is determined by

$$w_1 = \sigma_o \int_V \varepsilon_i dV \tag{16.291}$$

The elementary volume separated by two spherical surfaces with radii R and R+dR is

$$dV = 2\pi(1 - \cos\gamma)R^2 dR \tag{16.292}$$

By Eqs. (16.291) and (16.293) we get

$$w_1 = \sigma_o 4\pi w R_2^2 (1 - \cos\gamma) \int_{R_1}^{R_2} \frac{dR}{R} \tag{16.293}$$

After integrating we get

$$w_1 = \sigma_o 4\pi w R_2^2 (1 - \cos\gamma) \ln \frac{R_2}{R_1} \tag{16.294}$$

The work of friction forces on a conical die is

$$w_2 = \iint_A \tau_k u_{kR} dA \tag{16.295}$$

where τ_k is the tangent stress on the contact surface and u_{kR} is the displacement on the contact surface

$$u_{kR} = u_R = \frac{wR_2^2}{R} \tag{16.296}$$

The elementary surface dA is

$$dA = 2\pi R \sin\gamma dR \tag{16.297}$$

Assume, that the tangent stress τ_k is constant on the contact surface. By Eqs. (16.296), (16.297) and (16.295) we get

$$w_2 = \tau_k 2\pi w R_z^2 \sin\gamma \int_{R_1}^{R_2} \frac{dR}{R} \tag{16.298}$$

After integrating

$$w_2 = \tau_k 2\pi w R_2^2 \sin\gamma \ln\frac{R_2}{R_1} \tag{16.299}$$

The work of the friction forces in the sizing part is

$$w_3 = p_1 \frac{\pi d^2}{4} w_1 \tag{16.300}$$

By the continuity of process

$$w_1 = w\frac{R_2^2}{R_1^2} \tag{16.301}$$

Hence

$$w_3 = p_2 \frac{\pi d^2}{4} w \frac{R_2^2}{R_1^2} \tag{16.302}$$

By Eqs. (16.294), (16.299), (16.302) and (16.283) we get

$$\begin{aligned} P_2 w = {} & \sigma_o 4\pi w R_2^2 (1-\cos\gamma)\ln\frac{R_2}{R_1} \\ & + \tau_k 2\pi w R_2^2 \sin\gamma \ln\frac{R_2}{R_1} + p_1 \frac{\pi d^2}{4} w \frac{R_2^2}{R_1^2} \end{aligned} \tag{16.303}$$

Hence

$$\begin{aligned} P_2 = {} & \sigma_o 4\pi R_2^2 (1-\cos\gamma)\ln\frac{R_2}{R_1} \\ & + \tau_k 2\pi R_2^2 \sin\gamma \ln\frac{R_2}{R_1} + p_1 \frac{\pi d^2}{4} \frac{R_2^2}{R_1^2} \end{aligned} \tag{16.304}$$

By the relations

$$R_2^2 = \frac{D_o^2}{4\sin^2\gamma} = \frac{D_o^2}{4\left(1-\cos^2\gamma\right)} \tag{16.305}$$

$$\frac{R_2^2}{R_1^2} = \frac{D_o^2}{d^2} = \frac{A_o}{A} \tag{16.306}$$

we express the mean stress on the surface between the conical and cylindrical surfaces

$$p_2 = \left(\frac{\tau_k}{\sin\gamma} + \frac{2\sigma_o}{1+\cos\gamma}\right)\ln\frac{A_o}{A} + p_1 \tag{16.307}$$

In extrusion we assume that the tangent stress on a conical surface attains the maximal values

$$\tau_k = \frac{\sigma_o}{2} \tag{16.308}$$

The distribution of metal velocities in extrusion depends on the conditions of the process. Three approaches to metal flow in extrusion can be considered. The first case is when the material domain **3** is in an elastic state (Fig. 16.35) and moves rigidly with the punch. This case appears if the friction coefficient is low and the material is sufficiently uniform. The second case is when the material intensively flows near the axis, and the flow of the outer layers is strongly hindered by friction forces. Such a flow occurs at high resistance of friction and if the material is of low uniformity. The third case is characterized by intensive flow of material in the entire volume. The layers near the walls flow in an opposite direction to the punch direction, near the punch they change the direction of flow in to the direction of the punch movement. Such a case occurs at high friction resistance and high uniformity of material.

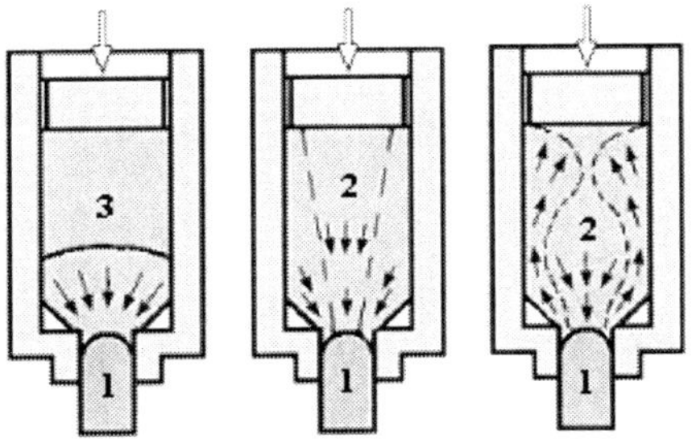

Fig. 16.35. Metal flow in extrusion

In the majority of cases the first kind of approach is observed. On the surface between conical and cylindrical parts the radial stress as to absolute value is higher than stress along the extrusion axis having the value of yield stress of the material. The high degree of deformation in this process and the high friction resistance cause the radial stress to attain a maximal value equal to half of the yield stress.

16.13 Strain Rates

16.13.1 Strain Rates in Rolling

Strain rates at any point are the function of location. The velocity of metal flow in rolling direction is determined from the continuity condition (Fig. 16.36)

$$A_o V_o = A_x V_x = A_l V_l \tag{16.309}$$

where subscripts o refer to initial parameters, x to parameters in the cross section at the distance x and the subscripts 1 refer to the exit zone. Hence

$$V_x = \frac{A_l}{A_x}, \qquad V_l = \frac{A_l}{A_o}, \qquad \frac{A_o}{A_x} V_l = \frac{\lambda_x}{\lambda} V_l \tag{16.310}$$

where λ is the elongation.

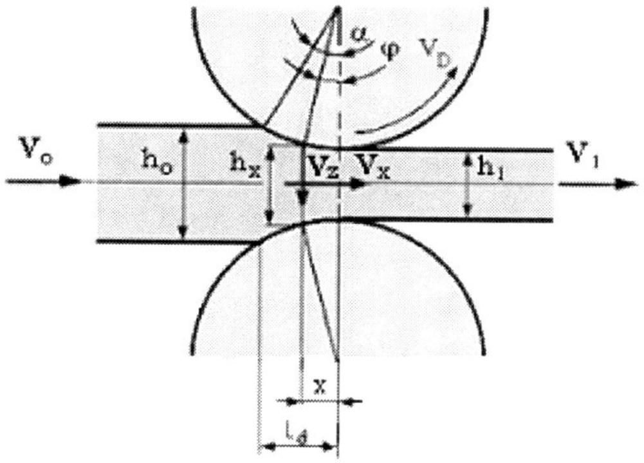

Fig. 16.36. Velocities of metal in rolling

Assume

$$\lambda_x = \gamma_x^{k-1} \tag{16.311}$$

where

$$\gamma_x = \frac{h_x}{h_o} \qquad k = \frac{\ln\beta}{\ln \frac{1}{\gamma}} \tag{16.312}$$

The distribution of material velocities in the rolling direction is

$$V_x = \frac{V_1}{\lambda}\gamma_x^{k-1} = V_1\left(\frac{\gamma_x}{\gamma}\right)^{k-1} \tag{16.313}$$

By Eq. (16.312) we get

$$V_x = V_1\left(\frac{h_x}{h_1}\right)^{k-1} \tag{16.314}$$

The height h_x at the cross-section x can be determined as

$$h_x = h_1 + D(1-\cos\varphi) \tag{16.315}$$

Finally we get

$$V_x = V_1\left[1+\frac{D}{h_1}(1-\cos\varphi)\right]^{k-1} \tag{16.316}$$

The strain rate in the direction of the height is

$$\dot{\varepsilon}_z = \frac{2V_z}{h_x} \tag{16.317}$$

The vertical components of the material velocity (Fig. 16.30) are

$$V_z = V_x \,\mathrm{tg}\varphi \tag{16.318}$$

By Eqs. (16.316) and (16.318) we get

$$\dot{\varepsilon}_z = 2V_1 \frac{\left[1+\frac{D}{h_1}(1-\cos\varphi)\right]^{k-1}}{h_1 + D(1-\cos\varphi)}\,\mathrm{tg}\varphi \tag{16.319}$$

16.13.2 Strain Rates in Wire Drawing

By continuity of the process we have

$$A_o V_o = A_x V_x = A_1 V_1 \tag{16.320}$$

where subscripts o refer to initial parameters

Lehmann Th (1983a) General frame for the definition of constitutive laws for large non-isothermic elastic-plastic and elastic-viscoplastic deformations. In: The constitutive law in thermoplasticity. CISM Course and lectures. Springer, Berlin Heidelberg New York

Lehmann Th (1983b) Einige Aspecte der Thermoplastizität. ZAMM 63 (3/13)

Lehmann Th (1984) Some considerations on the constitutive law in thermoplasticity. Mechanika Teoretyczna i Stosowana 1-2 (22): 3-20

Malinin NN (1975) The theory of plasticity and creep (in Russian). Maszinostrojenije, Moscow

Marciniak Z (1961) Mechanics of sheet forming. Polish Scientific Publishers, Warsaw (in Polish)

Mieldniczuk J (2000) Plasticity of porous materials. Technical University of Poznan Press

Mielniczuk J (1988) Plasticity of porous metal sinters. IFTR reports 15: 1-92

Mielniczuk J (1993) Plasticity of porous metals. ed J P Boehler Failure criteria of structured media 331-338. Balkema Rotterdam

Mises R (1913) Mechanik der festen Körper in plastisch deformablem Zustand. Götinger Nachrichten

Oyane M, Shima S, Kono Y (1973) Theory of plasticity of porous metals. Bull JSME 16: 1254-1262

Perzyna P (1961) Propagation of shock waves in non-homogeneous elastic-visco-plastic bodies. Arch Mech Stos 13: 851-867

Perzyna P (1962) Propagation of shock waves in an elastic-visco-plastic medium of a definite non-homogeneity type. Arch Mech Stos 14: 93-111

Perzyna P (1963) The constitutive equations for rate sensitivite plastic materials. Quart Appl Math 20: 321-332

Perzyna P (1963) The study of the dynamical behaviour of rate sensitive plastic materials. Arch Mech Stos 15: 113-130; Bull Acad Polon Sci Sér Sci Tech 12: 207-216

Perzyna P (1963) On the propagation of stress waves in a rate sensitivite plastic medium. ZAMP 14: 241-261

Perzyna P (1964) On a nonlinear boundary-value problem for a linear hyperbolic differential equation. Bull Acad Polon Série Sci Tech 12: 589-594

Perzyna P, Wierzbicki T (1964) Temperature dependent and strain rate sensitivite plastic materials. Arch Mech Stos 16: 135-143; (1964)

Perzyna P (1966) Theory of viscoplasticity (in Polish). Polish Scientific Publishers, Warsaw

Perzyna P (1978) Thermodynamics of non-elastic materials. Polish Scientific Publishers Warsaw

Sawczuk A (1989) Mechanics and plasticity of structures. Polish Scientific Publishers and Ellis Horwood Limited

Shima S, Oyane M (1976) Plasticity theory of porous metals. Int J Mech Sci 18: 285-291

Shima S, Tabata T, Oyane M, Kawakami T (1976) Upper bound theory for deformation of porous materials. Memoris of the faculty of engineering 38/3: 117-137 Kyoto University

Skrzypek J (1986) Plasticity and Creep. Polish Scientific Publishers, Warsaw

Slater RAC (1977) Engineering plasticity. Wiley, New York – Toronto

Służalec A (1984) A preliminary analysis of heat flow within roll-forging dies using a finite element method. Int J Mach Tool Des Res 24, 3: 171-179

Służalec A (1988) An analysis of thermal effects of coupled thermoplasticity in metal forming processes. Comm Appl Num Meth 4: 675-685
Służalec A (1989) An application of numerical simulation in the design of metal forging process (in Polish). Mechanika i Komputer 10: 103-118
Służalec A (1989) Yield function in simulation of the powder metal forming. Proceeding 2nd Int. Conference Computational Plasticity, Models, Software and Applications, ed. D.R.J. Owen, E. Hinton, E. Onate, Part II: 995-1007
Służalec A (1990) An evaluation of the internal dissipation factor in coupled thermoplasticity. Int J Nonlinear Mechanics 25, 4: 395-403
Służalec A (1991) An analysis of dead zones in the process of direct extrusion trough single-hole flat die. Comm Appl Num Meth 7: 281-287
Służalec A (1992) Temperature rise in elastic-plastic metal. Comp Meth Appl Mech Eng 96: 293-302
Służalec A (1993) Elastic stresses in porous material undergoing thermal loading. Int J Eng Sci 31, 3: 475-482
Służalec A (1997) Thermo-elastic analysis of spatial structure in random conditions. Proceedings of the International Colloquium on Computation of Shell & Spatial Structures, ICCSS'97: 403-408
Służalec A (1998) Heat transfer described in terms of random variables. Fourth World Congress on Computational Mechanics, Buenos Aires, Argentina
Służalec A (1999) Numerical simulation of stochastic metal forming process for rigid - viscoplastic material. Constitutive and Damage Modeling of Inelastic Deformation and Phase Transformation. VII International Symposium on Plasticity and its Current Applications, Cancun: 421-424
Służalec A (1999) Thermo-elastic-plastic porous material undergoing thermal loading. Int Journal of Engineering Science 37: 1985-2005
Służalec A (2000) Rigid – viscoplasticity described in terms of stochastic finite elements. Advances in Computational Engineering & Sciences, vol. II: 1156-1161
Służalec A (2000) Simulation of stochastic metal-forming process for rigid-viscoplastic material. Int J of Mechanical Sciences 42: 1935-1946
Służalec A (2000) Thermoelastic analysis in random conditions. Journal of Thermal Stresses 23:131-141
Służalec A (2001) Numerical simulation of stochastic thermo - rigid -viscoplastic processes. Proceeding Fourth International Congress on Thermal Stresses, Osaka: 223-226
Służalec A, Bruhns OT (1993) Densification of powder metals with assumed ellipsoidal yield surface. Int J Mech Sci 35, 9: 731-740
Służalec A, Grzywiński M (2000) Rigid - thermo - viscoplasticity in metal forming process described by stochastic finite elements. European Congress on Computational Methods in Applied Sciences and Engineering ECCOMAS 2000, Barcelona
Służalec A, Grzywiński M (2000) Stochastic convective heat transfer equations in finite differences method. Int J of Heat & Mass Transfer 43: 4003-4008
Służalec A, Grzywiński M (2002) Stochastic equations of rigid - thermo - viscoplasticity in metal forming process. Int J Eng Sci vol. 40/4: 367-383
Służalec A, Kubicki K (2002) Sensitivity analysis in thermo-elastic-plastic problems. J Thermal Stresses 25: 705-718
Storożew MW, Popow EA (1957) Teorija obrabotki mietałłow dawleniem. Maszgiz, Moskwa

Storożew MW, Popow EA (1963) Teorija obrabotki mietałłow dawleniem. Wysszaja Szkoła, Moskwa
Stüwe HP (1965) Flow curves of policrystalline metals and their application in the theory of plasticity (in German). Z Metallkd. 56: 633-642
Szopman LA (1964) Teorija i rasczoty processow chołodnoj sztampowki. Maszgiz, Moskwa
Tarnowskij II, Pozdiew AA, Mieandrow LW, Hasin GA (1960) Miechaniczeskije swojstwa stali pri goriaczej obrabotkie dawleniem. Mietałłurgizdat, Swierdłowsk.
Tarnowskij II, Pozdiew AA, Ganago OA, Kołmogorow WL, Trubin WN, Bajsburd RA, Tarnowskij WJ (1963) Teorija obrabotki mietałłow dawleniem. Mietałłurgizdat, Moskwa
Tayler GJ, Quinney H (1931) The plastic distortion of metals. Phil Trans Roy Soc A230
Washizu K (1975) Variational methods in elasticity and plasticity. Pergamon Press

Subject Index

Printed in the United Kingdom by
Lightning Source UK Ltd., Milton Keynes
140027UK00007B/150/P

9 783540 406488